Local Energy Markets

Meng Song · Ciwei Gao · Mingyu Yan ·
Yunting Yao · Tao Chen

Local Energy Markets

Paving the Path Toward the Low-Carbon Digital Power Distribution System

Springer

Meng Song
Southeast University
Nanjing, Jiangsu, China

Ciwei Gao
Southeast University
Nanjing, Jiangsu, China

Mingyu Yan
Huazhong University of Science and Technology
Wuhan, Hubei, China

Yunting Yao
Nanjing Normal University
Nanjing, Jiangsu, China

Tao Chen
Southeast University
Nanjing, Jiangsu, China

ISBN 978-981-97-9749-3 ISBN 978-981-97-9750-9 (eBook)
https://doi.org/10.1007/978-981-97-9750-9

This work was supported by National Natural Science Foundation of China (52007030, 52107079).

This Springer imprint is published by the registered company Springer Nature Singapore Pte Ltd.
The registered company address is: 152 Beach Road, #21-01/04 Gateway East, Singapore 189721, Singapore

Preface

The transition toward a low-carbon digital power distribution system is becoming a global priority, driven by the urgent need to mitigate climate change and achieve sustainable development. The local energy market (LEM) is emerging as a promising solution to address the supply-demand imbalance problem caused by the increasing penetration of distributed energy resources (DERs). LEM primarily focuses on energy transactions within power distribution systems, closely linked to human behaviors and social activities, and can extend to multi-energy carrier systems. It fosters greater flexibility, efficiency, and resilience for the energy system, enabling a more effective integration of renewable energy.

This book tries to introduce to researchers, policymakers, and regulators the history, ongoing developments, and future trends of LEM. This book consists of 12 chapters. Chapter 1 provides an overview of LEM. Chapters 2–4 model the trading behaviors of prosumers in LEM. Classical psychological theories, e.g., mental accounting, are employed to characterize prosumers' trading choices and preferences. New prosumers, i.e., networked data centers, are also discussed. Operation uncertainties of LEM are analyzed in Chaps. 5 and 6. Meanwhile, the distribution network constraints should be satisfied for LEM, and Chaps. 7 and 8 discuss this issue. Chapters 9 and 10 design green LEM to guide prosumers to trade in a green way. Chapters 11 and 12 develop cutting-edge technologies (blockchain and reinforcement learning) to operate LEM. The detailed introductions of the 12 chapters are given as follows.

- Chapter 1 comprehensively reviews LEM conceptual development in power distribution systems, including its fundamental features, structure, key techniques, and transactive control methodology. The integration of LEM with artificial intelligence (AI), multi-energy carriers, and social and human impacts is also discussed.
- Chapter 2 provides a trading decision-making method for prosumers based on the mental accounting theory. We construct the decision-making accounts to evaluate the trading behaviors of prosumers, including risk attitude and subjective value.

The multi-round bidding market mechanism has also been developed to improve social welfare and market efficiency.

- Chapter 3 proposes a trading strategy for prosumers based on the prospect theory and adaptive learning. Blockchain technology is utilized in LEM for higher trading efficiency through smart contract-based matching and settlement.
- Chapter 4 presents a bi-level optimization model for networked data centers (DCs) among multiple LEMs. The LEM model embedded with DCs is proposed, which is cleared by the alternating direction method of multipliers (ADMM). The differential evolution algorithm with the Kriging meta-model is utilized to solve the bi-level optimization problem.
- Chapter 5 offers a LEM mechanism to allocate cooperative surplus based on the market contribution of each participant for social welfare maximization, individual rationality, cost recovery, and incentive compatibility. An adaptive robustness optimization (ARO) method is developed to solve the uncertainty problems caused by renewable energy.
- Chapter 6 presents a peer-to-peer (P2P) trading strategy to mitigate the uncertainty brought by renewable energy. The distributionally robust optimization (DRO) method is leveraged to handle the power fluctuation issues from P2P trading. Based on dual theory, the nonlinear model is converted into a linear convex programming problem, and the ADMM approach is utilized to clear the market for prosumers' privacy preservation.
- Chapter 7 designs a P2P trading within a bi-level structure for multi-microgrid (MG) systems. The dynamic P2P interactions among MGs are captured at the lower level, which is modeled as a multi-leader multi-follower (MLMF) Stackelberg game. The distributed system operator (DSO) employs the reconfiguration at the upper level to adjust the distribution network.
- Chapter 8 introduces a sequential clearing method of the local energy and flexibility markets to decouple energy trading and congestion management. A local energy market based on continuous double auctions is proposed for efficient market clearing. Dynamic calculation of power loss and network fees is implemented to incentivize prosumer trading within neighborhoods. Furthermore, a flexibility market is established to encourage actively exploring prosumers' flexibility for congestion management.
- Chapter 9 proposes an energy-sharing strategy for green communities considering carbon credit and power traceability. Carbon credit is utilized to monetize the environmental cost of heterogeneous electricity. An electricity-carbon joint clearing model and traceability-based pricing method are developed to lower the carbon emission of prosumers.
- Chapter 10 incorporates the nodal carbon intensity (NCI) model into a P2P trading mechanism for electricity and gas to prevent data privacy leakage. The information gap decision theory is employed to construct a robust optimization problem. We adopt the normal boundary intersection (NBI) approach to cope with the multi-objective problem while maximizing social welfare.

- Chapter 11 discusses the joint energy and carbon allowance trading for MGs based on blockchain technology. The transaction behavior of MGs is modeled by a cooperative game, which is formulated as a two-stage problem to distribute the inner payoff rationally. The column-and-constraint generation (C&CG) method and Karush-Kuhn-Tucker (KKT) conditions are combined to solve the problem.
- Chapter 12 studies the indirect customer-to-customer energy trading in a local event-driven energy market. The trading process is modeled as a Markov decision process (MDP) and solved by incorporating reinforcement learning and data-driven approaches.

Nanjing, China
August 2024

Meng Song
Ciwei Gao
Mingyu Yan
Yunting Yao
Tao Chen

Acknowledgements

This book summarizes our research on local energy market from 2020 to 2024. These studies were carried out at three different universities/labs, including Southeast University, Huazhong University of Science and Technology, and Nanjing Normal University. Many people contributed to this book. We offer our tons of thanks to Prof. Mohammad Shahidehpour from Illinois Institute of Technology, Dr. Xingyu Yan, Mr. Yunfeng Cai, Ms. Xinyi Jing, Mr. Xudong Wang, and Mr. Jiacheng Cao from Southeast University. We also thank Mr. Tianyi Zhang and Mr. Chen Zhao for their careful examination of the entire book. Finally, we appreciate the assistance and help of the staff at Springer.

This book is supported by National Natural Science Foundation of China (Grant Nos. 52007030, 52107079).

Contents

Chapter 1
State of the Art of the Local Energy Market

Abstract The local energy market (LEM) is emerging as a novel market mechanism to balance supply and demand as more distributed energy resources (DERs) and flexible loads integrate with power systems. LEM primarily focuses on energy transactions within distribution systems, which are closely linked to human behaviors and social activities and can extend to multi-energy carrier systems. Additionally, artificial intelligence (AI) advancements facilitate more automatic and intelligent participation in various LEMs. Consequently, LEM can pave the way to the cyber-physical-society system and price signal-defined open-access energy facilities. This chapter comprehensively reviews the conceptual development of LEM in power distribution systems, encompassing its fundamental features, structure, key implementation techniques, and transactive control methodology. Furthermore, the integration of LEM with the cyber-physical-society system and AI, multi-energy carriers, and social and human impacts is discussed.

1.1 Introduction

The electricity market is experiencing an unprecedented change toward a high liberalization direction driven by the fast and sustainable penetration of renewable energy resources (RES). Unlike power generation from the supply side, which has high costs and unavoidable environmental pollution, demand-side RESs have natural advantages. The generation of RESs can achieve zero-carbon emission and slight generation expenditure, which could be regarded as a promising and economical way to meet the electricity requirement of end-users [1, 2]. However, the substantial integration of RES into the electricity market has caused severe uncertainties and significant challenges in maintaining the real-time power balance [3]. Notably, such uncertainty will exacerbate the need for corrective dispatch actions at the real-time level. Achieving this goal requires not only tremendous costs but also overprovisioning measures to sustain the robustness of system operation. Under such circumstances, improving the flexibility of the power system comes as an imperative paradigm, replacing previous system overprovision.

M. Song et al., *Local Energy Markets*, https://doi.org/10.1007/978-981-97-9750-9_1

Another major challenge is that the traditional market is incompatible with the increasing presence of RES, which poses an urgent signal for active energy management. On the one hand, the power system is structured as a top–bottom hierarchy, where end-users, usually at the bottom position, are only forced to accept the price signal from the top layer [4]. The price signal cannot incentivize users to exploit their capacity potential to smooth fluctuations or serve the network grid. Besides, the ownership of various distributed energy resources (DERs) enables electricity consumers to become "prosumers" [5, 6] with the ability to generate and consume electricity power. Usually, prosumers have a proactive willingness to manage their energy source and share the flexibility for the safe operation of the distribution network. On the other hand, the significant challenge is the growing presence of RES and other energy resources in low-voltage grids, which requires active energy management at the network's edge. Ensuring the safe operation of the distribution network is crucial while integrating these DERs into the broader network energy management system. Tracking globally efficient solutions becomes computationally challenging in a system with a high penetration of small DERs. This complexity arises due to the large number of small resources, the increased uncertainty they introduce, and the non-convexities resulting from distribution network flow constraints and DER models.

The sharing economy is flourishing in various sectors, enabling small buyers and sellers to negotiate and trade goods or services with each other [7, 8]. E-commerce platforms like eBay allow individuals to share and sell second-hand goods online [9]. Airbnb, a service for booking accommodations, connects travelers with homeowners, offering vacant rooms, benefiting both parties economically [10]. Uber exemplifies the sharing economy in transportation by matching passengers with nearby drivers who can fulfill ride requests, facilitating a direct peer-to-peer (P2P) exchange that benefits passengers and car owners [11]. These sharing economy models have transformed traditional large-scale marketing paradigms by removing barriers and enabling small-scale sellers and buyers to connect and trade more directly. As a result, idle commodities and services are fully utilized, and the creativity and enthusiasm of small-scale entities are unleashed. This transformation leads to a novel economic model with greater transaction efficiency and higher productivity.

Given the above-ensured condition, prosumers with their distributed generation can autonomously trade energy to pursue self-interest maximization [12]. Local energy markets (LEMs) emphasize local production and consumption [13], addressing issues in current electricity markets by considering the local context [14]. Developing new business models for local power supply can create favorable conditions for small community generators [15]. Local energy trade enhances microgrids' autonomy by reducing their dependence on the main grid [16]. Additionally, LEM can benefit the overall grid [17]. Some grid investments can be avoided by utilizing local flexibility sources, as additional capacity would only be needed for a few hours annually. Distribution system operators (DSOs) can circumvent the need for extra grid capacity by locally contracting flexibility sources as part of their planning activities [18].

As a promising methodology, LEM offers significant advantages for optimally scheduling DERs to balance supply and demand in real time [19–21]. Beyond electricity, LEM can be applied in other energy fields (i.e., gas) and is closely intertwined with information/data technology and human behaviors. The future LEM system will evolve into a cyber-physical-society system, incorporating cross-disciplinary knowledge, including data science [22, 23], multi-carrier energy technology [24], and social science [25]. Compared to existing reviews on LEM [26–29], the contributions of this chapter are as follows:

(1) LEM paves the path toward the cyber-physical-society system, which is inevitable and promising. LEM is reviewed from the perspective of the cyber-physical-society system;
(2) As a market paradigm at the power distribution system level, this chapter compares LEM and demand response (DR), capturing and highlighting crucial characteristics of LEM, which gives readers a comprehensive overview of LEM;
(3) The chapter reviews transactive control, which uses market mechanisms to control loads in a distributed manner. This emerging technology for regulating loads on the demand side is seldom reviewed.

1.2 Basic Features of LEM

With the reconstruction of the power system, LEMs have the potential to provide a suitable and adaptable negotiation environment to integrate the small-sized electricity end-users into the energy supply system. Besides, the emergence of LEM facilitates energy transactions at a local level. It emphasizes direct transactions between prosumers and active control of DERs, resulting in various participants and market business models. Although there is no consensus regarding the definition of LEM, we compare LEM and traditional electricity markets, DR to highlight the key features of the LEM and help provide a big picture of LEM.

1.2.1 Participators in LEM

This section is focused on the LEM at the level of power distribution systems, and the typical LEM participants can be summarized as follows:

(1) **Prosumers** [22, 30, 31]: A prosumer combines consuming, producing, or energy storage at its premises and is treated as an end-user connected to distributed networks. It can engage the LEM to fully utilize its energy flexibility. Generally, these prosumers with several DERs become the most common participants in the LEM.
(2) **Utility company/Virtual power plants/Aggregators** [32–34]: These entities can aggregate DERs to buy/sell electricity in the whole market and act as an entity to trade with others in the LEM.

(3) **Utility network/System operator** [35, 36]: Electricity is delivered from the producers to customers via the network managed by the utility network/system operator. Network constraints are monitored to prevent violations, ensuring power systems' safe and reliable operation.

1.2.2 Comparisons of the LEM and DR

DR is described as "Changes in electric usage by end-use customers from their normal consumption patterns in response to changes in the price of electricity over time, or to incentive payments designed to induce lower electricity use at times of high wholesale market prices or when system reliability is jeopardized" [37]. There are two main types of DR: price-based and incentive-based. Price-based DR encourages customers to adjust their electricity consumption through mechanisms like real-time pricing (RTP) [38], time of use (TOU) [23], or critical peak pricing (CPP) [39]. Incentive-based DR includes several strategies, with direct load control (DLC) [40–42] being the most common. DLC directly manages customers' devices to shape their electricity profiles [43]. In DR programs, customers react to price/control signals from the grid, functioning as followers while the grid acts as the leader. Signals flow from the grid to customers without allowing feedback on customers' willingness or capacity. This setup limits the customers' ability to utilize their flexibility fully and does not help balance renewable energy generation fluctuations at the distribution system level.

On the other hand, LEM transforms this one-way leader–follower relationship into a two-way and P2P interaction [44–48]. In this model, the grid provides price signals indicating the current state of the power systems. At the same time, end-users can interact and adjust their power usage based on their operational behaviors. This model maximizes power adjustment flexibility and promotes local consumption of renewable energy generation.

The differences between LEM and DR are summarized in Table 1.1.

1.3 Framework of LEM

Three LEM structures can be identified based on the involvement of supervisory agents such as managers, aggregators, or operators: full P2P market, community-based market, and hybrid market [26, 49]. These structures differ as follows.

Table 1.1 Comparisons between DR and LEM

Type	DR	LEM
Participants	Flexible loads	Prosumers; smart devices (e.g., electric vehicles)
Main regulation methods	RTP/TOU/CPP-driven; direct load control	Transaction driven
Main control/scheduling methods	Centralized	Decentralized
Operation patterns	Leader (the grid) and followers (customers)	P2P and self-interest
Interaction	One way with bad effects	Two-way with good effects
Activeness	Bad	Good
Privacy	Good for price-based DR while bad for DLC-based DR	Good

1.3.1 Full P2P Market

In this structure, prosumers are the sole participants in the LEM, with no overseeing entity regulating market operations [28]. As shown in Fig. 1.1, prosumers engage in decentralized negotiations and trades, each striving to optimize their generation and consumption to maximize benefits from local resources, including renewable energy, flexible loads, and energy storage, after obtaining price information from other prosumers [50]. A transactive mechanism is introduced in [51] to facilitate the trading of schedule deviations between uncontrollable and elastic resources, aiming for maximal welfare in a distributed manner. Reference [52] employs a non-cooperative game to model the competitive trading behaviors between buildings, enabling direct energy exchanges under specific energy and balance constraints. This market structure ensures the privacy, and preferences of prosumers are maintained, as only price information is shared [27]. However, omitting power system network constraints can result in issues such as line overloading or voltage violations. To mitigate this, Ref. [53] integrates optimal power flow with a bilateral energy trading mechanism, addressing both economic and technical challenges in the LEM. The primary problem is divided into two subproblems, solved in a distributed manner to enhance market efficiency. The application of full P2P trading across the entire power system is complicated by the need for distributed algorithms and iterative processes, which can significantly increase computational complexity and reduce efficiency.

1.3.2 Community-Based Market

The community-based market structure can be categorized into centralized and semi-centralized architectures, depending on the level of direct interaction between

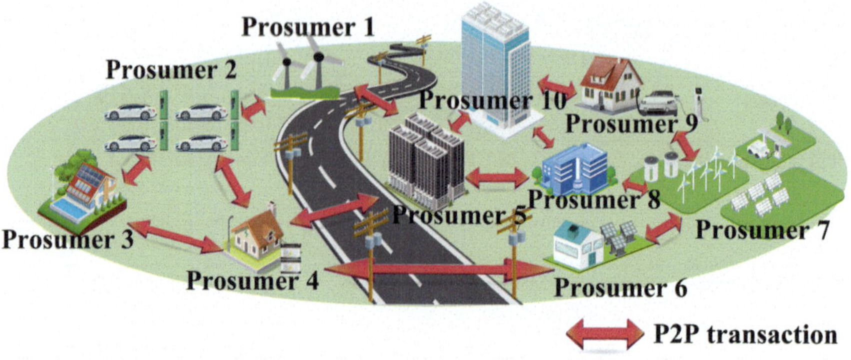

Fig. 1.1 Full P2P market structure

prosumers, as depicted in Fig. 1.2. In the centralized model, prosumers interact exclusively with the community manager without direct trading information exchange among themselves. The supervisory agent collects offers and bids from prosumers, manages the market, and clears transactions. For instance, Ref. [54] presents a modified auction-based LEM mechanism facilitating energy storage sharing between sellers and buyers within a residential community, where an auctioneer controls the process without information exchange between the participants.

In the semi-centralized model, prosumers can communicate with the supervisory agent and each other. For example, Ref. [55] describes a two-level framework where microgrids within the same distribution system trade with one another while adhering to network constraints. The trading outcomes at the microgrid level are submitted to the DSO, which ensures safe operation by reconfiguring the network. Reference [56] introduces a game-theoretic approach enabling direct interactions between buyers and sellers in a community microgrid, with support from the market operator.

By employing a supervisory agent, the trading process among prosumers is better managed, and connections to other markets or system operators are facilitated, allowing for quicker market clearing and more robust consideration of power system constraints. This structure enhances the power system's safe, efficient, and economical operation. However, the community-based P2P market is vulnerable to failures or attacks on the supervisory agent, highlighting the need for increased robustness in its design.

1.3.3 Hybrid Market Structure

This market type combines elements of the full P2P and community-based markets, as illustrated in Fig. 1.3. Structurally, it tends to be hierarchical by forming LEMs at different levels. For instance, in [50], a full P2P market is established to enable aggregators to trade with each other at an upper level, while a community-based market

(a) Centralized Community-based Market

(b) Semi-centralized Community-based Market

Fig. 1.2 Community-based market structure

manages the flexible loads. Similarly, Refs. [57, 58] propose a federated trading system via two-level P2P markets. At the same level, different types of peers can directly trade. The hybrid market offers several benefits by integrating the advantages of both full P2P and community-based markets. These include protecting customer privacy, enhancing peer cooperation, and improving scalability. The hybrid market model is considered the most promising LEM type compared to its predecessors [27]. It effectively combines the decentralized and privacy-preserving nature of full P2P markets with the organized and reliable structure of community-based markets, making it a robust and adaptable solution for modern energy trading.

A more detailed comparison of the three LEM organizations is summarized in Table 1.2.

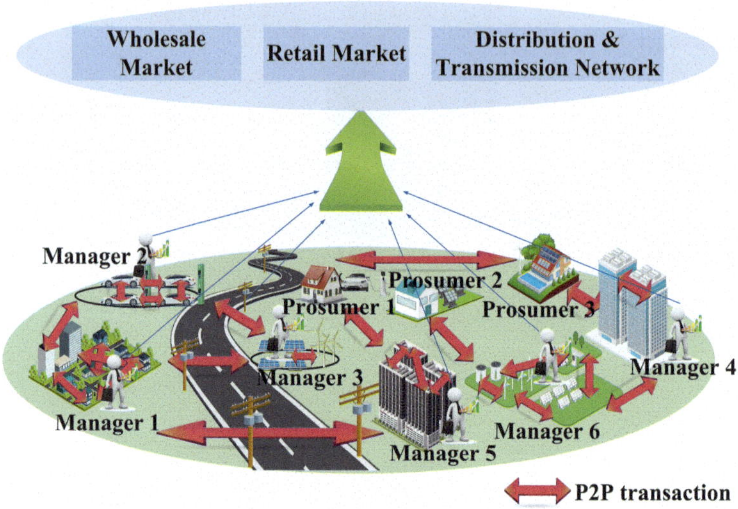

Fig. 1.3 Hybrid market structure

Table 1.2 Structure comparison for LEM

Framework	Full P2P market structure	Community-based market structure	Hybrid market structure
Control	Decentralized	Centralized	Centralized and decentralized
Flexibility	Low	Moderate	High
Social engagement	Very high	High	Moderate
Market	LEM	LEM/wholesale	LEM/ancillary/wholesale
Participants	End-users	End-users and local market operator	End-users, aggregators, DSO, and so on
Trading products	Energy	Energy and capacity	Energy, capacity, and ancillary service

1.4 Key Technologies

1.4.1 Modeling Technologies

1. **Game theory**

Game theory is employed to analyze the decision-making process of market players in a competitive surrounding. Regarding the relationship between market players, game theory is divided into several categories, which are introduced as follows.

(1) **Non-cooperative game**: Market participants often have conflicting interests, leading them to make independent decisions without direct communication. In this context, static and dynamic games, both forms of non-cooperative games, are typically utilized to formulate trading strategies. In static games, each player makes a single decision without considering time constraints, whereas in dynamic games, time plays a crucial role, and players engage in multiple actions [26]. The Stackelberg game, a well-known non-cooperative game, is particularly relevant in the LEM, as depicted in Fig. 1.4.

In the Stackelberg game, some participants act as leaders, while others are designated as followers. The leaders first make decisions and communicate them to the followers, who then optimize their actions in response. The leaders subsequently adjust their actions based on the followers' responses. This iterative

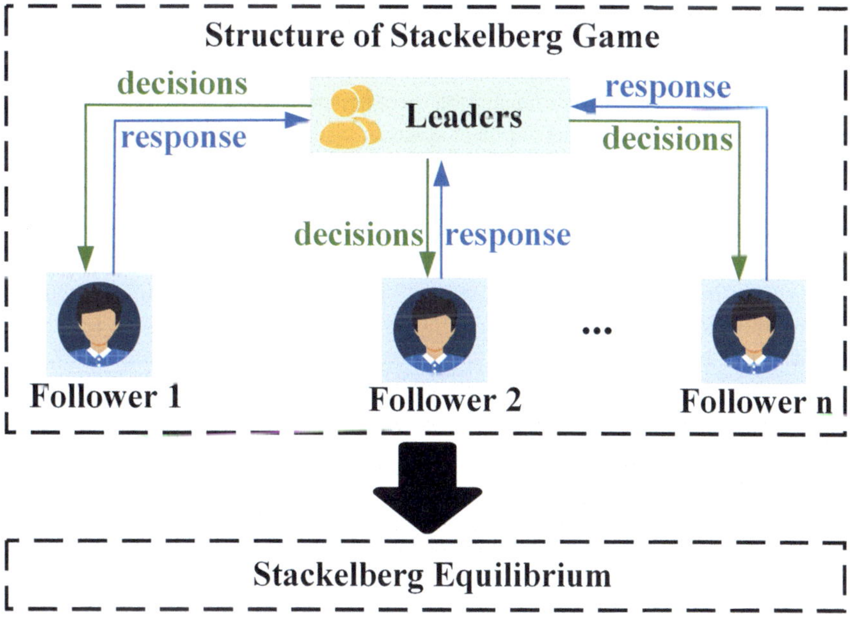

Fig. 1.4 Structure of Stackelberg game

process continues until a Stackelberg equilibrium is reached, wherein no player can improve their outcome by unilaterally altering their strategy. An example of this is demonstrated in Ref. [59], where the Stackelberg game determines the trading behaviors of sellers and buyers within a microgrid.

(2) **Cooperative game**: Market participants collaborate to enhance their benefits through incentives, utilizing the cooperative or coalitional game framework. This framework includes three main types: canonical coalition games, coalition formation games, and coalitional graph games [17]. The primary objective in canonical coalition games is to establish a stable grand coalition and distribute the resulting revenue equitably among participants [28]. Common methods for designing such distribution schemes include the Shapley value, the nucleolus, the Kernel, and the strong epsilon core [60]. In coalition formation games, the focus shifts to the network coalitional structure, adapting to environmental changes. Meanwhile, coalitional graph games aim to develop concise distributed algorithms for market participants with interconnected relationships, while also analyzing the properties of the network graph among players. For example, in the cooperative game context, an energy storage system is optimized to minimize coalitional energy costs as demonstrated in [61]. Additionally, Ref. [62] contrasts two different energy storage scenarios using cooperative game theory to identify a more effective approach for power systems.

2. **Double auction**

As shown in Fig. 1.5, in a double auction-based market, participants typically submit their energy volume and price to a third party for trading. The third party then ranks the bids and offers based on price, identifying the intersection point to clear the market. Various studies have examined the design of auction mechanisms within the LEM. For instance, Ref. [63] presents an auction mechanism for the P2P market that incorporates the physical constraints of power systems, with a procedure aimed at keeping power losses within acceptable limits. The auction mechanism is further analyzed in [64] using market supply and demand metrics. Additionally, Ref. [7] introduces an innovative optimization model based on a prediction-integration method within a continuous auction market to maximize operational efficiency and profits.

Given the high frequency of transactions and many prosumers in distribution systems, the LEM requires rapid market clearing within short time slots. In double auctions, distribution network operation constraints are generally overlooked, and no direct information exchange occurs between prosumers, making this approach suitable for LEMs with many participants. However, for the safe and reliable operation of distribution systems, the DSO must verify network constraints to prevent violations. This may lead to reassessing prosumer trading results, complicating the calculation of power loss and grid transit fees. Consequently, trading results from double auctions often serve as financial contracts rather than operational schedules for distribution systems.

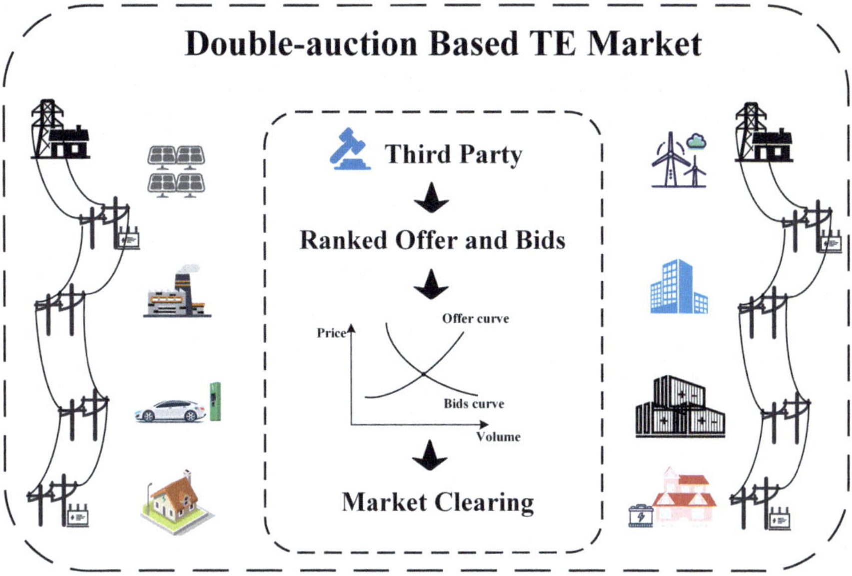

Fig. 1.5 Diagram of double auction

1.4.2 Solution Technologies

LEM clearing is an optimization problem that requires careful consideration of constraints related to the network structure of distribution systems, prosumer behavior, and electrical distances. Solving this problem centrally would compromise the privacy and preferences of market participants. To maintain privacy, the primary problem is decomposed into subproblems for each prosumer, enabling the application of distributed or consensus optimization algorithms. Among these, Lagrangian relaxation (LR), the alternating direction method of multipliers (ADMM), and consensus + innovation (CI) are prominent approaches [27].

(1) **LR**: In the context of LR, the optimization problem is transformed by relocating complex constraints into the objective function, thereby improving solution efficiency. Lagrangian multipliers, which act as prices, are iteratively exchanged between prosumers. The optimal solution is reached by reducing the gap between upper and lower bounds. LR is widely used in distributed algorithms for LEM clearing, as demonstrated in Ref. [50], which addresses private information leakage in a LEM involving aggregators. However, LR often suffers from slow convergence, necessitating enhancements in efficiency [13].

(2) **ADMM**: ADMM, a variant of the augmented Lagrangian method, is well-suited for addressing the complexities of LEM clearing through distributed algorithms. Reference [10] introduces the concept of "energy classes" to manage the heterogeneity of energy products, with ADMM handling the management

of these multi-class energy systems. The scalability of ADMM allows for integration with other methods to improve performance. For instance, Ref. [65] discusses a modified ADMM with an inner loop to alleviate the computational burden on prosumers, especially when dealing with many participants. Additionally, Ref. [15] proposes a consensus-based version of ADMM to enhance the computational efficiency of transactive energy trading.

(3) **CI**: CI distinguishes itself from other distributed algorithms by solving the first-order optimality conditions based on the Karush–Kuhn–Tucker framework iteratively [67]. CI safeguards prosumer privacy and enables the transparent operation of the LEM without the need for a centralized agent. Reference [53] presents a distributed multi-agent management approach utilizing CI, while a relaxed CI method is employed in Ref. [67] to address the multi-bilateral economic dispatch problem in a fully decentralized manner.

Despite the extensive use of decomposition techniques in the literature, the LEM remains challenging to clear swiftly due to the large number of prosumers and the iterative communication required. These factors hinder the timely scheduling of prosumers through market transactions in practical applications. Consequently, advancing decomposition methods that enable faster and more efficient LEM clearing is crucial.

1.4.3 Blockchain Technologies

With the advent of cryptocurrency technology and decentralization ideal [68], the traditional top-down vertical structure of the power system has been sharply shifted into a decentralized P2P transaction process for small-capacity prosumers without the participation of any centralized third party [69, 70]. To support effective data exchange and communication, the corresponding basic infrastructure is in the rapid development stage, which serves as the premise for realizing local energy trading. Of course, the decentralized platform blockchain can provide data storage and verification services to different trading kinds, like flexible load [71], demand response [72], and distributed energy. The main advantages of blockchain are containing data with time stamp storage, and decentralized authorization [73]. Besides, with the help of Internet of Things (IoT) technology, we can leverage the blockchain to construct the automatic execution transaction circumstances to promote local energy trading for different prosumers [74]. Furthermore, to defend the malicious cheating or attacks from agents, Refs. [74, 75] introduce a novel algorithm that combinates the blockchain and optimization method to promote the robustness and resilience of the system.

In addition, with the employment of smart contracts, blockchain can relate the trading components in LEM to achieve the desirable functionalities, like transparent public information, high-level privacy protection, and dynamic matching methods [10]. To illustrate different performance scenarios, various blockchain examples

Table 1.3 Some typical blockchain-enabled LEM

Reference	Blockchain platform	Trading commodity	Transactive mechanism	Key functions
[63]	Power ledger	Solar energy	Vickrey-Clarke-Groves (VCG)	Enable bidding strategy
[72]	DeTrade	Electricity	Optimization matching	Decentralized market clearing
[76]	Hyperledger fabric	Electricity, Demand response	Electron volt exchange	Smart contracts with multi-layer verification
[78]	Quorum	Electricity	Proposed decentralized trading	Bi-level cost minimization
[77]	Hyperledger fabric	Solar/wind energy	Agent coordination	Support for voltage regulation
[89]	Ethereum	Electricity	Multi-stakeholder optimization	Trading synchronization
[80]	MGMC blockchain	Microgrid energy	Upper-level and lower-level optimization	Trading adjustment and smart contract

are selected for their dominant application contexts. For instance, Hyperledger Fabric offers cross-platform access with a limited authorization process compared to Ethereum. According to Refs. [76, 77], utilizing Hyperledger for heterogeneous customers allows different agents to have diverse physical characteristics but similar trading logic, such as demand response or solar energy at different voltage levels. On the other hand, modern blockchain platforms like DeTrade and Quorum provide advanced functionalities, supporting more complex trading logic. These platforms incorporate optimization-oriented proof-of-work (PoW) or proof-of-stake (PoS) mechanisms, as discussed in [72, 78], and include carefully designed privacy protection features (Table 1.3).

1.4.4 Operation Frameworks

It is important to recognize that these techniques are not mutually exclusive; they are frequently combined to suit various scenarios and market structures. Table 1.4 presents LEM operation frameworks as discussed in representative literature. The Stackelberg game is commonly employed in centralized markets, with virtual power plants or aggregators serving as leaders and prosumers as followers. Decomposition techniques are often utilized to address privacy concerns. The cooperative game is another strategy applied in centralized markets. Additionally, blockchain technology can be integrated with any of the techniques mentioned above, serving as a ledger to enhance transparency and security.

Table 1.4 Typical operation frameworks of LEM

Reference	Market structure	Modeling technique	Solution technique	Blockchain
[10]	Semi-centralized community-based	Auction	ADMM	No
[50]	Hybrid market structure	Auction	LR	No
[81]	Full P2P market structure	Auction	Centralized optimization	No
[82]	Centralized community-based	Stackelberg game	Q-learning	No
[80]	Centralized community-based	Cooperative game	Centralized optimization	Yes
[75]	Centralized community-based	Auction	Proposed blockchain-based algorithm	Yes

1.5 Transactive Control

Transactive control, also referred to as market-based control, is defined as "a distributed control strategy that uses market mechanisms to engage self-interested responsive loads to provide services to the grid" [83]. This approach employs market prices to regulate end-users while preserving privacy and accommodating different preferences. Essentially, transactive control is used to implement LEM [84], requiring a platform for end-users to exchange price and quantity information. Transactive control primarily operates at the smart device level, which is this section's focus. These smart devices can bid based on their preferences and adjust their behaviors according to price changes. The main devices, market types, and time scales are introduced as follows.

(1) **Devices**: Heating, ventilation, and air conditioning (HVACs) [85–87], thermostatically controlled loads (TCLs) [43, 88], and EVs are the primary smart devices managed by transactive control. These devices are flexible loads that adjust power consumption based on price changes. As smart devices participate as entities in the LEM, their operating characteristics are fully considered. For HVACs and TCLs, internal temperature, affecting user comfort and satisfaction, is crucial for transactive control. Reference [88] proposes a market mechanism for coordinating TCLs, incorporating their internal temperatures. The relationship between internal temperatures and bidding prices is depicted in Fig. 1.6. If the internal temperature stays within the comfort range [$T_{\min}$, $T_{\max}$], the bidding price increases with rising temperature. Additionally, switch times are included in TCL bidding prices in [50] to reduce compressor strain. For EVs, arrival/departure times and EV battery constraints are considered. Reference [89] designs a LEM to coordinate EVs, mitigating their negative impact on

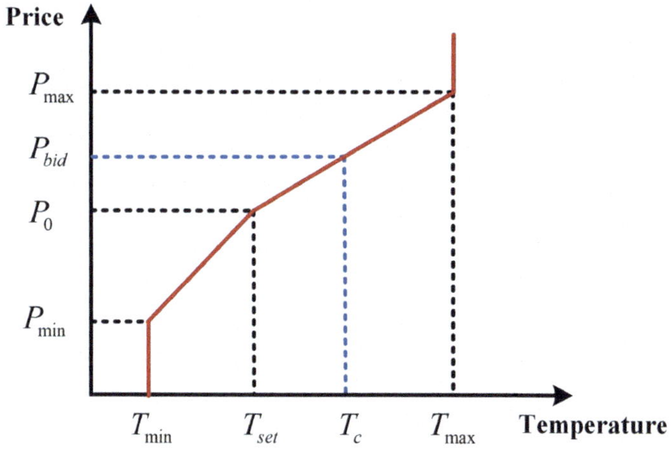

Fig. 1.6 Relationship between the bidding price and internal temperature

power systems and ensuring the desired state of charge (SoC) levels at departure time.

(2) **Market types**: Transactive control allows smart devices to act according to the price in either real or virtual LEM. In a real LEM, smart devices engage in financial transactions, receiving or paying money to other participants. This integration of economic benefits with technical functions is the focus of most research. In a virtual LEM, however, devices respond to price signals without actual financial transactions, using the price only to guide user behavior. Reference [84] presents a virtual LEM structure involving commercial HVACs for power balance. Each HVAC component participates as an entity in this virtual market, taking actions based on the clearing price to optimize overall HVAC operation. A similar virtual market concept is explored in [50].

(3) **Time scales**: DR includes price-based and DLC-based DR, as illustrated in Fig. 1.7. Price-based DR requires predicting end-user power consumption and analyzing their price sensitivity to set appropriate prices that guide behavior. Due to the difficulty in accurately quantifying and forecasting user habits and behaviors, price-based DR generally lacks precise power or time resolution and is suited for shaping consumption over hours or longer periods. DLC-based DR, managed by aggregators, virtual power plants, or operators, controls end-user power consumption directly, offering high power and time resolution suitable for minute- or second-scale services. However, this approach can compromise user privacy and autonomy. Transactive control merges the advantages of both price-based and DLC-based DR. Users develop bid/offer strategies and respond to prices based on their operating characteristics in the LEM, achieving high power and time resolution while maintaining privacy and autonomy. Typically, transactive control enables hour- and minute-scale services, respecting each user's privacy and free will.

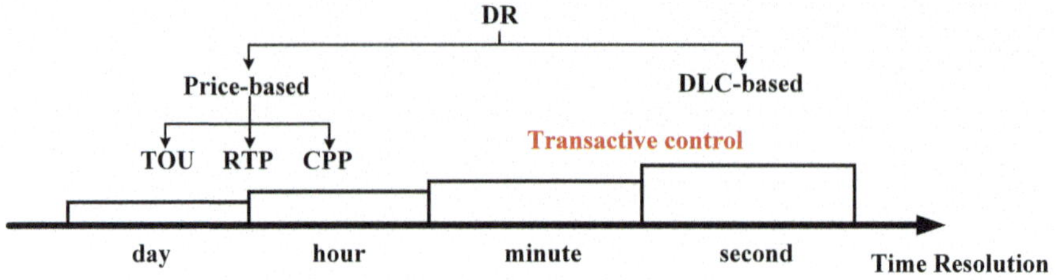

Fig. 1.7 Different time scales of DR and transactive control

1.6 Existing Pilot Projects of LEM

Numerous engineering explorations and practical implementations of LEM based on P2P transactions have been conducted, as summarized in Table 1.5. For instance, the Rotterdam Innovation Dock in the Netherlands developed the Distro power trading platform, leveraging artificial intelligence (AI) and blockchain technology. This platform disseminated dynamic market signals for demand and pricing, achieving a supply–demand balance. Remarkably, the project facilitated 20 million transactions within two months, significantly lowering energy costs and promoting the local consumption of renewable energy [90]. In the USA, the Pacific Northwest Smart Grid Demonstration Project (PNW-SGDP) established a LEM involving 60,000 users across five states. This project utilized locational marginal pricing (LMP) to implement distributed power management and demand response, effectively aligning grid supply with demand [91]. Similarly, the Shekou Energy Blockchain Project in China involved 100 community volunteers in green power virtual transactions. Smart contracts, automatically generated via blockchain technology, enabled direct P2P transactions between users [93].

Despite these advancements, participation in LEM projects remains limited, as noted in Table V. As the integration of DERs with power distribution systems increases, challenges such as limited data storage capacity and network latency are likely to impede the accommodation of a large number of DERs. Moreover, most LEM applications are currently in the demonstration phase. Therefore, substantial efforts are required to develop appropriate regulatory policies and market mechanisms that ensure the operational viability of distribution networks while safeguarding the interests of all stakeholders.

Table 1.5 Existing projects of LEM

Project name	Region	Time	Project content	Superiority	Shortcoming
Rotterdam Innovation Dock [90]	Netherlands	2020	Tool of trading agency, P2P trading platform	Renewable energy consumption, delaying grid investment	Only used in one port with a small scale
PNW-SGDP [91]	US	2012	Distributed generation management, demand response	On a large scale, Supply and demand matching mechanism based on LMP	No local transactive energy market between prosumers
TransActive Grid [92]	US	2016	P2P trading platform, transaction in microgrid	Smart contracts based on blockchain technology	Only used in microgrids with a small scale
Sun Exchange [85]	South Africa	2020	Rental of photovoltaic equipment, Sale of photovoltaic power	Innovation in business model, high return on investment	No local transactive energy market between prosumers
Shekou Energy Blockchain [93]	China	2018	P2P trading between the power station and users	Smart contracts based on blockchain technology	Only 100 users with a small scale
Scanergy [86, 87]	EU	2016	P2P trading platform, Transactive market between DSOs, suppliers, and users	Use highly trusted virtual currency for transactions	Regulatory policies are not yet clear
Share & Energy [94]	Germany	2017	Shared and P2P platform for private charging piles	High charging pile utilization	Regulatory policies are not yet clear
De La Salle University Project [87]	Philippines	2018	P2P trading platform, Transaction in school microgrid	High reliability of campus microgrid, Clean electricity	Only used in one school with a small scale

1.7 Discussions

1.7.1 AI Technology in LEM

LEM integrated with AI and data-driven methods can be designed for fully autonomous and intelligent operation, minimizing the need for human intervention. Ideally, every smart device within a LEM and energy ecosystem can autonomously adjust its energy consumption in response to internal or external price signals, seamlessly connecting low-level kW control mechanisms with high-level MW system requirements [95]. Furthermore, data generated across the distribution network (e.g., SCADA, micro-PMU) and on the end-user side (e.g., EMS, smart meters) can be harnessed for instant, localized analysis using advanced AI techniques, such as deep learning, deep reinforcement learning, and federated learning [96].

Several factors are driving the integration of LEM systems with AI. One key driver is the vast amount of operational and trading data collected from LEM systems and customers, enabling real-time, data-driven analysis to monitor LEM modules with closed-loop feedback. Another critical factor is the need to manage the simultaneous decision-making of numerous customers, each with limited knowledge of the overall LEM system and trading processes. AI-based trading techniques can automate decision-making, freeing customers from manual processes without requiring them to exchange or rely on external information.

A notable trend is the shift from focusing primarily on estimation or identification tasks, akin to traditional pattern recognition, to employing reinforcement learning (RL) and deep reinforcement learning (DRL) to tackle complex decision-making challenges more directly related to LEM system operations. These techniques, now prevalent in the latest LEM and P2P local energy trading research [97], leverage RL and DRL's efficient data sampling capabilities to address the non-convex energy management problems [98]. This trend suggests that AI advancements could significantly enhance LEM systems by simulating real-world trading behaviors, potentially incorporating psychological factors.

1.7.2 LEM Integrated with Multi-energy System

The diversification of energy forms—such as electricity, heat, and cold—presents new challenges for balancing energy supply and demand [99]. Current research into multi-energy integration with LEM explores various models, including energy hubs [100], which integrate generation and consumption as a unified entity for energy management, and multi-energy networks connected by energy conversion devices [101]. In multi-energy systems, LEM can function as both a provider and consumer, ensuring optimal energy allocation and network balance.

Research in this area often focuses on decentralized models that allow independent market member decisions while maintaining social welfare. However, the diversity of equipment in multi-energy markets and the differing characteristics of energy transmission infrastructure create complex variables in market-clearing models, leading to significant computational demands and efficiency issues. Future research could focus on improving solution speeds through methods such as convexification and linearization of complex energy network models [102].

1.7.3 Impacts of Society and Humans on the LEM

Traditional economic models often assume that market participants are perfectly rational. However, behavioral economics suggests that decision biases can influence traders, particularly under risk [103]. In LEM research, human behavior is typically modeled based on degrees of rationality, with prospect theory being a prominent framework [104]. This theory can optimize prosumer bidding strategies in distributed energy trading [105] and mitigate market price uncertainty [106].

Additionally, social factors significantly influence the market, particularly through government regulation of market transactions [107]. However, modeling LEM markets with social attributes remains challenging due to inherent subjectivity. While these models can simulate human-like decision-making, they often fall short of predicting actual trader preferences and market outcomes. Therefore, future research could focus on incorporating machine learning, big data interpolations, and data-driven approaches to develop more realistic predictions and optimizations of market transactions.

1.7.4 Future LEM

The future version of LEM is depicted in Fig. 1.8. The advanced cyber system equips the energy system with extensive data on energy consumption and related parameters, facilitating transactions in the LEM and informing the bidding strategies of prosumers. Future LEM will extend beyond electricity to include multiple types of energy, allowing prosumers in the distribution network to trade electricity, gas, heat, and other energy forms. Additionally, the integration of LEM mechanisms with social networks will enable optimization of the energy network's operation from a societal perspective. Understanding human energy consumption and trading preferences will aid in designing suitable LEM mechanisms and formulating effective energy policies. This approach will enable the provision of personalized and customized energy products. Ultimately, a coordinated operation among the cyber system, energy system, and social network will be achieved, enhancing the overall efficiency and effectiveness of the LEM.

1.8 Conclusions

This section reviews the state of the art of LEM conceptual development in power distribution systems, encompassing fundamental features of LEM, the structure of the LEM, key techniques, and transactive control. LEM, as a promising technology, integrates AI, energy systems, and social behaviors, paving the way toward a cyber-physical-society system. LEM shifts the modern power system from a grid-centric

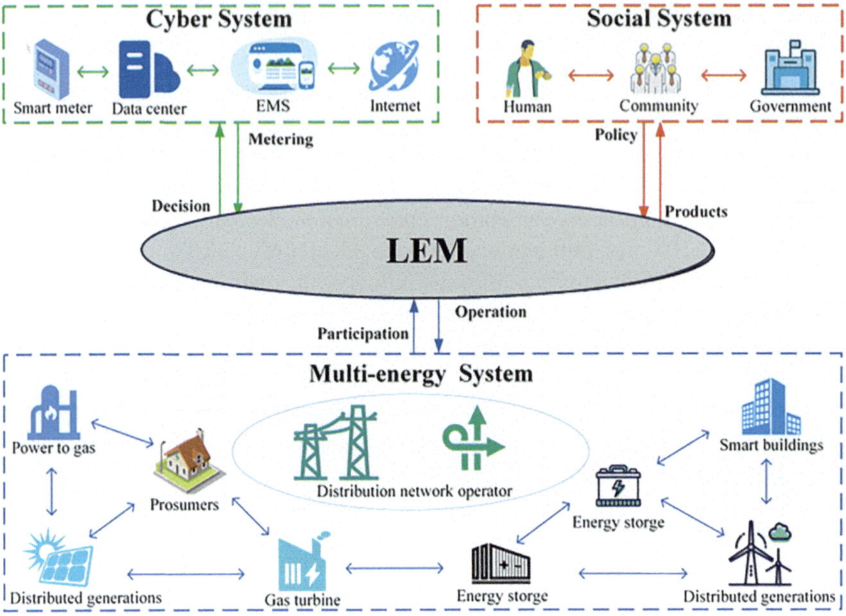

Fig. 1.8 Diagram of future LEM (cyber-physical-society)

to a customer-centric paradigm and has garnered increasing attention in recent years. By enabling end-users to access and respond to price signals directly, LEM stimulates customer engagement and provides greater flexibility to manage power fluctuations caused by the rise of DERs.

The future trend of LEM is the evolution toward a cyber-physical-society system, where AI empowers LEM to make informed and automated decisions, enhancing flexibility and speed in various energy transactions. Additionally, the accurate coupling of social and human activities with actual LEM operations will further refine and optimize the system's performance.

References

1. Bedoya, J. C., Ostadijafari, M., Liu, C. C., & Dubey, A. (2020). Decentralized transactive energy for flexible resources in distribution systems. *IEEE Transactions on Sustainable Energy, 12*(2), 1009–1019.
2. Chen, M., Gao, C., Song, M., Chen, S., Li, D., & Liu, Q. (2020). Internet data centers participating in demand response: A comprehensive review. *Renewable and Sustainable Energy Reviews, 117*, 109466.
3. Song, M., Gao, C., Yan, H., & Yang, J. (2017). Thermal battery modeling of inverter air conditioning for demand response. *IEEE Transactions on Smart Grid, 9*(6), 5522–5534.
4. Rahimi, F. A., & Ipakchi, A. (2012). Transactive energy techniques: Closing the gap between wholesale and retail markets. *The Electricity Journal, 25*(8), 29–35.

5. Cui, S., Wang, Y. W., Li, C., & Xiao, J. W. (2019). Prosumer community: A risk aversion energy sharing model. *IEEE Transactions on Sustainable Energy, 11*(2), 828–838.
6. Chen, K., Lin, J., & Song, Y. (2019). Trading strategy optimization for a prosumer in the continuous double auction-based peer-to-peer market: A prediction-integration model. *Applied Energy, 242*, 1121–1133.
7. Zhang, J., Choi, T. M., & Cai, Y. J. (2021). Consumer-to-consumer product trading with strategic consumer behaviors in the sharing economy. *IEEE Transactions on Engineering Management, 70*(5), 1708–1723.
8. Einav, L., Farronato, C., & Levin, J. (2016). Peer-to-peer markets. *Annual Review of Economics, 8*(1), 615–635.
9. Morstyn, T., & McCulloch, M. D. (2018). Multiclass energy management for peer-to-peer energy trading driven by prosumer preferences. *IEEE Transactions on Power Systems, 34*(5), 4005–4014.
10. Zekanovic-Korona, L., & Grzunov, J. (2014). Evaluation of shared digital economy adoption: Case of Airbnb. In *2014 37th International Convention on Information and Communication Technology, Electronics and Microelectronics* (pp. 1574–1579).
11. Rocky Mountain Institute. (2014). *An Airbnb or Uber for the electricity grid*. RMI Outlet.
12. Mendes, G., Nylund, J., Annala, S., Honkapuro, S., Kilkki, O., & Segerstam, J. (2018). Local energy markets: Opportunities, benefits, and barriers.
13. Lüth, A., Zepter, J. M., Del Granado, P. C., & Egging, R. (2018). Local electricity market designs for peer-to-peer trading: The role of battery flexibility. *Applied Energy, 229*, 1233–1243.
14. Teotia, F., & Bhakar, R. (2016). Local energy markets: Concept, design and operation. In *2016 National Power Systems Conference* (pp. 1–6).
15. Hall, S., & Roelich, K. E. (2015). Local electricity supply: Opportunities, archetypes and outcomes.
16. Mustafa, M. A., Cleemput, S., & Abidin, A. (2016). A local electricity trading market: Security analysis. In *2016 IEEE PES Innovative Smart Grid Technologies Conference Europe* (pp. 1–6).
17. Saad, W., Han, Z., Poor, H. V., & Basar, T. (2012). Game-theoretic methods for the smart grid: An overview of microgrid systems, demand-side management, and smart grid communications. *IEEE Signal Processing Magazine, 29*(5), 86–105.
18. Ramos, A., De Jonghe, C., Gómez, V., & Belmans, R. (2016). Realizing the smart grid's potential: Defining local markets for flexibility. *Utilities Policy, 40*, 26–35.
19. Rahimi, F., Ipakchi, A., & Fletcher, F. (2016). The changing electrical landscape: End-to-end power system operation under the transactive energy paradigm. *IEEE Power and Energy Magazine, 14*(3), 52–62.
20. Ambrosio, R. (2016). Transactive energy systems. *IEEE Electrification Magazine, 4*(4), 4–7.
21. Melton, R., & Fuller, J. (2016). Transactive energy: Envisioning the future. *IEEE Electrification Magazine, 4*(4), 2–3.
22. Kok, K., & Widergren, S. (2016). A society of devices: Integrating intelligent distributed resources with transactive energy. *IEEE Power and Energy Magazine, 14*(3), 34–45.
23. Lu, T., Wang, Z., Wang, J., Ai, Q., & Wang, C. (2018). A data-driven Stackelberg market strategy for demand response-enabled distribution systems. *IEEE Transactions on Smart Grid, 10*(3), 2345–2357.
24. Yang, Z., Hu, J., Ai, X., Wu, J., & Yang, G. (2020). Transactive energy supported economic operation for multi-energy complementary microgrids. *IEEE Transactions on Smart Grid, 12*(1), 4–17.
25. Chen, L., Liu, N., Li, C., & Wang, J. (2020). Peer-to-peer energy sharing with social attributes: A stochastic leader–follower game approach. *IEEE Transactions on Industrial Informatics, 17*(4), 2545–2556.
26. Tushar, W., Saha, T. K., Yuen, C., Smith, D., & Poor, H. V. (2020). Peer-to-peer trading in electricity networks: An overview. *IEEE Transactions on Smart Grid, 11*(4), 3185–3200.
27. Sousa, T., Soares, T., Pinson, P., Moret, F., Baroche, T., & Sorin, E. (2019). Peer-to-peer and community-based markets: A comprehensive review. *Renewable and Sustainable Energy Reviews, 104*, 367–378.

28. Adeyemi, A., Yan, M., Shahidehpour, M., Bahramirad, S., & Paaso, A. (2020). Transactive energy markets for managing energy exchanges in power distribution systems. *The Electricity Journal, 33*(9), 106868.
29. Zhang, C., Wu, J., Long, C., & Cheng, M. (2017). Review of existing peer-to-peer energy trading projects. *Energy Procedia, 105*, 2563–2568.
30. Pouttu, A., et al. (2017). P2P model for distributed energy trading, grid control and ICT for local smart grids. In *EuCNC 2017—European Conference on Networks and Communications*, July 2017, no. EuCNC 2017-Eur. Conf. Networks Commun. https://doi.org/10.1109/EuCNC.2017.7980652
31. Lee, J., Guo, J., Choi, J. K., & Zukerman, M. (2015). Distributed energy trading in microgrids: A game-theoretic model and its equilibrium analysis. *IEEE Transactions on Industrial Electronics, 62*(6), 3524–3533. https://doi.org/10.1109/TIE.2014.2387340
32. Pan, Z., et al. (2020). Stochastic transactive control for electric vehicle aggregators coordination: A decentralized approximate dynamic programming approach. *IEEE Transactions on Smart Grid, 11*(5), 4261–4277. https://doi.org/10.1109/TSG.2020.2992863
33. Qiu, J., Meng, K., Zheng, Y., & Dong, Z. Y. (2017). Optimal scheduling of distributed energy resources as a virtual power plant in a transactive energy framework. *IET Generation, Transmission & Distribution, 11*(13), 3417 [Online]. Available: https://search.ebscohost.com/login.aspx?direct=true&db=edsbl&AN=vdc.100114551190.0x000001&lang=zh-cn&site=eds-live
34. Qiu, J., Zhao, J., Yang, H., & Dong, Z. Y. (2018). Optimal scheduling for prosumers in coupled transactive power and gas systems. *IEEE Transactions on Power Systems, 33*(2), 1970–1980. https://doi.org/10.1109/TPWRS.2017.2715983
35. Renani, Y. K., Ehsan, M., & Shahidehpour, M. (2018). Optimal transactive market operations with distribution system operators. *IEEE Transactions on Smart Grid, 9*(6), 6692–6701. https://doi.org/10.1109/TSG.2017.2718546
36. Chen, S., et al. (2018). Forming bidding curves for a distribution system operator. *IEEE Transactions on Power Systems, 33*(5), 5389–5400. https://doi.org/10.1109/TPWRS.2018.2821673
37. Federal Energy Regulatory Commission. (2007). 09-07-Demand-Response. Pdf (Application/Pdf Objeto). [Online]. Available: https://www.ferc.gov/sites/default/files/2020-04/09-07-demand-response.pdf
38. Yoon, J. H., Baldick, R., & Novoselac, A. (2014). Dynamic demand response controller based on real-time retail price for residential buildings. *IEEE Transactions on Smart Grid, 5*(1), 121–129.
39. Somasundaram, S., Pratt, R. G., Akyol, B., Fernandez, N., Foster, N., Katipamula, S., & Taylor, Z. (2014). *Reference guide for a transaction-based building controls framework*. Pacific Northwest National Laboratory.
40. Song, M., Gao, C., Shahidehpour, M., Li, Z., Yang, J., & Yan, H. (2018). State space modeling and control of aggregated TCLs for regulation services in power grids. *IEEE Transactions on Smart Grid, 10*(4), 4095–4106.
41. Haring, T. W., Mathieu, J. L., & Andersson, G. (2015). Comparing centralized and decentralized contract design enabling direct load control for reserves. *IEEE Transactions on Power Systems, 31*(3), 2044–2054.
42. Ruiz, N., Cobelo, I., & Oyarzabal, J. (2009). A direct load control model for virtual power plant management. *IEEE Transactions on Power Systems, 24*(2), 959–966.
43. Song, M., Ciwei, G., Yang, J., Liu, Y., & Cui, G. (2017). Novel aggregate control model of air conditioning loads for fast regulation service. *IET Generation, Transmission & Distribution, 11*(17), 4391–4401.
44. Liu, Y., Wu, L., & Li, J. (2019). Peer-to-peer (P2P) electricity trading in distribution systems of the future. *The Electricity Journal, 32*(4), 2–6.
45. Long, C., Wu, J., Zhou, Y., & Jenkins, N. (2018). Peer-to-peer energy sharing through a two-stage aggregated battery control in a community microgrid. *Applied Energy, 226*, 261–276.

46. Paudel, A., Sampath, L. P. M. I., Yang, J., & Gooi, H. B. (2020). Peer-to-peer energy trading in smart grid considering power losses and network fees. *IEEE Transactions on Smart Grid, 11*(6), 4727–4737.
47. Guo, Z., Pinson, P., Chen, S., Yang, Q., & Yang, Z. (2020). Chance-constrained peer-to-peer joint energy and reserve market considering renewable generation uncertainty. *IEEE Transactions on Smart Grid, 12*(1), 798–809.
48. Zhang, Z., Li, R., & Li, F. (2019). A novel peer-to-peer local electricity market for joint trading of energy and uncertainty. *IEEE Transactions on Smart Grid, 11*(2), 1205–1215.
49. Azizi, A., Aminifar, F., Moeini-Aghtaie, M., & Alizadeh, A. (2020). Transactive energy market mechanism with loss implication. *IEEE Transactions on Smart Grid, 12*(2), 1215–1223.
50. Song, M., Sun, W., Wang, Y., Shahidehpour, M., Li, Z., & Gao, C. (2019). Hierarchical scheduling of aggregated TCL flexibility for transactive energy in power systems. *IEEE Transactions on Smart Grid, 11*(3), 2452–2463.
51. Le, X., Chen, S., Yan, Z., Ping, J., Zhao, X., Zhang, N., & Xi, J. (2018). Enabling a transactive distribution system via real-time distributed optimization. *IEEE Transactions on Smart Grid, 10*(5), 4907–4917.
52. Cui, S., Wang, Y. W., Shi, Y., & Xiao, J. W. (2020). A new and fair peer-to-peer energy sharing framework for energy buildings. *IEEE Transactions on Smart Grid, 11*(5), 3817–3826.
53. Moret, F., & Pinson, P. (2018). Energy collectives: A community and fairness based approach to future electricity markets. *IEEE Transactions on Power Systems, 34*(5), 3994–4004.
54. Tao, Z., & Chen, J. (2015). *Energy storage technologies in smart grid.*
55. Yan, M., Shahidehpour, M., Paaso, A., Zhang, L., Alabdulwahab, A., & Abusorrah, A. (2020). Distribution network-constrained optimization of peer-to-peer transactive energy trading among multi-microgrids. *IEEE Transactions on Smart Grid, 12*(2), 1033–1047.
56. Long, C., Wu, J., Zhang, C., Thomas, L., Cheng, M., & Jenkins, N. (2017). Peer-to-peer energy trading in a community microgrid. In *2017 IEEE Power & Energy Society General Meeting* (pp. 1–5). IEEE.
57. Petri, I., Alzahrani, A., Reynolds, J., & Rezgui, Y. (2020). Federating smart cluster energy grids for peer-to-peer energy sharing and trading. *IEEE Access, 8*, 102419–102435.
58. Morstyn, T., Farrell, N., Darby, S. J., & McCulloch, M. D. (2018). Using peer-to-peer energy-trading platforms to incentivize prosumers to form federated power plants. *Nature Energy, 3*(2), 94–101.
59. Anoh, K., Maharjan, S., Ikpehai, A., Zhang, Y., & Adebisi, B. (2019). Energy peer-to-peer trading in virtual microgrids in smart grids: A game-theoretic approach. *IEEE Transactions on Smart Grid, 11*(2), 1264–1275.
60. Chen, Y., Liu, F., Wei, W. E. I., & Mei, S. (2021). Energy sharing at demand side: Concept, mechanism and prospect. *Automation of Electric Power Systems, 45*(2), 1–11.
61. Han, L., Morstyn, T., & McCulloch, M. (2018). Incentivizing prosumer coalitions with energy management using cooperative game theory. *IEEE Transactions on Power Systems, 34*(1), 303–313.
62. Chakraborty, P., Baeyens, E., Poolla, K., Khargonekar, P. P., & Varaiya, P. (2018). Sharing storage in a smart grid: A coalitional game approach. *IEEE Transactions on Smart Grid, 10*(4), 4379–4390.
63. Leong, C. H., Gu, C., & Li, F. (2019). Auction mechanism for P2P local energy trading considering physical constraints. *Energy Procedia, 158*, 6613–6618.
64. Lin, J., Pipattanasomporn, M., & Rahman, S. (2019). Comparative analysis of auction mechanisms and bidding strategies for P2P solar transactive energy markets. *Applied Energy, 255*, 113687.
65. Cui, S., Wang, Y. W., Shi, Y., & Xiao, J. W. (2019). An efficient peer-to-peer energy-sharing framework for numerous community prosumers. *IEEE Transactions on Industrial Informatics, 16*(12), 7402–7412.
66. Hug, G., Kar, S., & Wu, C. (2015). Consensus+ innovations approach for distributed multiagent coordination in a microgrid. *IEEE Transactions on Smart Grid, 6*(4), 1893–1903.

67. Sorin, E., Bobo, L., & Pinson, P. (2018). Consensus-based approach to peer-to-peer electricity markets with product differentiation. *IEEE Transactions on Power Systems, 34*(2), 994–1004.
68. Narayanan, A., Bonneau, J., Felten, E., Miller, A., & Goldfeder, S. (2016). *Bitcoin and cryptocurrency technologies: A comprehensive introduction*. Princeton University Press.
69. Su, W. (2014). The role of customers in the us electricity market: Past, present and future. *The Electricity Journal, 27*(7), 112–125.
70. Chen, T., Alsafasfeh, Q., Pourbabak, H., & Su, W. (2017). The next-generation US retail electricity market with customers and prosumers—A bibliographical survey. *Energies, 11*(1), 8.
71. Li, M., Hu, D., Lal, C., Conti, M., & Zhang, Z. (2020). Blockchain-enabled secure energy trading with verifiable fairness in industrial Internet of Things. *IEEE Transactions on Industrial Informatics, 16*(10), 6564–6574.
72. Esmat, A., de Vos, M., Ghiassi-Farrokhfal, Y., Palensky, P., & Epema, D. (2021). A novel decentralized platform for peer-to-peer energy trading market with blockchain technology. *Applied Energy, 282*, 116123.
73. Zhao, C., Han, D., Li, C., & Wang, H. (2024). A blockchain consensus mechanism to optimize reputation-based distributed energy trading in urban energy system. *IEEE Access, 12*, 53698–53712. https://doi.org/10.1109/ACCESS.2024.3387715
74. Chen, S., Shen, Z., Zhang, L., Yan, Z., Li, C., Zhang, N., & Wu, J. (2021). A trusted energy trading framework by marrying blockchain and optimization. *Advances in Applied Energy, 2*, 100029.
75. Chen, S., Zhang, L., Yan, Z., & Shen, Z. (2021). A distributed and robust security-constrained economic dispatch algorithm based on blockchain. *IEEE Transactions on Power Systems, 37*(1), 691–700.
76. Saha, S., Ravi, N., Hreinsson, K., Baek, J., Scaglione, A., & Johnson, N. G. (2021). A secure distributed ledger for transactive energy: The Electron Volt Exchange (EVE) blockchain. *Applied Energy, 282*, 116208.
77. Saxena, S., Farag, H. E., Turesson, H., & Kim, H. (2020). Blockchain based transactive energy systems for voltage regulation in active distribution networks. *IET Smart Grid, 3*(5), 646–656.
78. Yang, Q., & Wang, H. (2020). Blockchain-empowered socially optimal transactive energy system: Framework and implementation. *IEEE Transactions on Industrial Informatics, 17*(5), 3122–3132.
79. Eisele, S., Laszka, A., Schmidt, D. C., & Dubey, A. (2020). The role of blockchains in multi-stakeholder transactive energy systems. *Frontiers in Blockchain, 3*, 593471.
80. Shahidehpour, M., Yan, M., Shikhar, P., Bahramirad, S., & Paaso, A. (2020). Blockchain for peer-to-peer transactive energy trading in networked microgrids: Providing an effective and decentralized strategy. *IEEE Electrification Magazine, 8*(4), 80–90.
81. Haggi, H., & Sun, W. (2021). Multi-round double auction-enabled peer-to-peer energy exchange in active distribution networks. *IEEE Transactions on Smart Grid, 12*(5), 4403–4414.
82. Xu, X., Xu, Y., Wang, M. H., Li, J., Xu, Z., Chai, S., & He, Y. (2020). Data-driven game-based pricing for sharing rooftop photovoltaic generation and energy storage in the residential building cluster under uncertainties. *IEEE Transactions on Industrial Informatics, 17*(7), 4480–4491.
83. Hao, H., Corbin, C. D., Kalsi, K., & Pratt, R. G. (2016). Transactive control of commercial buildings for demand response. *IEEE Transactions on Power Systems, 32*(1), 774–783.
84. Hu, J., Yang, G., Kok, K., Xue, Y., & Bindner, H. W. (2017). Transactive control: A framework for operating power systems characterized by high penetration of distributed energy resources. *Journal of Modern Power Systems and Clean Energy, 5*(3), 451–464.
85. Sun Exchange. Earn with purpose: Offset your carbon footprint selling solar energy to the world. [Online]. Available: https://thesunexchange.com/
86. Mihaylov, M., Jurado, S., Avellana, N., Van Moffaert, K., de Abril, I. M., & Nowé, A. (2014, May). NRGcoin: Virtual currency for trading of renewable energy in smart grids. In *11th International Conference on the European Energy Market (EEM14)* (pp. 1–6). IEEE.

87. Mihaylov, M., Jurado, S., Avellana, N., Razo-Zapata, I. S., Van Moffaert, K., Arco, L., & Nowé, A. (2015, May). SCANERGY: A scalable and modular system for energy trading between prosumers. In *AAMAS* (Vol. 15, pp. 1917–1918).
88. News Desk. (2018). Energo boost clean energy production through Qtum Blockchain implementation in Philippines. [Online]. Available: https://www.geospatialworld.net/news/energo-clean-energy-production-qtum-blockchain/
89. Saber, H., Ehsan, M., Moeini-Aghtaie, M., Fotuhi-Firuzabad, M., & Lehtonen, M. (2020). Network-constrained transactive coordination for plug-in electric vehicles participation in real-time retail electricity markets. *IEEE Transactions on Sustainable Energy, 12*(2), 1439–1448.
90. S&P Global Platts. (2020). *World's first high-frequency decentralized energy market helps drive Port of Rotterdam's energy transition*. https://www.spglobal.com/platts/en/about-platts/media-center/press-releases/2020/051020-world-s-first-high-frequency-decentralized-energy-market-driveport-of-rotterdam-energy-transition.
91. Huang, P., Kalagnanam, J., Natarajan, R., Hammerstrom, D., Melton, R., Sharma, M., & Ambrosio, R. (2010, October). Analytics and transactive control design for the Pacific Northwest smart grid demonstration project. In *2010 First IEEE International Conference on Smart Grid Communications* (pp. 449–454). IEEE.
92. Power Technology. (2017). The Brooklyn microgrid: Blockchain-enabled community power. [Online]. Available: https://www.power-technology.com/features/featurethe-brooklyn-microgrid-blockchain-enabled-community-power-5783564/
93. Zhu, S., Song, M., Lim, M. K., Wang, J., & Zhao, J. (2020). The development of energy blockchain and its implications for China's energy sector. *Resources Policy, 66*, 101595.
94. Share&Charge. Open charging network: The next level of OCPI-based e-roaming. [Online]. Available: https://shareandcharge.com/
95. Wu, Y., Wu, Y., Guerrero, J. M., & Vasquez, J. C. (2021). Digitalization and decentralization driving transactive energy Internet: Key technologies and infrastructures. *International Journal of Electrical Power & Energy Systems, 126*, 106593.
96. Chen, T., & Su, W. (2018). Indirect customer-to-customer energy trading with reinforcement learning. *IEEE Transactions on Smart Grid, 10*(4), 4338–4348.
97. Zhang, Z., Zhang, D., & Qiu, R. C. (2019). Deep reinforcement learning for power system applications: An overview. *CSEE Journal of Power and Energy Systems, 6*(1), 213–225.
98. Mohammad, A., Zamora, R., & Lie, T. T. (2020). Transactive energy management of PV-based EV integrated parking lots. *IEEE Systems Journal, 15*(4), 5674–5682.
99. Yao, Y., Gao, C., Lai, K., Chen, T., & Yang, J. (2021). An incentive-compatible distributed integrated energy market mechanism design with adaptive robust approach. *Applied Energy, 282*, 116155.
100. Huang, W., Du, E., Capuder, T., Zhang, X., Zhang, N., Strbac, G., & Kang, C. (2021). Reliability and vulnerability assessment of multi-energy systems: An energy hub based method. *IEEE Transactions on Power Systems, 36*(5), 3948–3959.
101. Zhao, B., Zlotnik, A., Conejo, A. J., Sioshansi, R., & Rudkevich, A. M. (2018). Shadow price-based co-ordination of natural gas and electric power systems. *IEEE Transactions on Power Systems, 34*(3), 1942–1954.
102. Chen, R., Wang, J., & Sun, H. (2017). Clearing and pricing for coordinated gas and electricity day-ahead markets considering wind power uncertainty. *IEEE Transactions on Power Systems, 33*(3), 2496–2508.
103. Kai-Ineman, D. A. N. I. E. L., & Tversky, A. (1979). Prospect theory: An analysis of decision under risk. *Econometrica, 47*(2), 363–391.
104. Tversky, A., & Kahneman, D. (1992). Advances in prospect theory: Cumulative representation of uncertainty. *Journal of Risk and Uncertainty, 5*, 297–323.
105. Yao, Y., Gao, C., Chen, T., Yang, J., & Chen, S. (2021). Distributed electric energy trading model and strategy analysis based on prospect theory. *International Journal of Electrical Power & Energy Systems, 131*, 106865.

106. El Rahi, G., Etesami, S. R., Saad, W., Mandayam, N. B., & Poor, H. V. (2017). Managing price uncertainty in prosumer-centric energy trading: A prospect-theoretic Stackelberg game approach. *IEEE Transactions on Smart Grid, 10*(1), 702–713.
107. Yao, Y., Gao, C., Tian, H., & Zhang, H. (2020). Review of mid-to long-term trading mechanism for renewable electricity consumption in Ningxia, China. *Renewable and Sustainable Energy Reviews, 134*, 110325.

Part I
Trading Behaviors of Prosumers in Local Energy Markets

Chapter 2
Mental Accounting Theory-Based Trading Strategy of Prosumers

Abstract In the local energy market (LEM), prosumers, participants who produce and consume energy, are typically considered rational. Nevertheless, their trading behaviors are impacted by multiple factors, such as social relations, personal preferences, and habits. Each of them holds different trading demands regarding benefits, comfort levels, and environmental consciousness, demonstrating bounded rationality. This chapter presents a mental accounting theory-based trading strategy for prosumers. This trading strategy accounts for the trading behaviors in reality via quantifying prosumers' motivations and psychological changes, which are evaluated by the risk attitude and subjective value models during the multi-account management process. Furthermore, to increase social welfare and market efficiency, this chapter introduces the updated rules of trading decisions and the multi-round bidding market mechanism. Case studies are conducted and following conclusions are verified: (1) the proposed trading strategy can effectively reflect the prosumers' intentional and limited rationality during the LEM; (2) the proposed market mechanism can effectively improve social welfare and market efficiency.

2.1 Introduction

Renewable energy is now the critical solution to global climate change and energy resource depletion [1]. However, renewable energy generation is often intermittent and varies abruptly. Therefore, the high penetration level of renewable energy in the whole energy supply challenges the operation of power systems. Local Energy Market (LEM) [2, 3] was proposed to solve this issue. LEM is a set of economic and control mechanisms that allow the dynamic balance of supply and demand across the entire electrical infrastructure using value as a key operational parameter' [4]. Under the LEM mechanism, customers on a small scale can trade directly with other participants at the distribution system level. In this way, exploiting the flexibility provision from LEM customers becomes essential to enhance the local utilization of renewable energy [5–7].

M. Song et al., *Local Energy Markets*, https://doi.org/10.1007/978-981-97-9750-9_2

Among the LEM participants, customers possessing local generation facilities, such as rooftop photovoltaic panels or wind turbines, are both energy suppliers and consumers. They are regarded as prosumers in the LEM. Most existing studies established the decision-making models of the prosumers based on utility theory [8, 9], assuming full rationality and sufficient cognitive capabilities of each prosumer [10]. However, in reality, the decision-making of prosumers depends on factors such as individual preferences and consumption patterns [11, 12] (as shown in Fig. 2.1). For example, some prosumers may prefer trading the same transaction objects to avoid risks; some prosumers may incline to increase renewable energy usage to cut down negative environmental impacts. As LEM participants, prosumers need to weigh benefits against risks with limited availability of market information. Utility theory-based decision-making models fail to account for prosumers' diverse preferences and restricted access to market information. This leads to less accurately representing prosumers' trading behaviors in the LEM.

Some studies have assumed that prosumers' trading decision-making process is subjective and not fully rational, and they have incorporated non-economic factors, such as people's inclinations and social connections [13], into the modeling of prosumers' trading strategies. Reference [14] introduced the concept of comfort gain, demonstrating the importance of suitable price incentives in encouraging prosumers' participation in intelligent electricity allocation. Reference [15] utilized a piece-level function to represent prosumers' sensitivity to price. Some studies developed classification models that extract factors. Reference [16] categorized prosumers by behavioral characteristics and established differentiation models. Reference [17] utilized a linear function to evaluate the additional value of specific transactions under individual preferences. Reference [18] developed four transaction models based on resource allocation, representing the suburb, residential, commercial, and industrial

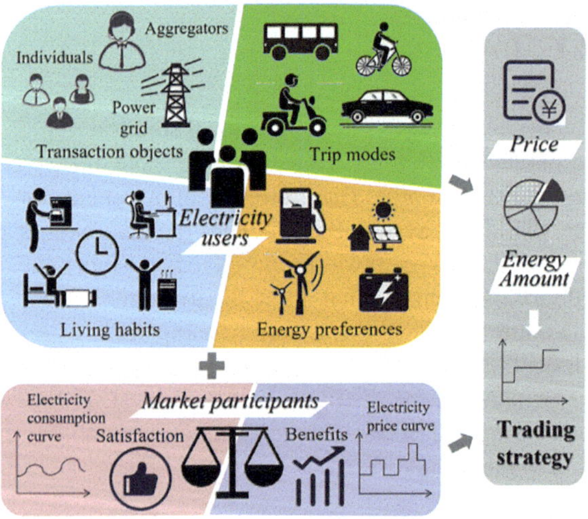

Fig. 2.1 Factors that influence prosumers' trading decisions

prosumers. Furthermore, behavioral economics theories have examined psychological and social factors during decision-making. Reference [19] proposed a distributed trading strategy using prospect theory and the adaptive learning process. Reference [20] developed models describing prosumers' subjective decision-making process when faced with uncertain future prices, which were delivered using the framing effect. Reference [21] proposed a subjective utility model incorporating risk-utility and social factors. Reference [22] developed a data-driven random game model to evaluate how social factors affect the performance of peer-to-peer (P2P) energy sharing.

In the studies above, prosumers were assumed to show constant preferences with one single trading objective. However, prosumers have varying demands, including benefits, comfort levels, and environmental awareness. They are also likely to pursue multiple goals and have different attitudes toward each one. Therefore, the trading methods that account for one objective cannot capture all the trading preferences. It is important to investigate the decision-making process of prosumers with more than one trading demand.

Mental accounting, an important concept in behavioral economics, states that people tend to manage funds in classified mental accounts based on subjective goals, and these mental accounts are categorized by different preferences and are managed independently. Mental accounting theory has been used in multiple areas, such as consumption analysis [23], asset allocation [24], and portfolio optimum [25]. Reference [26] revealed that discounts have a stronger encouraging effect on consumer purchases than bonuses from the mental accounting perspective. Reference [27] developed a portfolio selection model consisting of two mental accounts, one for safety and one for self-actualization. This model provides better average returns than the conventional models for small datasets. While mental accounting has been broadly utilized in decision-making, it has not yet been applied to electricity transactions. Given prosumers' autonomous behaviors in the LEM, mental accounting theory is a compatible modeling tool for the prosumers' trading behaviors.

LEM is designed to host simultaneous small-size transactions between numerous prosumers, necessitating an efficient trading mechanism. Double auction, a well-established mechanism, has been widely used in the LEM. Reference [28] developed an adaptive dynamic strategy within the continuous double auction framework to achieve optimal trading solutions. Reference [29] proposed an optimal scheduling method based on the double auction mechanism applied to microgrids while accounting for complex physical constraints. Reference [30] established a prediction model to help prosumers seek profits via continuous double auctions. However, the existing auction mechanisms do not account for prosumers' limited level of rationality, and the trading preferences and behaviors are often neglected. The improvement of social welfare and transaction efficiency should be in line with prosumers' willingness.

To address the aforementioned issue, this chapter utilizes the mental accounting theory to model the bounded-rationality trading behaviors of prosumers in the LEM [31]. The main contents of this chapter are:

(1) Contrary to the existing research, which assumed the full rationality of prosumers and used single-objective optimization models deviating from real-life willingness, this chapter introduces a multi-objective trading decision-making method based on mental accounting theory. The proposed method can capture prosumers' diverse subjectiveness, with multiple mental accounts representing prosumers' expectations of basic electricity demand, economic benefit, and attitude toward local renewable energy. The risk attitude and subjective value models are used to evaluate prosumers' risk attitude and paranoia psychology, characterizing LEM prosumers' bounded-rationality trading behaviors.

(2) This chapter demonstrates a multiple-round bidding market mechanism and trading decision updating rules to enhance social welfare and market efficiency. The market payment rules utilize the generalized second-price auction mechanism for fairness. Accordingly, prosumers can dynamically adapt their trading strategies to their preferences and goals.

2.2 Mental Accounting Theory

2.2.1 Mental Accounting Theory and Electricity Trading

Mental accounting theory is an important concept in behavioral economics. It states that people allocate their funds into different mental accounts based on the fund sources, destinations, and purposes. In Fig. 2.1, people's behaviors are grouped into three psychological effects: non-fungibility effect, sunk cost effect, and transaction utility effect [32], which demonstrate how mental accounts shape the decision-making behaviors of an individual (Fig. 2.2).

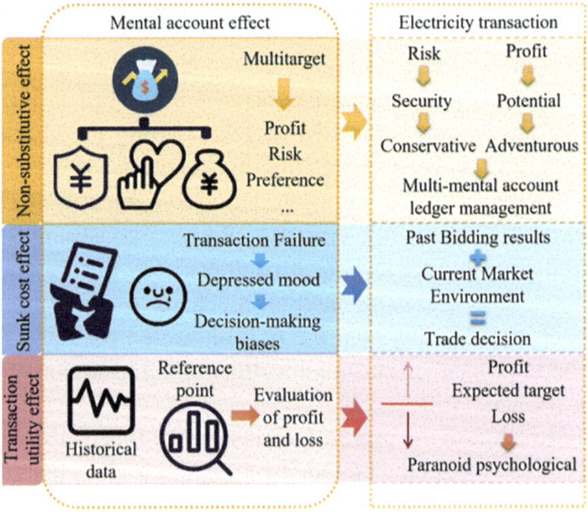

Fig. 2.2 Relationship between the mental accounts and electricity transaction

The non-fungibility effect occurs during consumption or investment activities. People unintentionally develop a mental account system and assign different incomes and expenditures to distinct mental accounts. These accounts are regarded as independent and non-interchangeable.

The sunk cost effect refers to the tendency of people to stick with the existing behavior modes to avoid the negative emotions caused by previous losses. Influenced by the sunk cost effect, people tend to consider the consequences of past investments, current income, and future investments simultaneously.

The transaction utility effect refers to the behavior in which people tend to assess their gains and losses based on the relative value of income compared to expectations rather than the absolute worth of revenue. When the actual income is lower than expected, people are inclined to experience a sense of loss, which may lead to a paranoid mindset. In general, people exhibit risk-avoiding behaviors when faced with potential losses while being more risk-seeking when there are potential rewards. Consequently, people are more likely to stay forceful when anticipating positive incomes and vice versa.

The non-fungibility effect in electricity trading suggests that prosumers are likely to manage multiple mental accounts such as risk, revenue, and personal preference. Studies indicate that some attributes of electricity, such as production technology, location, and trading objects, are valuable. Prosumers may tend to pay a premium for renewable energy that is generated locally [33–35] and even be in favor of specific prosumers [36, 37]. As suggested by the sunk cost effect, prosumers tend to make quotations according to the outcomes of previous transactions, so their attitudes toward risk will change along with the bidding process. Influenced by the transaction utility effect, prosumers' reference price may be dependent on the anticipation and market environment. Each mental account holds a distinct reference price for gain and loss that aligns with its corresponding goals.

Three mental accounts are proposed to pursue prosumers' trading goals related to the generation/consumption basic demand, economic benefits, and preference for local renewable energy, which are demonstrated in Fig. 2.3.

M1: a low-risk account that ensures the basic electricity demand is satisfied or low-risk profit is achieved;
M2: a high-risk account that intends to reduce electricity costs or seek arbitrage;
M3: a local green transaction account designated for renewable energy trading, which assumes some prosumers are willing to sacrifice a certain portion of their revenues to reduce carbon emissions.

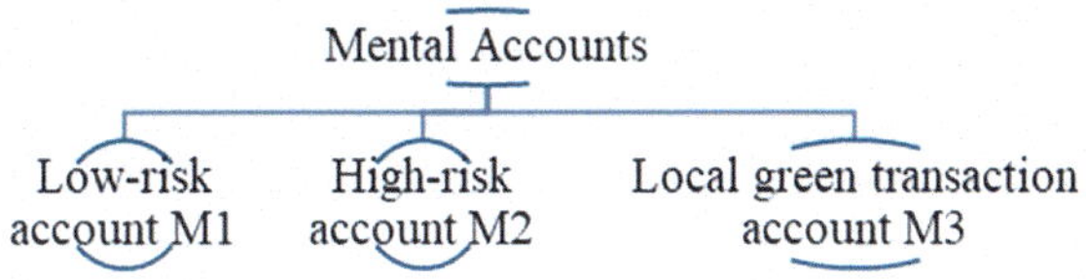

Fig. 2.3 Three mental accounts

2.2.2 Risk Attitude Models

The LEM-participating prosumers are confident to seek higher benefits by trading with other participants. A convex quadratic function is used to represent the probability of successful trading $f(\pi_i)$ $(0 \le f(\pi_i) \le 1)$. The profit space of prosumers comes from the difference between the Feed-in tariff (FIT) and the Time-of-use electricity price (ToU). Prosumer behaviors are also incentivized by ToU [38]. Therefore, prosumers' quotations fluctuate between FiT and the ToU price.

For a seller, a lower price can increase their chances of succeeding in the asking process. Accordingly, the proximity of its price to the FIT is used to measure the probability of successful trading:

$$f(\pi_i) = 1 - \left[\frac{\pi_i - \pi_{\text{sell}}}{\mu(t) - \pi_{\text{sell}}}\right]^2, \quad i = 1, 2, 3 \tag{2.1}$$

where i represents the index of mental accounts, π_i is a quotation of M1, π_{sell} is FiT price, $\mu(t)$ is time-of-use electricity price.

For buyers, a higher price can increase their chances of succeeding in the bidding process. Accordingly, the proximity between the price and the ToU electricity price is used to quantify the probability of successful trading:

$$f(\pi_i) = 1 - \left[\frac{\mu(t) - \pi_i}{\mu(t) - \pi_{\text{sell}}}\right]^2, \quad i = 1, 2, 3 \tag{2.2}$$

In accordance with "safety, potential, and expectation theory" by Lopes [39], the prosumers are motivated by the prospective interest to focus on the potential success of their bids or asks and have a sense of fluke. At the same time, the prosumers are still cautious and conservative as they are aware of the probability of failure. The prosumers exhibit different risk attitudes toward different mental accounts based on the purpose of each account. Therefore, M1, M2, and M3 risk attitudes are conservative, speculative, and neutral, respectively. This is represented by the weight functions $\omega(\pi_i)$ of $f(\pi_i)$ [40], as demonstrated in Fig. 2.4.

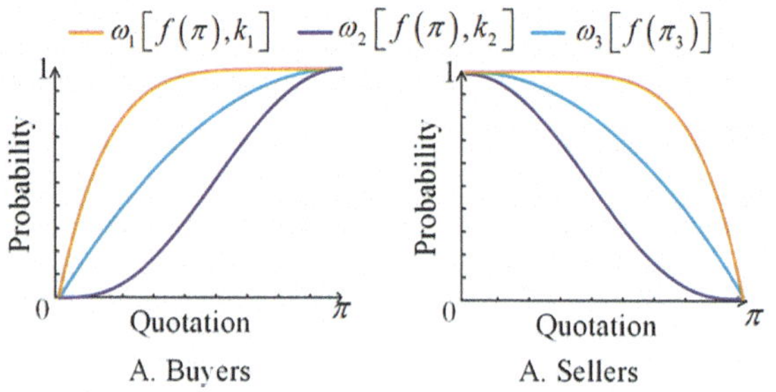

Fig. 2.4 Weight functions of three mental accounts

For the mental account M1 with a conservative attitude, prosumers emphasize the security of the trading decisions as fear of loss takes a dominant position. The prosumers keep lower expectations and aim for successful bids or ask by offering more competitive prices. Consequently, prosumers are inclined to underestimate the successful probability of a bid/ask of M1. This is represented by a concave weight function:

$$\omega_1(\pi_1) = \left[f(\pi_1)\right]^{1+k_1} \tag{2.3}$$

where the larger the k_1 is, the stronger the prosumers' risk-averse attitude.

For the mental account M2 with a speculative attitude, prosumers emphasize the potential benefits, with optimism as the dominant influence. The prosumers expect to achieve greater economic returns. Consequently, prosumers are inclined to overestimate the successful probability of a bid/ask of M2. This is represented by a convex weight function:

$$\omega_2(\pi_2) = 1 - \left[1 - f(\pi_2)\right]^{1+k_2} \tag{2.4}$$

where the larger the k_2 is, the stronger the prosumers' risk-pursuing attitude.

For the mental account M3, prosumers aim to trade with local renewable energy suppliers at a reasonable price while balancing security and risk during the decision-making process. Therefore, prosumers take the successful probability of a bid/ask more rationally and objectively. In this way, the weight function for M3 is the same as the probability function.

$$\omega_3(\pi_3) = f(\pi_3) \tag{2.5}$$

2.2.3 Subjective Value Models

Influenced by the transaction utility effect and different intentions of M1 to M3, prosumers assess the economic benefits of each mental account by using different reference points. Consequently, prosumers may attribute distinct subjective values to the same bid/ask price. The subjective value of a bid/ask is influenced by the expected target $R_{A,i}$, which is calculated based on the highest, lowest, and average electricity price of the previous day and the fluctuation of the ongoing market. When trading outcomes fall short of expectations, prosumers consider this as a loss. Stimulated by the desire to reach the ideal target, prosumers may become paranoid, causing a subjective value deviation.

(1) Subjective value model for a successful bid/ask

When the quotation reaches the expected target, prosumers experience self-satisfaction and refrain from excessive greed due to risk aversion. In this state, the subjective value remains a linear relationship with the bid/ask. However, when the quotation is below the expected target, prosumers exhibit risk-seeking behaviors and adjust the bid/ask to meet the expected target. Therefore, the subjective value function becomes concave as it reflects an aggressive attitude. The subjective value model for a successful bid/ask $V(\pi_i, R_{A,i}, k_{A,i})$ is set as below.

For sellers:

$$V(\pi_i, R_{A,i}, k_{A,i}) = \begin{cases} \pi_i & \pi_i \geq R_{A,i} \\ R_{A,i} \times \left(\frac{\pi_i}{R_{A,i}}\right)^{1+k_{A,i}} & \pi_i < R_{A,i} \end{cases} \quad i = 1, 2, 3 \tag{2.6}$$

For buyers:

$$V(\pi_i, R_{A,i}, k_{A,i}) = \begin{cases} \frac{R_{A,i}-\mu(t)}{R_{A,i}}\pi_i + \mu(t) & \pi_i \leq R_{A,i} \\ R_{A,i} \times \left(\frac{\pi_i}{R_{A,i}}\right)^{-(1+k_{A,i})} & \pi_i > R_{A,i} \end{cases} \quad i = 1, 2, 3 \tag{2.7}$$

where the larger the $k_{A,i}$ is, the more determined the prosumer is to attain the desired goal.

In terms of benefit assessment, prosumers exhibit greater caution in M1 and more tolerance in M2. If the trading outcome deviates from the expected target, prosumers are expected to be more sensitive to M1 rather than M2. Additionally, as economic benefit is not the main goal of M3, prosumers are more inclined to make relatively rational decisions.

$$k_{A,1} \geq k_{A,2} \geq k_{A,3} \tag{2.8}$$

(2) Subjective value model for an unsuccessful bid/ask

If the bid/ask is unsuccessful, the only option for prosumers is to trade with the grid, which would reduce income or increase cost. The LEM doesn't provide any value increment. The subjective function of unsuccessful bis/ask $V(\pi_i, R_{A,i}, k_{A,i})$ is calculated as below.

$$V\_(\pi_i, R_{A,i}, k_{A,i}) = 0, \quad i = 1, 2, 3 \tag{2.9}$$

2.3 Trading Decision-Making Method of Prosumers Based on Multi-mental Accounts

---

Algorithm 1: Quotation decision solving algorithm

Step 1: Determine the identify (Seller or Buyer) and the type of account $M_{i(i=1,2,3)}$
Step 2: Initialize maximum expected value $E_{max}=0$
Step 3: Initialize quotation parameters $R_{A,i}$, $k_{A,i}$, k_i
for Each π_i from the FIT π_{sell} to the TOU price $\mu(t)$ with a fixed step length ρ **do**
  while $\pi_{sell} \leq \pi \leq \mu(t)$ **do**
    a. Calculate the probability of successful bidding $f(\pi_i)$;
    b. Calculate the weight probability $\omega(\pi_i)$;
    c. Calculate the subjective value $V(\pi_i)$;
    d. Calculate the expected value $E(\pi_i)$;
    if $E(\pi_i) \geq E_{max}$ **then**
      $E_{max}=E(\pi_i)$
    else
      $\pi_i=\pi_i+\rho$
    end
  end
end
Step 4: Get the decision quotation π_i of $M_{i(i=1,2,3)}$, whose expected value is E_{max}

During the previous sections, the prosumers' expectation value $E_i(\pi)$ for a specific quotation π_i is determined by the weight functions $\omega(\pi_i)$ and the corresponding subjective value function $V(\pi_i, R_{A,i}, k_{A,i})$ and $V(\pi_i, R_{A,i}, k_{A,i})$. As prosumers have more than one decision-making goal simultaneously, the trading decision-making model is established based on multiple mental accounts, as demonstrated by Algorithm 1.

The bid/ask price of M1 to M3 are calculated as below.

$$\max M_i : E(\pi_i) = V(\pi_i, R_{A,i}, k_{A,i}) \times \omega(\pi_i) + V\_(\pi_i, R_{A,i}, k_{A,i}) \times \{1 - \omega(\pi_i)\} \quad i = 1, 2, 3 \tag{2.10}$$

Once the bid/ask price of three accounts in each period is obtained, the allocated electricity quantity is calculated by solving the optimization model as below. When the prosumer sells electricity, the allocated quantity is positive; when the prosumer purchases electricity, the allocated quantity is negative.

$$\max \sum_{t=1}^{24} \left\{ \begin{array}{l} \pi_1(t)Q_1(t) + \pi_2(t)Q_2(t) + \pi_3(t)Q_3(t) - u_{\mathrm{AC}}(t) \\ -C_{\mathrm{dep}}[b_{\mathrm{ch}}(t) + b_{\mathrm{dis}}(t)]\Delta t - \alpha Q_2^2(t) - \beta Q_3^2(t) \end{array} \right\} \tag{2.11}$$

where $Q_i(t)$ denotes electricity quantity of M_i, C_{dep} is the depreciation cost per energy storage charging and discharging unit, b_{ch} and b_{dis} are charging power and discharging power of energy storage, α is the opportunity cost coefficient of M2, β is the transaction risk coefficient of M3.

The constraints include:

(1) Power balance constraint

Throughout the optimization time horizon, the total power generation and consumption should be balanced.

$$Q_1(t) + Q_2(t) + Q_3(t) = \mathrm{y}(t) + b_{\mathrm{dis}}(t) - l(t) - z(t) - b_{\mathrm{ch}}(t) \tag{2.12}$$

where y is photovoltaic power generation, z is the fixed loads.

(2) Air conditioning constraints

The air conditioning is modeled as below [41].

$$H(t) = H_{\mathrm{out}}(t) - [H_{\mathrm{out}}(t) - H(t-1)]\mathrm{e}^{-1/RC} + \rho l(t-1) \tag{2.13}$$

where H_{out} is outdoor temperature. l denotes the electricity consumption of air conditioning, C is the equivalent thermal capacity of the air condition, R is the resistance of the air conditioning, ρ is the operating mode of the air conditioning.

When a prosumer with air conditioning participates in the LEM, the indoor temperature deviates from the preferred temperature h. The prosumer's declined comfort could be evaluated by a utility function $u_{\mathrm{AC}}(t)$:

$$u_{\mathrm{AC}}(t) = a_{\mathrm{AC}}[H(t) - h]^2 \tag{2.14}$$

where a_{AC} is the sensitivity level of prosumers, H represents the indoor temperature, h is preferred temperature.

The indoor temperature is limited within a temperature fluctuation range.

$$H_{\mathrm{AC}}^{\min} \leq H(t) \leq H_{\mathrm{AC}}^{\max} \quad \forall t \in T \tag{2.15}$$

where $H_{\mathrm{AC}}^{\min}$ and $H_{\mathrm{AC}}^{\max}$ are minimum and maximum acceptable indoor temperatures.

(3) Photovoltaic constraints

The photovoltaic power generation is influenced by illumination and ambient temperature. The relationship between the actual and forecast values of photovoltaic power generation $y(t)$ and $y^{pred}(t)$ is represented by:

$$y(t) = (1 + \delta)y^{pred}(t) \tag{2.16}$$

where y^{pred} is predicted generation power of PV, $\delta \in [-0.005, 0.005]$ follows uniform distribution.

(4) Energy storage constraints

The modeling of energy storage operation includes a state of charge (SoC) constraint, and the upper/down range constraints of SoC and charging/discharging power.

$$s(t+1) = s(t) + \frac{\eta_{ch} b_{ch}(t)\Delta t}{E_{bi}} - \frac{b_{dis}(t)\Delta t}{\eta_{dis} E_{bi}} \tag{2.17}$$

$$s^{min} \leq s(t) \leq s^{max} \tag{2.18}$$

$$b^{min} \leq b_{ch}(t)/b_{dis}(t) \leq b^{max} \tag{2.19}$$

$$0 \leq \Delta t \leq 1 \tag{2.20}$$

where s is the state of charge, s^{min} and s^{max} represent the maximum state of charge and the minimum state of charge, η_{ch} and η_{dis} are efficiencies of charging and efficiencies of discharging, E_{bi} is the capacity of the battery, b^{min} and b^{max} are the maximum and minimum charging/discharging power.

In this model, decision variables include $Q_1(t)$, $Q_2(t)$, $Q_3(t)$ of M1, M2, and M3, $l(t)$ of air conditioning, and $b_{ch}(t)$ and $b_{dis}(t)$ of energy storage. This model is a convex optimization problem, which can be solved using commercial solvers, such as Gurobi.

2.4 LEM Mechanism and Trading Decision Updating Method

2.4.1 Market Entities

LEM comprises prosumers, the trading center, and the distribution network operators (DNOs). Figure 2.5 demonstrates its structure.

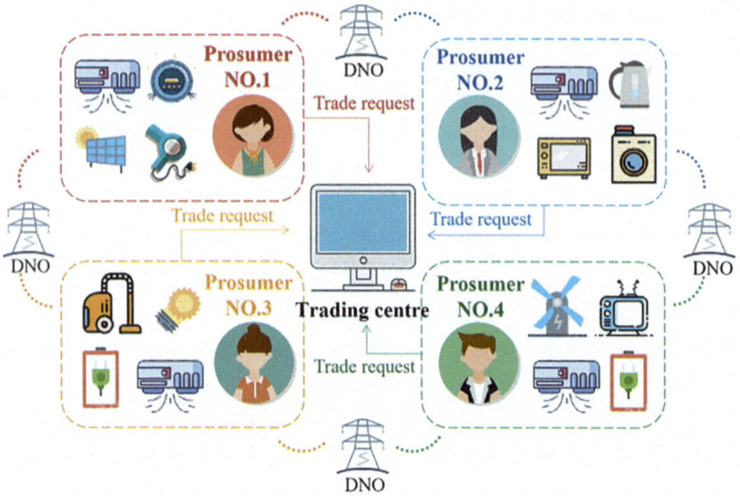

Fig. 2.5 Market entities

The prosumers from LEM can be classified into two types: consumption-type and production-type. Consumption-type prosumers are electricity users who have not installed equipment for distributed generation. They can lower electricity costs by purchasing a lower LEM electricity price. Production-type prosumers are electricity users with equipment for distributed power generation, such as photovoltaic equipment. They can enhance their revenues by selling excess electricity when the price exceeds FIT. The trading center is in charge of providing market information, collecting demand information from prosumers, making trading matches, and organizing the settlements of completed transactions. DNO offers power transmission services to prosumers and balances the excess electricity supply or usage that LEM couldn't address.

2.4.2 *Trading Process*

A multi-round auction mechanism is adopted for market clearing to stimulate prosumers' engagement and enhance social welfare. Each day is divided into 24 trading periods lasting one hour. The trading process is as below.

(1) Parameter setting

For each trading period, the prosumers choose one role: electricity seller or buyer.

(2) Submission of the price and quantity

LEM participants submit their declaration on the bid/ask price and the quantity for each trading period.

(3) Matching and clearing

The declaration of price and quantity made by buyers and sellers from the order sequence:

$$\{(\pi_i, Q_i)|i \in \Omega_\mathrm{B}\} \tag{2.21}$$

$$\{(\pi_j, Q_j)|j \in \Omega_\mathrm{S}\} \tag{2.22}$$

where Ω_B and Ω_S represent set of unmatched buyers and unmatched sellers.

The orders to be matched satisfy:

$$\{\pi_b^{\mathrm{match}}|\pi_b^{\mathrm{match}} \geq \pi_i, \quad i \in \Omega_B\} \tag{2.23}$$

$$\{\pi_s^{\mathrm{match}}|\pi_s^{\mathrm{match}} \leq \pi_j, \quad j \in \Omega_S\} \tag{2.24}$$

When $\pi_b^{\mathrm{match}} \geq \pi_s^{\mathrm{match}}$, a transaction is confirmed with the trading price executed aligned with the payment rules. The trading quantity is:

$$Q_{\mathrm{deal}} = \min(Q_b^{\mathrm{match}}, Q_s^{\mathrm{match}}) \tag{2.25}$$

Then the order sequence is updated using:

$$\Omega_\mathrm{B} = \Omega_\mathrm{B} - \{b\} \tag{2.26}$$

$$\Omega_\mathrm{S} = \Omega_\mathrm{S} - \{s\} \tag{2.27}$$

Otherwise, the orders are not matched. Prosumers need to update their quotations based on the feedback from the trading center and return to Step (2) for another round of bids/asks.

The market clearing process concludes when no electricity margin (the margin for buyers or sellers reaches zero) or the final round of bids/asks is completed. The sellers then sell any unsold electricity to the power grid at the FIT. The buyers purchase electricity from the power grid at the ToU price.

2.4.3 Payment Rules

To encourage the prosumers to bid or ask for actual prices in line with their trading objectives, payment rules are designed by adopting the Generalized Second Price Auction (GSP). Within the GSP mechanism, an extension of the Vickrey-Clarke-Groves (VCG) auction mechanism, the highest bidder wins and pays the second-highest price. This approach encourages each prosumer to stick to a constant optimal

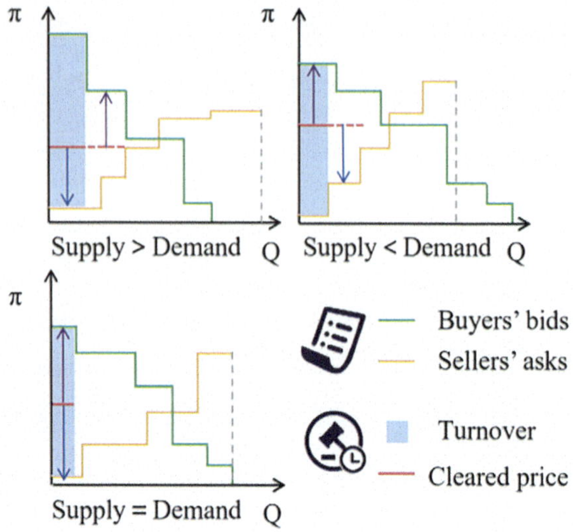

Fig. 2.6 Payment rules

strategy, with the bid/ask price equal to its evaluation [42]. GSP meets the personal rationality and incentive compatibility [43], with the benefits of the "sealed second price" to the maximum. Additionally, GSP mitigates the budget deficit issue in VCG, allowing it to reach a stable equilibrium.

Assume that the orders (π_b^{match}, Q_b^{match}) and (π_s^{match}, Q_s^{match}) are to be matched and the condition $\pi_b^{\text{match}} \geq \pi_s^{\text{match}}$ is met, the clear price can be calculated by the following steps (demonstrated in Fig. 2.6):

(1) Buyer-dominated market

When the electricity supply surpasses the demand, all buyers' demands can be met through reasonable bidding. Still, some sellers have to sell their electricity to the power grid at a lower price.

$$\pi_{\text{deal}} = \frac{\pi_{b-1}^{\text{match}} + \pi_s^{\text{match}}}{2} \tag{2.28}$$

where π_{b-1}^{match} represents the buyer's highest quotation (excluding π_b^{match}) within the order sequence. In this case, the buyers take a dominating role in the market. GSP prevents excessive speculation by reducing electricity prices, potentially reducing total market turnover.

(2) Seller-dominated market

When the electricity supply cannot meet the demand, sellers can sell their electricity by reasonably asking, but some buyers have to purchase electricity from the grid at a higher price.

$$\pi_{\text{deal}} = \frac{\pi_b^{\text{match}} + \pi_{s-1}^{\text{match}}}{2} \tag{2.29}$$

where, π_{s-1}^{match} represents the lowest sellers' lowest quotation (excluding π_s^{match}) within the order sequence. In this case, sellers take a dominating role in the market. GSP prevents excessive pursuit of profits by increasing the electricity price.

(3) Equilibrium market

In this case, the electricity supply equals the demand.

$$\pi_{\text{deal}} = \frac{\pi_b^{\text{match}} + \pi_s^{\text{match}}}{2} \tag{2.30}$$

Buyers and sellers participate in the market with the same level of competitive intensity. Even without GSP, market participants would still make their bid/ask reasonably by weighing their revenues.

(4) Special circumstances

The trading price equals to the equilibrium market case under these special circumstances: when no sub-high-price buyer or sub-low-price seller is within the order sequence; the cleared price from a one-party-dominated case exceeds the buyers' quotations or falls below the sellers' quotations.

2.4.4 Automatic Updating Rules of Prosumers' Quotation Parameters

When LEM trading starts, prosumers will likely be optimistic, probing the price based on their expectations of ample trading demand. However, after a few bid/ask rounds, the market margin decreases, and prosumers become cautious due to the pressure of consecutive failures. To conclude the LEM transactions, the remaining prosumers will tune their quotations based on the wait-and-see outcomes from the previous rounds and balance the benefits against the probability of successful transactions. This is conducted by updating the two quotation parameters below based on the feedback from the trading center.

(1) Risk attitude coefficient k_1 and k_2

Based on the previous bid/ask failures, prosumers will likely stay conservative. M1 account becomes more risk-avoiding (k_1 value increases) while M2 account becomes less risk-pursuing (k_2 value decreases). Their attitude toward risk is calibrated according to the price from the previous round of market clearing.

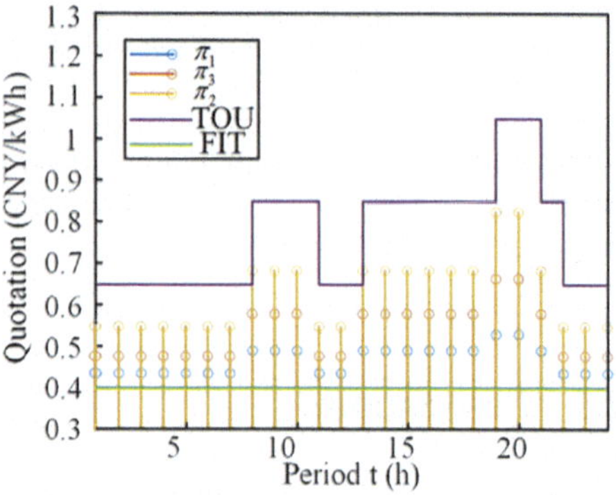

Fig. 2.8 Seller's quotation curve

Figure 2.9 demonstrates an example of the decision-making quotation process, which is the buyer's quotation results between 9:00 and 10:00. The optimal quotations for M1, M2, and M3 are 0.803 CNY/kWh, 0.580 CNY/kWh, and 0.711 CNY/kWh, respectively. The curves in Fig. 2.9 reflect the risk attitude of each account. For M1, the weight probability curve lies below the probability distribution curve, which shows that the prosumers hold a conservative attitude and underestimate the bidding success rate. On the contrary, M2's weight probability curve lies over the probability distribution curve, showing that the prosumer holds an aggressive attitude and overestimates the bidding success rate. As for M3, the probability curve overlaps with the probability distribution curve, which shows that the prosumer is objective and neutral. Each subjective value curve is concave when the quotation doesn't meet the expected target, showing the prosumers' different levels of paranoia. Each expected value curve rises first and then falls. The optimal quotation value is the maximum of the expected value.

Figures 2.10, 2.11 and 2.12 illustrate the details of the seller's quotation between 9:00 and 10:00, including the detailed sensitivity analysis for each parameter. In Fig. 2.10, when k_1 and k_2 climb from 0 to 5, M1's quotation decreases while M2's quotation increases, with M3's quotation remaining constant. This is consistent with the risk attitude analysis. In Fig. 2.11, when the paranoia coefficients climb from 0 to 1, the prosumer is more eager to reach the expected target. The quotation only increases with $k_{A,1}$, $k_{A,2,}$ $k_{A,3}$ when the expectation is not achieved. This indicates that paranoid psychology has a great impact on the quotation decision-making process. Figure 2.12 shows that when the selling targets $R_{A,1}$, $R_{A,2}$, $R_{A,3}$ climb from FIT to ToU price by steps of 0.02, the revenue and the probability of bidding failure increase simultaneously. When $R_{A,i}$ increases, the sellers moderately adjust the quotation to a higher value for more profit. However, if $R_{A,i}$ becomes too high, the sellers would stop raising the quotation out of consideration on the bidding failure probability. The quotation ceases growing when the expected target is satisfied.

Fig. 2.9 Optimal quotation decision-making curve of M1/M2/M3

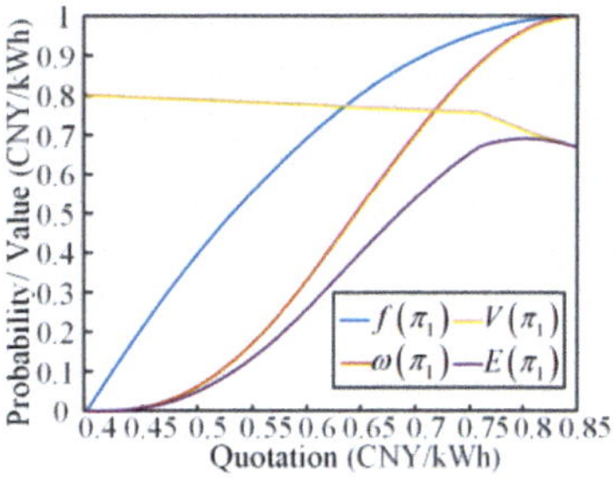

(a) M1

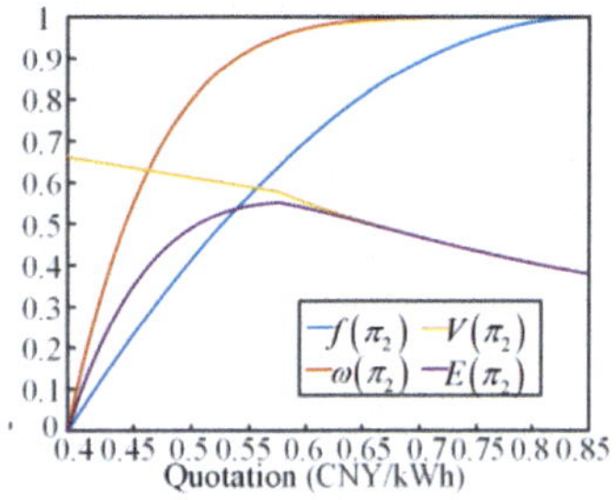

(b) M2

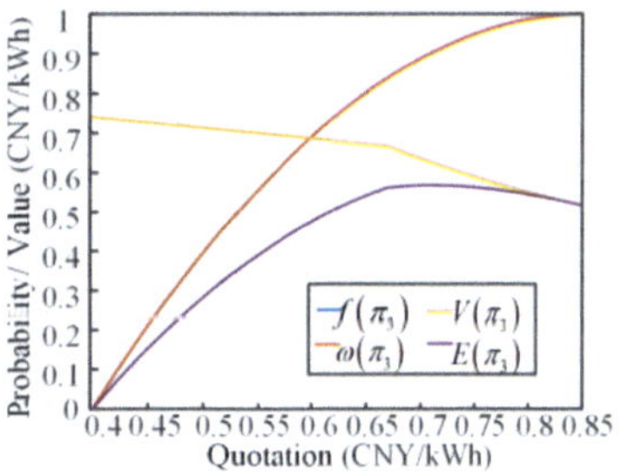

(c) M3

Fig. 2.10 Variation curve of each account's quotation with the risk coefficient *k*

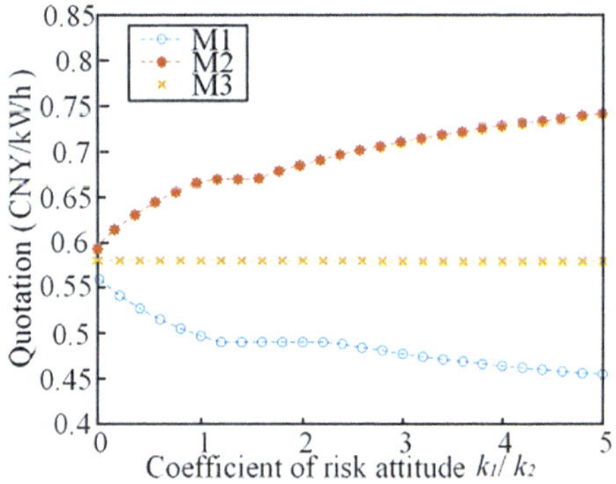

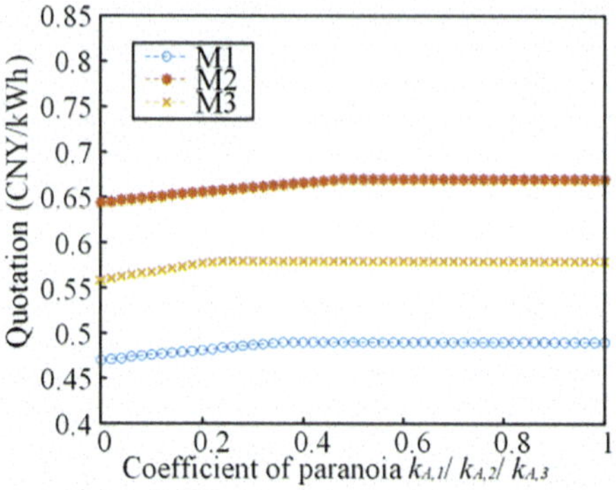

Fig. 2.11 Variation curve of each account's quotation with the paranoia coefficient k_A

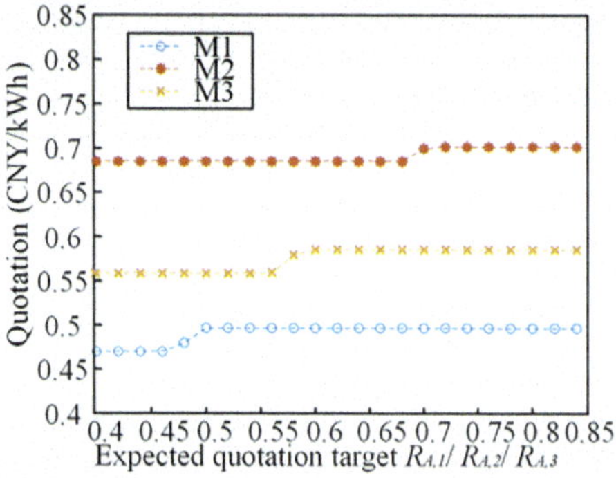

Fig. 2.12 Variation curve of each account's quotation with the expected target R_A

2.5.2 Results of Electricity Distribution in M1–M3

The resource parameters (air conditioning and energy storage) are given in Table 2.2. Market risk coefficients are given in Table 2.3. Decision-making parameters are given in Table 2.12, and the load-related profiles are presented in Figs. 2.24, 2.25 and 2.26.

Figure 2.13 demonstrates a production-type prosumer's electricity distribution over 24 h. Constrained by photovoltaic power generation availability, the prosumer can only sell excessive electricity in the LEM from 7:00 to 17:00 and buys electricity as a buyer during the rest of the day. The prosumer has to weigh profit against risk to reach maximal revenue. For example, from 9:00 to 10:00, the quotation and quantity

Table 2.2 Parameters of different resources

Resource	Parameters							
Air conditioning	a_{AC}	R	h	C	H_{AC}^{min}		ρ	H_{AC}^{max}
	1.1	1.35	20	3	16		– 0.185	24
Energy storage	η_{ch}	η_{dis}	b^{min}	b^{max}	s^{min}	s^{max}	C_{dep}	E_{bi}
	0.95	1.05	0	25	0	1	0.05	200

Table 2.3 Market risk coefficients

Prosumers type	Coefficients	
	α	β
Production-type	0.008	0.0013
Consumption-type	0.009	0.0015

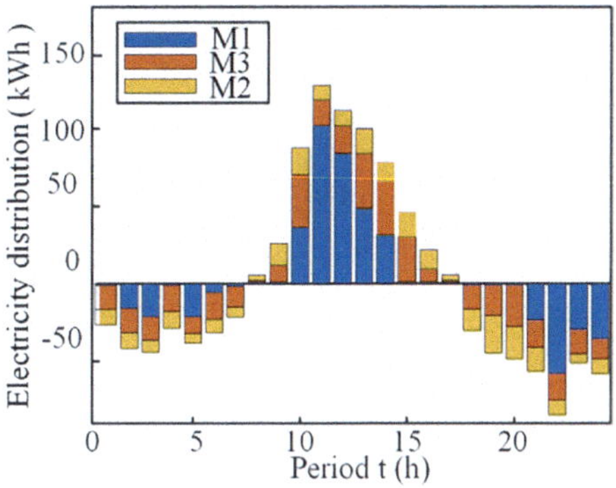

Fig. 2.13 Electricity distribution diagram of a production-type prosumer in a day

of M1, M2, and M3 are 0.47 CNY/kWh of 1 kWh, 0.759 CNY/kWh of 10.84 kW, and 0.559 CNY/kWh of 14.27 kWh, respectively. M1 has a limited role since its quotation is the lowest one. M2 has both the highest quotation and highest bidding risk; thus, its distributed quantity is smaller than that of M3, which is more balanced regarding the quotation and bidding risk. M1, M2, and M3's distributed quantities are influenced by prosumers' preferences, FIT, and ToU prices. This strategy enables the prosumers to gain more benefits while mitigating risks and accounting for their preferences.

Figure 2.14 presents the operating behaviors of two types of resources. As for energy storage, which has a limited capacity, it discharges from 1:00 to 6:00 and charges from 9:00 to 17:00 to fully utilize the photovoltaic power generation. It discharges again from 18:00 to 22:00 to alleviate the need for electricity at a high price.

2.5.3 *Results of Multi-round Market Clearing and Quotation Updating*

A simulation is conducted using a test case with 16 consumption-type prosumers and 5 production-type prosumers in the LEM. k_n is set as 5, $n(0)$ is set as 10. The bidding information of prosumers is given in Tables 2.13 and 2.14.

Figure 2.15 presents the LEM clearing result of the first round between 9:00 and 10:00. Upon market clearing, the total transaction electricity amount is 295.7032 kWh, with the lowest bid from the buyers as 0.579 CNY/kWh and the

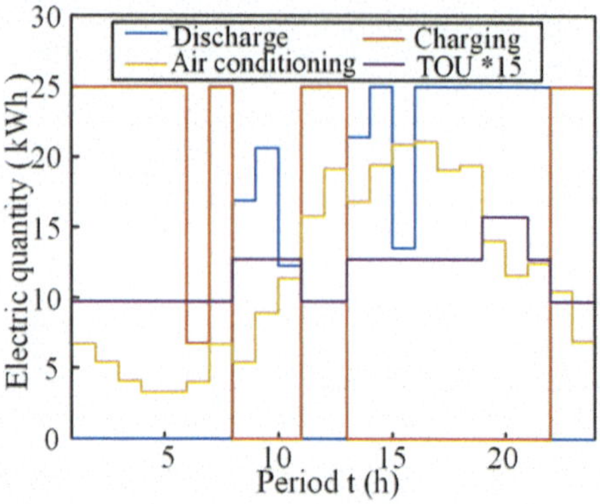

Fig. 2.14 Operating behaviors of energy storage and air conditioning for the production-type prosumer

highest bid from the sellers as 0.559 CNY/kWh. The cleared prices range from 0.55 to 0.65 CNY/kWh. With the surplus supply of electricity, the market is buyer-dominated. To restrain buyers' blind speculation, the cleared price is determined as the average value of the sublowest buyer's bid price and the current matching seller's ask. In this way, most orders are cleared at the price of around 0.6 CNY/kWh, which is below 0.625 CNY/kWh (the mean value of ToU price is 0.85 CNY/kWh and FIT 0.4 CNY/kWh). This approach guarantees a fair distribution of market dividends in the complex trading environment.

Figures 2.16 and 2.17 show the market clearing process during the second and third rounds, respectively. During the second round, no deal is made as the buyers' highest ask is below the sellers' lowest ask. After the buyers and the sellers revise their quotation based on the feedback on the market information, more transactions are made during the third round. With the trading process going forward, the prosumers prioritize the probability of successful transactions to make their quotations more competitive, which leads to a narrower price gap between the buyers and the sellers.

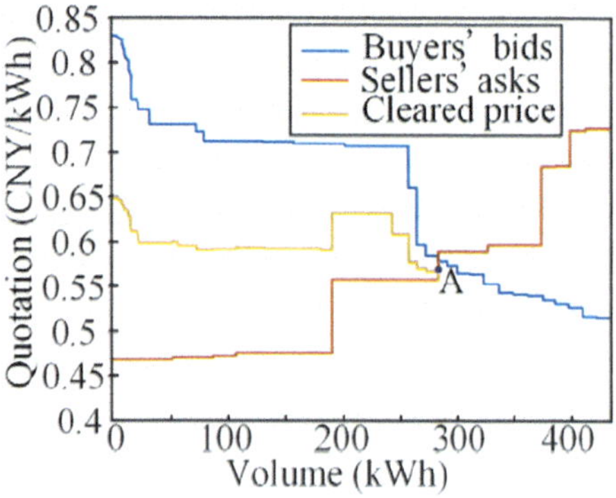

Fig. 2.15 Market clearing process diagram in the first round

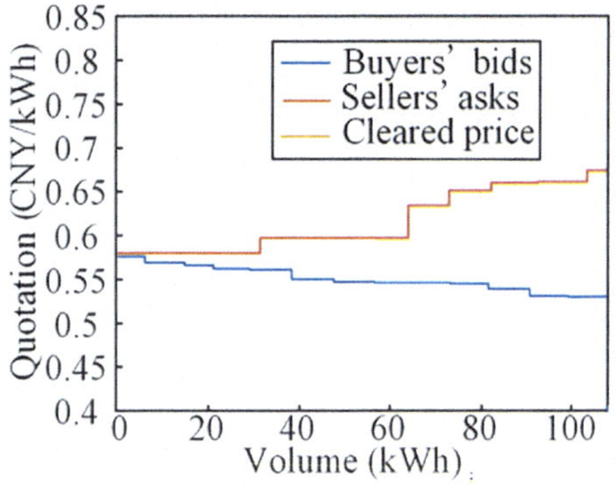

Fig. 2.16 Market clearing process diagram in the second round

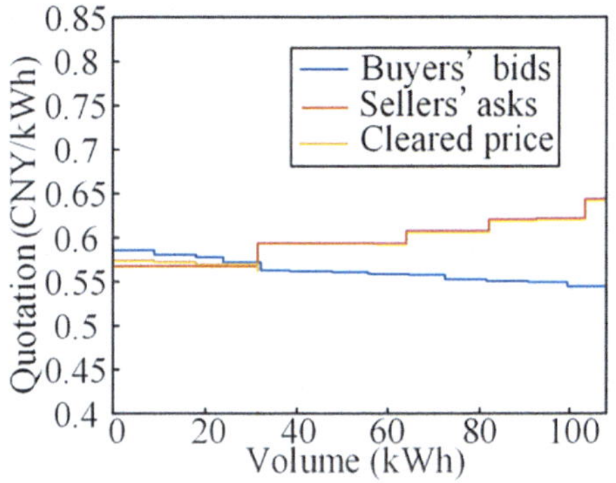

Fig. 2.17 Market clearing process diagram in the third round

Table 2.4 presents the transaction outcomes in each round (the detailed profiles of the market clearing process during the fourth and fifth rounds are given in Figs. 2.27 and 2.28). All the electricity is cleared after five rounds, promoting locally generated green energy consumption.

Figures 2.18 and 2.19 demonstrate how prosumers who fail in transaction clearing in the first round update their quotations during the following rounds. For most sellers (apart from No. 3 and No. 16), the untraded electricity comes from the M2 account due to the high-risk and high-profit attributes of M2. M2's electricity could only be cleared with a reasonable quotation, which shows that moderate speculation can enhance benefits, whereas overspeculation behavior raises the risk of bidding failure. Typically, as the market margin narrows, buyers raise their quotations. However, some specific buyers (No. 3 and No. 16) still reduce their quotations during the second

Table 2.4 Transaction quantity of each round without multi-mental accounts

Round k	1	2	3	4	5
Tradable quantity (kWh)	455.28	159.58	159.58	113.04	6.77
Turnover (kWh)	295.70	0.00	46.54	106.26	6.77

and third rounds as they can still bid in the following rounds. Some adventurous and speculative buyers continue to test the market environment. Nonetheless, with the final round approaching, prosumers feel stressed by the previous failure, and they would make comprehensive adjustments in their risk attitude, confidence function, and expected quotation target.

Table 2.5 and Fig. 2.20 present sellers' final transaction outcomes (buyers' final transaction outcomes are presented in Table 2.15 and Fig. 2.29). Prosumers can adapt their trading strategies based on the market information within the multi-round mechanism. This segmented quotation trading strategy reduces the likelihood of trading failure, helping both buyers and sellers to achieve substantial benefits.

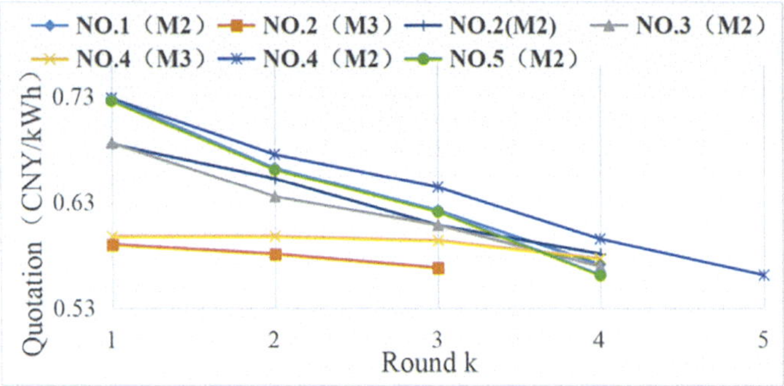

Fig. 2.18 Sellers' quotation update curve

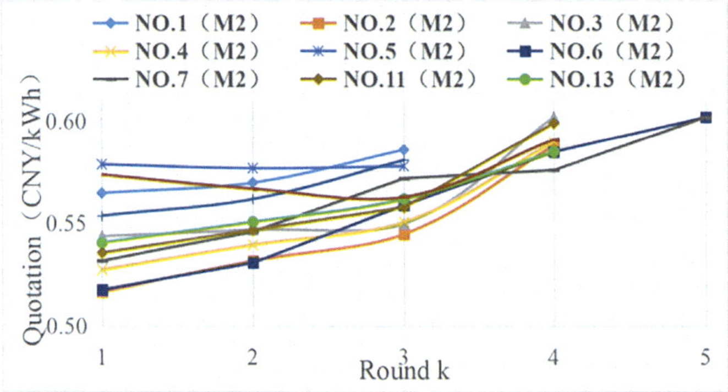

Fig. 2.19 Buyers' quotation update curve

Table 2.5 Trading results of each seller in one period

Seller's number	Cumulative trading quantity (kWh)	Untraded quantity (kWh)
1	87.3403	0
2	114.1420	0
3	84.7764	0
4	74.9885	9.1639
5	94.0325	0

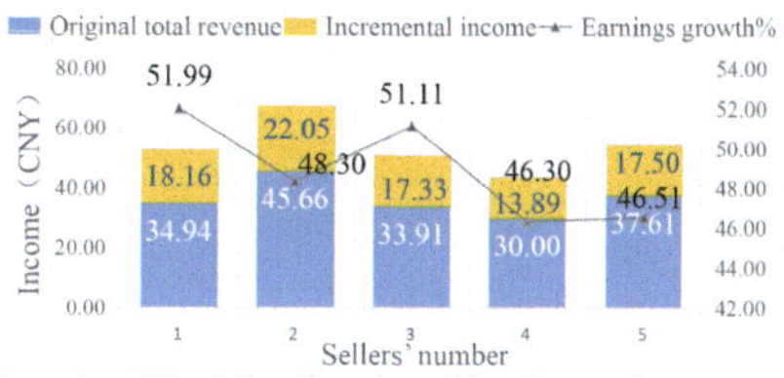

Fig. 2.20 Income of each seller in one period of bidding

Table 2.6 Trading results of each seller in the first round of bidding

Seller's number	Cumulative trading quantity (kWh)	Untraded quantity (kWh)
1	71.2778	16.0625
2	54.1035	60.0385
3	71.7139	13.0625
4	20.1380	64.0144
5	78.4700	15.5625

Table 2.6 and Fig. 2.21 present sellers' transaction outcomes during the first round (buyers' transaction outcomes are presented in Table 2.16 and Fig. 2.30). Compared to Fig. 2.20, the buyers' income decreases, as there is no market information feedback or interactive process in the single-round market mechanism, leading to a higher probability of transaction failure. As prosumers trade with bounded rationality and limited market information, the multi-round market mechanism can effectively improve their transaction behaviors through continuous information exchange. Consequently, the cleared price could accurately reflect the relationship between supply and demand, improving the market's fairness, efficiency, and vitality.

2.5.4 Impacts of Multi-mental Accounts

A comparison between the benchmark case and the proposed multi-mental account approach is made to validate the proposed method. The benchmark case assumes that the prosumers maintain a stable risk attitude. Prosumers submit quotations based

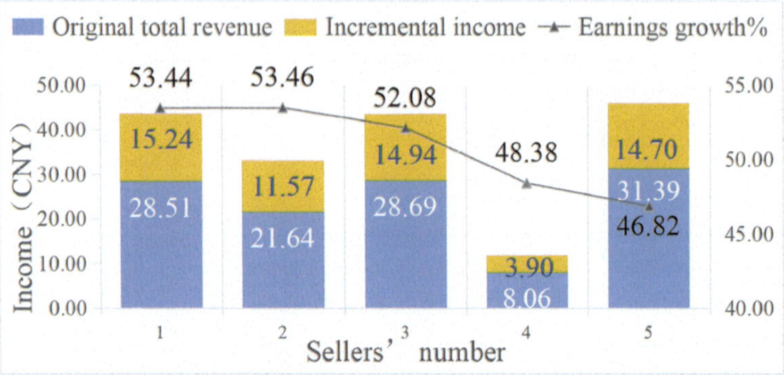

Fig. 2.21 Income of each seller in the first round of bidding

solely on a unified trading objective. The 16 consumption-type prosumers are categorized by risk attitude: four adventurous, eight neutrals, and four conservatives. The five production-type prosumers are categorized as one adventurous, three neutrals, and one conservative. There are five rounds of bidding.

Table 2.7 presents the clearing results. All transaction demands of conservative prosumers are fulfilled, as they prioritize the security of trading decisions and opt for a price advantage with low profit to increase the success rate. All transaction demands of neutral prosumers are fulfilled, too. They consider risks and benefits when making quotations and utilize the proposed updating strategy. Some adventurous prosumers' transaction demands are unsatisfied due to diagnosis bias and over-speculation behavior. Notably, no transactions are made in the second and third rounds. If the LEM only has one bidding round, prosumers with speculative and profit-seeking behavior would be less likely to achieve their goals in the LEM.

Table 2.8 and Fig. 2.22 present the transaction results of each seller (the buyers' transaction results are presented in Table 2.17 and Fig. 2.31). Both parties, buyers and sellers, gain certain benefits. Buyers' costs get reduced by approximately 25%, less than the 30% reduction by the multi-mental account approach. Sellers obtain approximately a 60% income raise, higher than the 48% increase in Fig. 2.20. The income growth fosters sellers' optimistic attitude. The cost savings for buyers are less significant, which may not offer strong incentives for them to continue participating in the LEM. The unbalanced participation levels of buyers and sellers exacerbate

Table 2.7 Transaction quantity of each round in the benchmark case

Round k	1	2	3	4	5
Tradable quantity (kWh)	497.042	93.364	93.364	93.364	65.461
Turnover (kWh)	403.678	0.000	0.000	27.903	54.596

Table 2.8 Trading results of each seller in the benchmark case

Seller's number	Cumulative trading volume (kWh)	Untraded volume (kWh)
1	96.6314	0
2	103.3106	0
3	131.2810	0
4	100.3584	0
5	54.5959	61.0753

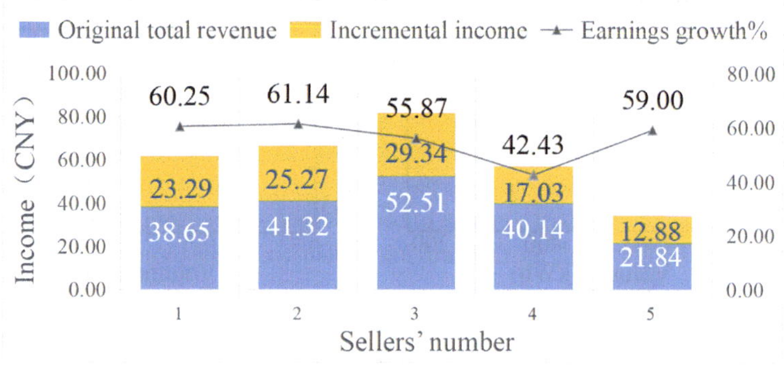

Fig. 2.22 Income of each seller in the benchmark case

the gap between supply and demand, obstructing the involvement of sustainable generation in the LEM.

The interaction mechanism is still evolving during the initial phase of the LEM. Though the failure of electricity trading might depress prosumers, the proposed trading decision method based on metal accounting theory could effectively mitigate this issue. Out of the three mental accounts, M1 guarantees a high successful trading probability, thereby improving prosumers' confidence while meeting basic income or demand. M2 and M3 offer opportunities to leverage local renewable energy or seek higher profits.

2.5.5 *Scalability and Practicality*

To assess the scalability and practicality of the proposed model, a simulation is carried out on a larger test case, which is a LEM with 120 consumption-type prosumers and 40 production-type prosumers. As the model is formulated in the form of linear programming problem, the computation time remains short even with a substantial increase in the number of prosumers. Table 2.9 and Fig. 2.23 present the simulation results consistent with the previous case study. This validates the scalability and practicality of the proposed method Table 2.10.

Table 2.14 Buyers' bidding information

Buyer's number	Quotation (CNY/kWh)	Quantity (kWh)	Account number
1	0.8290	1.0000	1
1	0.7080	30.7659	2
1	0.5650	13.0721	3
2	0.7920	1.0000	1
2	0.7140	10.2456	2
2	0.5170	12.6520	3
3	0.8050	1.0000	1
3	0.7600	5.5820	2
3	0.5440	12.9303	3
4	0.8160	1.0000	1
4	0.7140	19.8181	2
4	0.5280	13.6363	3
5	0.8310	1.0000	1
5	0.7080	11.9114	2
5	0.5790	9.1519	3
6	0.8060	1.0000	1
6	0.7080	14.6835	2
6	0.5180	13.0028	3
7	0.8090	1.0000	1
7	0.7110	13.1103	2
7	0.5320	12.1295	3
8	0.8270	1.0000	1
8	0.7320	25.6667	2
8	0.6610	8.2222	3
9	0.8270	1.0000	1
9	0.7480	10.6569	2
9	0.5660	11.8873	3
10	0.8200	1.0000	1
10	0.7320	16.9507	2
10	0.5420	13.3807	3
11	0.8310	1.0000	1
11	0.7110	11.2456	2
11	0.5360	11.5965	3
12	0.7880	1.0000	1
12	0.7240	6.7071	2
12	0.5980	8.1178	3
13	0.8280	1.0000	1

(continued)

Table 2.14 (continued)

Buyer's number	Quotation (CNY/kWh)	Quantity (kWh)	Account number
13	0.7140	24.0000	2
13	0.5410	13.6111	3
14	0.8310	1.0000	1
14	0.7130	27.8501	2
14	0.5540	13.4750	3
15	0.8300	1.0000	1
15	0.7110	22.2016	2
15	0.5850	10.7003	3
16	0.8030	1.0000	1
16	0.7570	1.0000	2
16	0.5740	9.3183	3

Table 2.15 Trading results of each buyer in one period

Buyer's number	Cumulative trading volume (kWh)	Untraded volume (kWh)
1	44.8380	0
2	23.8977	0
3	19.5123	0
4	34.4544	0
5	22.0634	0
6	28.6863	0
7	26.2398	0
8	34.8889	0
9	23.5441	0
10	31.3313	0
11	23.8421	0
12	15.8249	0
13	38.6111	0
14	42.3251	0
15	33.9019	0
16	11.3183	0

Case D

See Fig. 2.31 and Table 2.17.

Table 2.16 Trading results of each buyer in the first round of bidding

Buyer's number	Cumulative trading volume (kWh)	Untraded volume (kWh)
1	31.7659	13.0721
2	11.2456	12.6521
3	6.5820	12.9303
4	20.8181	13.6363
5	13.1788	8.8846
6	15.6835	13.0028
7	14.1103	12.1295
8	34.8889	0.0000
9	11.6569	11.8872
10	17.9507	13.3806
11	12.2456	11.5965
12	15.8249	0.0000
13	25.0000	13.6111
14	28.8501	13.4750
15	33.9019	0.0000
16	2.0000	9.3183

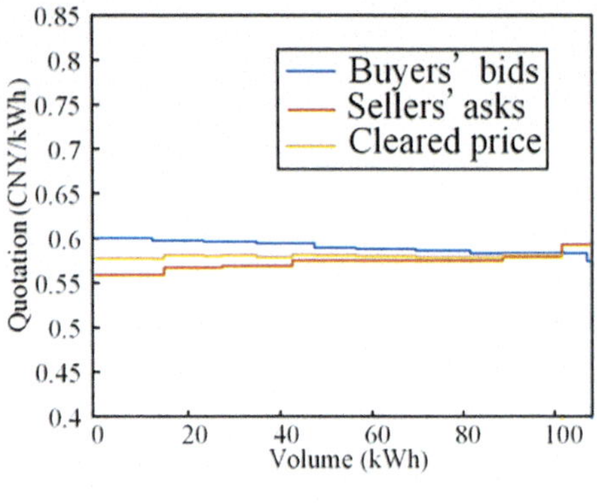

Fig. 2.27 Market clearing process diagram in the fourth round

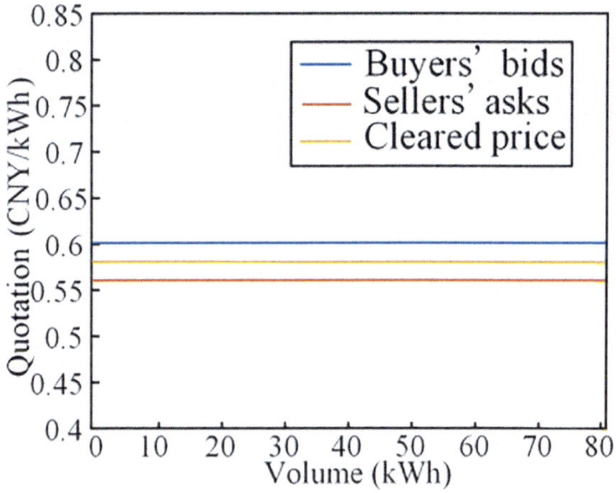

Fig. 2.28 Market clearing process diagram in the fifth round

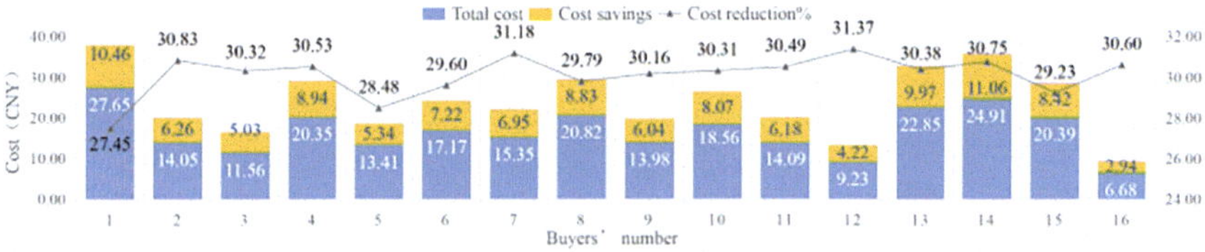

Fig. 2.29 Cost of each buyer in a single period of bidding

Fig. 2.30 Cost of each buyer in the first round of bidding

Table 2.17 Trading results of each buyer in the benchmark case

Buyer's number	Cumulative trading volume (kWh)	Untraded volume (kWh)
1	34.9651	0
2	23.6828	0
3	5.2700	0
4	40.6087	0
5	8.3879	0
6	62.4837	0
7	16.7586	0
8	16.2849	0
9	18.6224	0
10	13.6548	0
11	63.6229	0
12	23.1912	0
13	54.5959	0
14	58.3192	0
15	17.8262	0
16	27.9030	10.8649

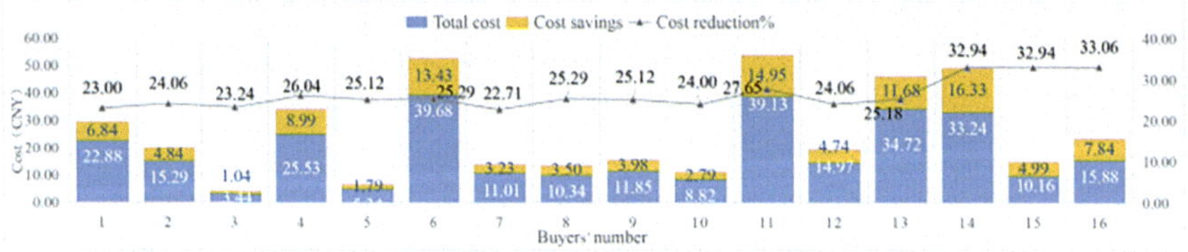

Fig. 2.31 Cost of each buyer in the benchmark case

References

1. Adeyemi, A., Yan, M., Shahidehpour, M., Bahramirad, S., & Paaso, A. (2020). Transactive energy markets for managing energy exchanges in power distribution systems. *The Electricity Journal, 33*(9), 106868.
2. Chen, Q., Wang, K., Chen, S., & Xia, Q. (2018). Transactive energy system for distributed agents: Architecture, mechanism design and key technologies. *Automation of Electric Power Systems, 42*(03), 1–7.
3. Song, M., & Gao, C. (2022). *Integration of distributed resources in smart grids for demand response and transactive energy: A case study of TCLs*. Springer Nature Press.
4. Melton, R. B. (2019). Gridwise transactive energy framework Ver. 1.1. Pacific Northwest National Lab. (PNNL), Richland, WA (United States).
5. Song, M., Sun, W., Wang, Y., Shahidehpour, M., Li, Z., & Gao, C. (2019). Hierarchical scheduling of aggregated TCL flexibility for transactive energy in power systems. *IEEE Transactions on Smart Grid, 11*(3), 2452–2463.

6. Song, M., Cai, Y., Gao, C., Chen, T., Yao, Y., & Ming, H. (2022). Transactive energy in power distribution systems: Paving the path towards cyber-physical-social system. *International Journal of Electrical Power & Energy Systems, 142*, 108289.
7. Bjarghov, S., Loschenbrand, M., Saif, A., et al. (2021). Developments and challenges in local electricity markets: A comprehensive review. *IEEE Access, 9*, 58910–58943.
8. Lu, K., Xu, H., & Zheng, Y. (2022). Distributed resource allocation via multi-agent systems under time-varying networks. *Automatica, 136*, 110059.
9. Yu, A., Zhang, C., & Zhang, Y. J. A. (2019). Optimal bidding strategy of prosumers in distribution-level energy markets. *IEEE Transactions on Power Systems, 35*(3), 1695–1706.
10. Ziras, C., Sousa, T., & Pinson, P. (2021). What do prosumer marginal utility functions look like? Derivation and analysis. *IEEE Transactions on Power Systems, 36*(5), 4322–4330.
11. Bai, X. I. A. O., Hanqi, C., Zhuo, J., Xu, Z., & Sun, L. (2021). Customized electricity price package design based on limited rational user selection behavior. *Power System Technology, 45*(03), 1050–1058.
12. Li, Y., et al. (2022). Decision-making model of electricity procurement and sale for electricity retailers considering multiple uncertain factors. *Automation of Electric Power System, 46*(7), 33–41.
13. Zhang, T. (2015). *Irrational consumption decision making in behavioral economics*. Yan' an University.
14. Cheng, L., Yang, R., et al. (2020). Multi-population asymmetric evolutionary game dynamics and its applications in power demand-side response in smart grid. *Proceedings of the CSEE, 40*(S1), 20–36.
15. Shen, Y., Li, Y., Zhang, Q., Li, F., & Wang, Z. (2020). Consumer psychology based optimal portfolio design for demand response aggregators. *Journal of Modern Power Systems and Clean Energy, 9*(2), 431–439.
16. Liu, Y., Gao, S., Wei, X., Huang, T., Zang, T., & Chen, K. (2021). Research on trading model of prosumers considering the social behaviors in the power Internet of Things. *Proceedings of the CSEE, 41*(9), 3173–3186.
17. Morstyn, T., & McCulloch, M. D. (2018). Multiclass energy management for peer-to-peer energy trading driven by prosumer preferences. *IEEE Transactions on Power Systems, 34*(5), 4005–4014.
18. Hongjun, G., Fan, Z., Junyong, L., et al. (2022). Design of P2P transaction mechanism considering differentiation characteristics of multiple prosumers in community microgrid system. *Proceedings of the CSEE, 42*(4), 1455–1470.
19. Yao, Y., Gao, C., Chen, T., Yang, J., & Chen, S. (2021). Distributed electric energy trading model and strategy analysis based on prospect theory. *International Journal of Electrical Power & Energy Systems, 131*, 106865.
20. El Rahi, G., Etesami, S. R., Saad, W., Mandayam, N. B., & Poor, H. V. (2017). Managing price uncertainty in prosumer-centric energy trading: A prospect-theoretic Stackelberg game approach. *IEEE Transactions on Smart Grid, 10*(1), 702–713.
21. Chen, L., Liu, N., Li, C., & Wang, J. (2020). Peer-to-peer energy sharing with social attributes: A stochastic leader–follower game approach. *IEEE Transactions on Industrial Informatics, 17*(4), 2545–2556.
22. Chen, L., Liu, N., Liu, L., Yu, X., & Xue, Y. (2021). Data-driven stochastic game with social attributes for peer-to-peer energy sharing. *IEEE Transactions on Smart Grid, 12*(6), 5158–5171.
23. Zhao, Y., & Gong, Z. (2021). Analysis on the influencing factors of customer consumption decision based on the theory of dual channel mental account. *Journal of Commercial Economics, 23*, 46–49.
24. Zhan, Z., & Wu, Z. (2022). Mental accounting, loss aversion and behavioral asset allocation empirical study. *Operations Research and Management Science, 31*(8), 177.
25. Zhang, P., & Ye, S. (2022). Random fuzzy mean semi-absolute deviation portfolio selection problem with mental accounts. *Fuzzy Systems and Mathematics, 36*(3), 154–163.
26. Qiu, S., Wu, L., Yang, Y., & Zeng, G. (2022). Offering the right incentive at the right time: Leveraging customer mental accounting to promote prepaid service. *Annals of Tourism Research, 93*, 103367.

27. Li, Z., Jiang, H., Chen, Z., & Wong, W. K. (2022). A mental account-based portfolio selection model with an application for data with smaller dimensions. *Computers & Operations Research, 144*, 105801.
28. Liang, J., & Tang, W. (2019). Optimal trading strategies in continuous double auctions for transactive energy. In *2019 North American Power Symposium (NAPS)*.
29. Jalali, M., Zare, K., & Tohidi, S. (2022). Impartial pricing approach in double auction transactive distribution systems. *International Journal of Electrical Power & Energy Systems, 135*, 107204.
30. Chen, K., Lin, J., & Song, Y. (2019). Trading strategy optimization for a prosumer in continuous double auction-based peer-to-peer market: A prediction-integration model. *Applied Energy, 242*, 1121–1133.
31. Thaler, R. H. (1999). Mental accounting matters. *Journal of Behavioral Decision Making, 12*(3), 183–206.
32. Li, M. (2021). Research on residents' consumption behavior based on the theory of mental accounts. *China Journal of Commerce, 17*, 35–37.
33. Kaenzig, J., Heinzle, S. L., & Wüstenhagen, R. (2013). Whatever the customer wants, the customer gets? Exploring the gap between consumer preferences and default electricity products in Germany. *Energy Policy, 53*, 311–322.
34. Da Silva, P. G., Karnouskos, S., & Ilic, D. (2012). A survey towards understanding residential prosumers in smart grid neighbourhoods. In *2012 3rd IEEE PES Innovative Smart Grid Technologies Europe (ISGT Europe)*.
35. Tabi, A., Hille, S. L., & Wüstenhagen, R. (2014). What makes people seal the green power deal?—Customer segmentation based on choice experiment in Germany. *Ecological Economics, 107*, 206–215.
36. Dagher, L., & Harajli, H. (2015). Willingness to pay for green power in an unreliable electricity sector: Part 1. The case of the Lebanese residential sector. *Renewable and Sustainable Energy Reviews, 50*, 1634–1642.
37. Zorić, J., & Hrovatin, N. (2012). Household willingness to pay for green electricity in Slovenia. *Energy Policy, 47*, 180–187.
38. Li, Y., Han, M., Yang, Z., & Li, G. (2021). Coordinating flexible demand response and renewable uncertainties for scheduling of community integrated energy systems with an electric vehicle charging station: A bi-level approach. *IEEE Transactions on Sustainable Energy, 12*(4), 2321–2331.
39. Lopes, L. L. (1987). Between hope and fear: The psychology of risk. In *Advances in experimental social psychology* (Vol. 20, pp. 255–295). Academic Press.
40. Lv, Q., Wu, Y., & Li, W. (2006). A novel bid decision-making model for generation companies based on multiple mental account. *Proceedings of the CSEE, 24*, 24–30.
41. Yang, Q., Wang, H., Wang, T., Zhang, S., Wu, X., & Wang, H. (2021). Blockchain-based decentralized energy management platform for residential distributed energy resources in a virtual power plant. *Applied Energy, 294*, 117026.
42. Doan, H. T., Cho, J., & Kim, D. (2021). Peer-to-peer energy trading in smart grid through blockchain: A double auction-based game theoretic approach. *IEEE Access, 9*, 49206–49218.
43. Lin, J., Pipattanasomporn, M., & Rahman, S. (2019). Comparative analysis of auction mechanisms and bidding strategies for P2P solar transactive energy markets. *Applied Energy, 255*, 113687.
44. Guo, L., Chen, X., Deng, H., He, Y., & Chen, Q. (2018). A framework of operating mechanism based on peer-to-peer transaction among distributed energy resources in community microgrid. *Electric Power Construction, 39*, 2–9.

Chapter 3
Prospect Theory-Based Trading Strategy and Analysis of Prosumers

Abstract Due to the increasing penetration of renewable energy (RE), existing traditional centralized clearing mechanisms are no longer suitable for the local energy market (LEM). Moreover, existing research often overlooks prosumers' subjective preferences, leading to suboptimal decisions that weaken social welfare and market efficiency. To tackle these challenges, we propose a trading strategy based on prospect theory (PT) and an adaptive learning method to optimize decisions for diverse prosumers and facilitate transactions in LEM. Using blockchain technology, we introduce a smart contract-based matching and settlement mechanism to ensure prosumers' peer-to-peer (P2P) and efficient transaction process. The advantages of the proposed trading strategy are highlighted through a comparison with the zero intelligence (ZI) method. We also adopted the Remix Ethereum platform to confirm its practicality.

3.1 Introduction

Prosumers who own renewable energy (RE) generation devices, serve both as producers and consumers of electricity [1, 2], marking a new position in the local energy market (LEM). With the continuous development of RE on the user side, massive prosumers in the distribution network are rising, bringing significant variations in their power consumption and behavior [3–5]. This growth has surpassed the traditional centralized clearing transaction model, which cannot meet the decentralized power generation structure requirement in LEM. Additionally, third-party trading institutions pose extra transaction costs, hindering market efficiency and reducing consumer participation [6]. A decentralized transaction model facilitates peer-to-peer (P2P) transactions among market participants, enhancing market efficiency and gradually replacing the centralized model [6]. The volatility and unpredictability of electricity prices also affect prosumers' subjective estimations and influence their transaction strategies [7]. Consequently, designing a trading mechanism that accommodates the different characteristics of prosumers is a crucial research focus in LEM.

M. Song et al., *Local Energy Markets*, https://doi.org/10.1007/978-981-97-9750-9_3

For previous research on LEM, prosumers' trading processes are based on rational decisions and the objective expected utility [8–10]. For example, [8] designs a P2P energy trading market that optimizes prosumers' energy consumption costs using rational rules, [9] proposes a P2P clearing mechanism to maximize the revenue rewards of users executed through incentive contracts, and [10] develops a bilateral contract network for P2P energy trading aimed at maximizing the utility of rational market participants. However, these above models fail to capture the subjective preferences of market participants accurately [11, 12]. As a result, in practice, human choices are often excluded from these objectives, such as user-agnostic decisions [13, 14]. Unlike the expected utility theory, prospect theory (PT) adopts the probability weights to convert objective probabilities into subjective probabilities, accounting for individual differences among market participants and addressing the limitations of overly rational assumptions [7]. Under PT, [7] designs a Stackelberg Game that incorporates each prosumer's valuation of gains and losses relative to a personal utility reference point. Additionally, [13] evaluates the impact of prosumers' subjectivity on micro-grid power trading results by assessing their objective weight, and [15] presents a stochastic game that considers prosumers' subjective behavior to analyze interactions in P2P trading. Building on these concepts, this chapter proposes a LEM trading mechanism based on PT to analyze prosumer decision-making behavior with varying subjective degrees as the objective probability of market prices changes.

Based on the inefficiencies of traditional centralized trading, it also leads to issues such as unmet real-time operation constraints in the distribution network [16], high costs associated with centralized organization, and significant information leakage risks and trust issues, as all market participants must offer essential transaction information to third-party platform [17].

We can leverage blockchain technology to address these issues and enable P2P management and market operations. The ability of Blockchain is to maintain a continuously growing and immutable data record eliminates the need for third-party organizations, ensuring the safe and stable execution of P2P power transactions [18]. The distribution nature of blockchain promotes P2P transactions among end-users and enhances the self-balancing capability of P2P transactions. Smart contracts, a key component of blockchain transactions, automatically execute pre-defined rules and terms, ensuring the fairness and efficiency of energy transactions. For instance, [16] proposes a distribution network transaction mechanism based on smart contracts that operate without a central agency, using active prosumers to mitigate deviations between electricity generation and consumption, thereby maintaining the dynamic balance of supply and demand in the distribution network. Similarly, [19] employs smart contracts for automatic demand response and energy scheduling for prosumers in regional power networks. However, recent research concerning smart contract technology's utilization for P2P transaction markets often assumes rational prosumers, neglecting the subjective characteristics and behavioral differences among them. This chapter addresses this gap by fully considering prosumer characteristics such as subjectivity, comfort requirements, price sensitivity, and transmission distance sensitivity. By applying PT to the P2P transaction, this study seeks to overcome the limitations of existing research and provide a more comprehensive framework for understanding and facilitating prosumer behavior in P2P transactions.

Unlike the wholesale market, where transactions are typically centralized, P2P trading is initiated by market participants based on their expected trading volume and price decisions. Continuous double auction (CDA) is a significant mechanism in P2P trading, and it is widely used to enable prosumers to make strategic decisions [20]. For example, Minghua et al. [21] proposes a fuzzy-logic-based bidding strategy to determine the best bid under various market conditions. Yinxin et al. [22] designs an optimal pricing strategy for generators within a CDA system, while [23] explores how prosumer operations and market trading strategies can be jointly optimized through a CDA market. However, these studies primarily focus on bidding price strategies in CDA, often neglecting the optimization of trading volume. This oversight can lead to suboptimal individual decisions, especially when considering the diverse characteristics of prosumers. An optimization method for trading volume decisions (TVD) is proposed to address this gap. This method integrates with the CDA process to help multi-featured prosumers make the best possible trading decisions.

In addition, after implementing the CDA, some market participants fail to achieve a transaction match due to incomplete market information, resulting in social welfare loss and reduced market efficiency [24, 25]. To mitigate this issue, an adaptive learning method combined with the CDA strategy is proposed to enhance bidding decisions for prosumers. This strategy makes market participants more competitive in the bidding process and increases their chances of completing transactions, thereby improving social welfare and market efficiency. Research on updating bidding strategies in CDA often focuses on estimating future market prices based on historical data using machine learning [23] or adjusting market participants' aggressiveness toward bidding prices [26]. For example, a P2P trading strategy driven by novel forecasting methods was proposed in [27]. Additionally, Gode and Sunder [28] introduced a zero intelligence (ZI) strategy, which assumes that market participants make profitable decisions based on bidding price constraints. Building on this, an advanced ZI plus strategy [29] was developed, offering higher market efficiency compared to [28]. However, these studies overlook the impact of market participants' subjectivity on their bidding decisions. This chapter proposes a P2P trading model that considers prosumers' characteristics to address these gaps. The model comprises two main components: a PT-based TVD model and an adaptive learning strategy-based bidding decision model. First, the optimal TVD is determined and validated through the model's analytical solution. Then, incorporating prosumers' subjectivity and the TVD results, adaptive learning is utilized to update and decide the prosumers' bidding prices. Furthermore, smart contracts utilizing blockchain technology are employed to manage the matching, transaction, and settlement processes among market participants in the P2P transaction market. A case study is conducted to verify the rationality and effectiveness of the proposed model. Finally, we analyze the impact of prosumers' subjectivity on market transaction outcomes and profits, and the feasibility of the designed smart contract is tested using the Ethereum platform. The major contributions of this chapter can be concluded below.

(1) To address the imprecise representation of prosumers' subjective preferences in existing research, this chapter proposes a distributed market trading strategy based on PT to accommodate prosumers with varying degrees of subjectivity. Furthermore, the feasibility and rationality of this transaction model are verified using smart contract architecture within the blockchain platform.
(2) Existing research on P2P trading with the CDA strategy often overlooks the optimization of prosumers' TVD. We bridge this gap by constructing an optimal TVD model integrated with the CDA strategy. The model optimizes prosumers' TVD considering variations in market price forecasts while accounting for factors such as comfort, price responsiveness, transmission distance sensitivity, and subjectivity. Additionally, the optimized values are obtained through analytical solutions, enhancing convenience for smart contract execution.
(3) This chapter adopts an adaptive learning strategy within the CDA model to update prosumers' bids based on changes in market information. Notably, prosumers' subjectivity influences their estimation of the Competitive Equilibrium Price (CEP) during the bidding process in our model. Case study results demonstrate that transactions between competitive market members can be effectively matched through the combination of adaptive learning. Moreover, the proposed trading model enhances social welfare and market efficiency compared to the ZI method.

3.2 Prosumers' Trading Model

3.2.1 Prosumers' Trading Strategy

The trading strategy framework for prosumers can be described using the Information, Knowledge, and Behavior (IKB) architecture, as illustrated in Fig. 3.1. This framework includes the information, knowledge, and behavior layers. Firstly, the information layer collects historical trading data from the LEM and the characteristics and preferences of prosumers, then transfers these necessary parameters to the knowledge layer. The knowledge and behavior layers include the adaptive aggressiveness (AA) strategy and PT, which update bids and calculate the optimal TVD. In the knowledge layer, PT uses the data from the information layer to calculate the subjective probability of the objective market price (OMP). The AA strategy then uses this subjective probability to estimate the CEP and develop an aggressiveness-adapting model.

(1) PT based TVD

According to the theory of PT, each prosumer i uses the probability weight to maximize the utility function Γ_i^{PT} as the objective for the decision-making of trading quantity and price. The objective function is:

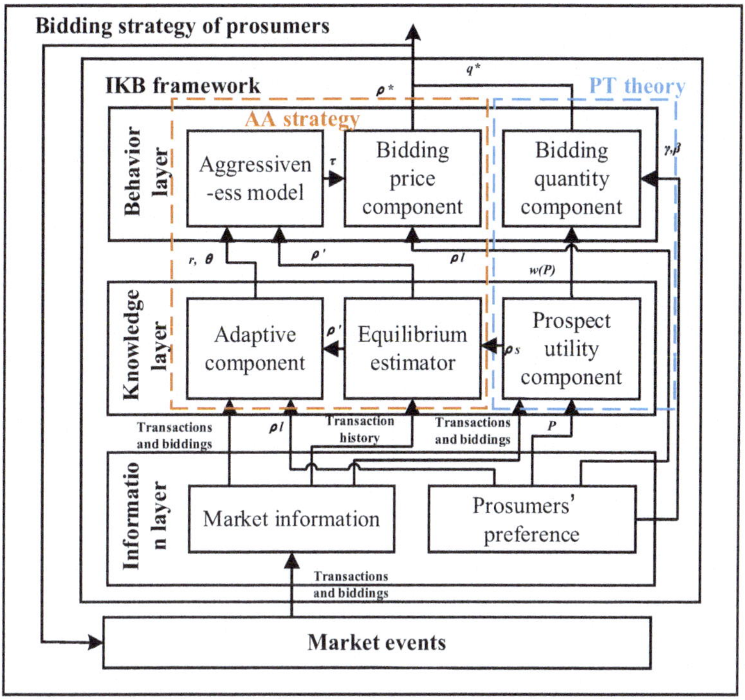

Fig. 3.1 Framework of prosumers' trading strategy

$$\max \Gamma_i^{\mathrm{PT}} = \sum_{m \in \Omega_p} u_i \omega_{i,m} \tag{3.1}$$

And the subjective probability $\omega_{i,m}$ can be formulated as [30, 31]

$$\omega_{i,m} = \exp(-(-\log P_m)^{\alpha_i}) \tag{3.2}$$

where α_i is a subjective factor used to measure the subjectivity degree of the prosumers on the objective probability. The higher the value, the more rational and the less subjective the prosumers, which ranges between (0, 1]. When $\alpha_i = 1$, the prosumer is completely rational, i.e., the optimization result of the Eq. (3.1) is an objective decision.

The u_i is expressed as

$$u_i = F_i^U - F_i^T - F_i^D \tag{3.3}$$

where F_i^U, F_i^T and F_i^D are the load benefit of the prosumer [32], the transaction expenditure, and the heat loss due to the transmission of electrical energy [33], which are calculated as

$$F_i^U = \beta_i \log(q_i + R_i + 1) \tag{3.4}$$

$$F_i^T = \rho_i q_i \tag{3.5}$$

$$F_i^D = \gamma_i q_i^2 \tag{3.6}$$

$$q_i > -R_i \tag{3.7}$$

where q_i is the amount of electricity purchased/sold by the prosumer. A positive value indicates the purchase of electricity, otherwise the sale of electricity. γ_i is the distance sensitivity coefficient. The larger the value, the higher the sensitivity of prosumers to transmission costs, i.e., they tend to accept short-distance electricity transactions.

It is worth noting that β_i and γ_i are constants with values between (0, 1). Generally, it keeps different values of β_i and γ_i. for prosumers. Therefore, we can divide the prosumers according to comfort requirements, price responsiveness, and transmission distance sensitivity to characterize them in the market.

(2) AA strategy-based bidding and asking decisions (BAD)

As the trading price ρ_i of the prosumer in the market undergoes adjustments in the analytical solution outlined in Eqs. (3.1–3.7) (detailed in Sect. 3.4.4), this chapter employs the AA strategy, leveraging short-term and long-term learning mechanisms to determine ρ_i, which denotes the bidding price for buyers and the asking price for sellers. The AA strategy implementation comprises three main steps: (a) CEP estimation; (b) aggressiveness model; (c) short-term and long-term adaptive learning.

(a) CEP estimation

In a perfectly competitive market, the equilibrium price is achieved when supply equals demand, known as competitive equilibrium [34]. This price can be forecasted by data prediction techniques. However, the estimation of the CEP for each prosumer is affected by individual subjectivity. Consequently, this section employs the OMP ρ_m algorithm along with each prosumer's subjective probability $\omega_{i,m}$ to estimate the CEP ρ_i', as demonstrated in Eq. (3.8):

$$\rho_i' = \frac{\sum_m (\rho_m \omega_{i,m})}{\sum_m \omega_{i,m}} \tag{3.8}$$

Since $\sum_m \omega_{i,m} = 1$, the above formula can be written as:

$$\rho_i' = \sum_m (\rho_m \omega_{i,m}) \tag{3.9}$$

(b) Aggressiveness model

Aggressiveness reflects the level of engagement of prosumers in market competition. When prosumers determine a target ask (for sellers) or bid (for buyers) based on their estimation of the CEP, their aggressiveness plays a significant role. Aggressive sellers (buyers) will lower (raise) their target ask (bid) to enhance competitiveness and increase trading opportunities. Conversely, less aggressive (passive) sellers (buyers) will raise (lower) their target ask (bid) to secure higher transaction profits. Additionally, when prosumers maintain a neutral stance, their target asks (bids) are aligned with their expected transaction prices. In other words, neutral prosumers set their target asks (bids) equal to their subjectively estimated CEP. According to the relationship between the prosumers' target asks (bids) and ρ_i', this section splits them into intra-marginal prosumer (IMP) and extra-marginal prosumer (EMP). As for prosumers who sell electricity, when $\rho_i < \rho_i'$, it is defined as the IMP, otherwise they are the EMPs.

We use the parameter r to represent the market competition initiative of prosumers. When $r \in [-1, 1]$ and $r \to 1$, it indicates that prosumers have strong aggressiveness. Conversely, $r \to -1$ means that the prosumers have weaker aggressiveness. When $r = 0$, prosumers are considered neutral market participants. Figure 3.2 illustrates the relationship between the target bidding price (asking price) and the aggressiveness of both IMP and EMP.

The function of the target bid (ask) of prosumers and their aggressiveness is as follows:

$$\rho^{\text{in},b} = \begin{cases} (\rho' - \rho_{\min})\left(1 - \frac{e^{-r\theta}-1}{e^{\theta}-1}\right) + \rho_{\min}, & r \in (-1, 0] \\ \rho' + (\rho_h - \rho')\frac{e^{r\theta}-1}{e^{\theta}-1}, & r \in (0, 1) \end{cases} \tag{3.10}$$

$$\rho^{\text{in},s} = \begin{cases} \rho' + (\rho_{\max} - \rho')\frac{e^{-r\theta}-1}{e^{\theta}-1}, & r \in (-1, 0] \\ \rho_l + (\rho' - \rho_l)\left(1 - \frac{e^{r\theta}-1}{e^{\theta}-1}\right), & r \in (0, 1) \end{cases} \tag{3.11}$$

$$\rho^{\text{ex},b} = \begin{cases} (\rho_h - \rho_{\min})\left(1 - \frac{e^{-r\theta}-1}{e^{\theta}-1}\right) + \rho_{\min}, & r \in (-1, 0] \\ \rho_h, & r \in (0, 1) \end{cases} \tag{3.12}$$

$$\rho^{\text{ex},s} = \begin{cases} \rho_l + (\rho_{\max} - \rho_l)\left(\frac{e^{-r\theta}-1}{e^{\theta}-1}\right), & r \in (-1, 0] \\ \rho_l, & r \in (0, 1) \end{cases} \tag{3.13}$$

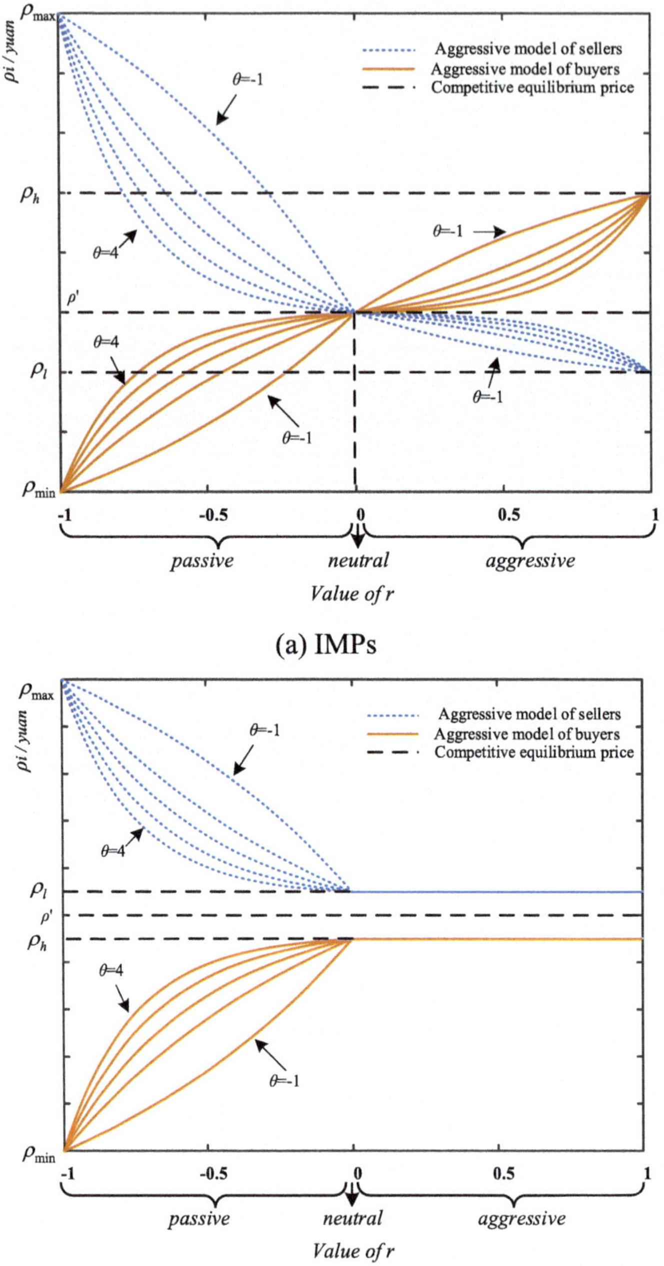

Fig. 3.2 Relationship curve between ρ_i and aggressiveness

According to Eqs. (3.10–3.13), the function curves depicted in Fig. 3.2 illustrate that when $r \to 1$, the target bids of buyers continue to rise, while the target asks of sellers decrease. This occurs because, as prosumers become more aggressive, they aim to secure market transactions by enhancing their competitiveness, thus adjusting their target bids (asks) to favor trading. For the buyers classified as EMP, because the maximum price of the target bid is lower than ρ_i', when $r > 0$ the target bid is maintained at the maximum price ρ_h. This is the same case with sellers of EMP: When $r > 0$, their ask is maintained at the minimum price ρ_l. We assume that the highest limit bidding price ρ_h for buyers equals the CEP, meaning the buyer's target bid does not exceed their unit load utility. The minimum asking price ρ_l for a seller is set to γ_i. This assumption is rational because the marginal cost for users generating electricity from RE in a LEM is approximately zero. However, the transmission loss incurred during the transaction needs to be considered, which can be represented by the unit loss cost γ_i. Therefore, the asking price for sellers should be greater than γ_i.

It is important to note that the prosumers' BAD is influenced by the variable parameter θ. For both IMPs and EMPs, when $r = 0$, the curve changes more slowly with a higher θ and more rapidly with a lower θ.

(c) Adaptive learning process

The adaptive learning process refers to updating and adjusting the parameters r and θ through short-term and long-term learning processes, respectively.

Short-term learning can be expressed as:

$$\begin{cases} r_{k+1} = r_k + \beta_1[\delta_k - r_k] \\ \delta_k = (1 + \lambda) r_k^{\text{op}} \\ \lambda = \{-0.05, 0.05\} \\ 0 < \beta_1 < 1 \end{cases} \tag{3.14}$$

when $r(k)=\delta(k)$, the prosumer makes a competitive bid within his bidding (asking) range, to strike a transaction. When $\lambda = -0.05$, the prosumer tends to improve his transaction profit by reducing r_k. Otherwise, the prosumer tends to strike a transaction in the market by increasing r_k. If prosumer i strikes none transactions in the round k, $r_{\text{op}}(k) = \rho^T$. Otherwise, $r_k^{\text{op}} = \rho_{i,k}$.

According to Eq. (3.14), the algorithm engages in an ongoing learning process, where the prosumer's level of aggressiveness in round $k + 1$ is updated based on the discrepancy between the expected and actual aggressiveness from the previous round. It is important to note that this aggressiveness varies with changes in $\delta(k)$, and the rate of this change is influenced by β_1. The steps of this short-term learning process are detailed in Algorithm 1.

Algorithm 1: Short-term learning process

If (prosumer is a buyer)
  If (the bidding price of the traded prosumer is q in round k)
    If ($\tau \geq q$) → reduce r in k+1
    Else increase r in k+1
  Else if (k=1, optimal bidding price is b)
    If ($\tau \leq b$) → increase r in k+1
    Else reduce r in k+1
Else if (prosumer is a seller)
  If (the asking price of the traded prosumer is q in round k)
    If ($\tau \leq q$) → reduce r in k+1
    Else increase r in k+1
  Else if (k=1, optimal asking price is s)
    If ($\tau \geq b$) → increase r in k+1
    Else reduce r in k+1

The rate at which a prosumer's bids converge to the CEP can be altered through long-term learning by adjusting the parameter θ, which is influenced by the variability of OMP ρ_m. For instance, if OMP shows considerable volatility, the long-term learning mechanism can modify the prosumer's bids to achieve rapid convergence by lowering θ. This process of long-term learning is depicted as:

$$\theta_{k+1} = \theta_k + \beta_2(\theta' - \theta_k) \tag{3.15}$$

$$p_v = \frac{\sqrt{\frac{1}{n}\sum_{i=m-n+1}^{n}(p_i - p')^2}}{p'} \tag{3.16}$$

$$\theta' = \theta'(p_v) = (\theta_{\max} - \theta_{\min})\left(1 - \frac{p_v - p_{v\ \min}}{p_{v\ \max} - p_{v\ \min}}\right)e^{\varepsilon\left(\frac{p_v - p_{v\ \min}}{p_{v\ \max} - p_{v\ \min}} - 1\right)} + \theta_{\min} \tag{3.17}$$

$$0 < \beta_2 < 1 \tag{3.18}$$

(d) BAD model

Based on the target bid and the optimal bid of the last round, prosumers make final BADs intending to strike a transaction. The BAD strategies for buyers p_k^b and sellers p_k^s are as follows:

$$p_k^b = \begin{cases} p_k^{\text{obid}} + \eta\left[\min\left\{v_i, p_k^{\text{oask}}\right\} - p_k^{\text{obid}}\right], & k = 1 \\ p_k^{\text{obid}} + \eta\left[\rho_k^b - p_k^{\text{obid}}\right], & k \geq 2 \end{cases} \tag{3.19}$$

$$p_k^s = \begin{cases} p_k^{\text{oask}} - \eta\left[p_k^{\text{oask}} - \max\left\{c_j, p_k^{\text{obid}}\right\}\right], & k = 1 \\ p_k^{\text{oask}} - \eta\left[p_k^{\text{oask}} - \rho_k^s\right], & k \geq 2 \end{cases} \tag{3.20}$$

$$0 < \eta \leq 1 \tag{3.21}$$

In this section, considering the TVD of prosumers, η is set to be proportional to the absolute value of q_i. This approach benefits competitive prosumers with higher trading volumes, enhancing their chances of completing transactions.

3.2.2 Smart Contract-Based Transaction Matching and Market Settlement

The designed smart contract encompasses a series of distributed energy transaction details, including transaction time, trading volume, and the addresses of buyers and sellers [18, 35]. The transaction matching and settlement process, based on the smart contract technology outlined in this section, is illustrated in Fig. 3.3.

(1) Calculation of Subjective TVD Based on PT: Utilizing PT to maximize the Γ_i^{PT} for each prosumer, prosumers solve the optimization problems specified in Eqs. (3.1–3.7). Consequently, the subjective TVD for each prosumer can be determined based on these solutions.
(2) Definition of the roles of prosumers. Based on the TVD of prosumers, the smart contract assigns them roles as buyers or sellers. When $q_i \geq 0$, the prosumer is defined as a buyer; otherwise, the prosumer is defined as a seller.
(3) The BAD update and transaction pair matching. The prosumers' bids (asks) are updated according to the AA strategy. When $k = 1$, in the absence of transaction information, all prosumers adjust their bidding (asking) strategies based on the optimal bidding (asking) price, which is the maximum value among buyers' bids and the minimum value among sellers' asks. When $k > 1$, adaptive

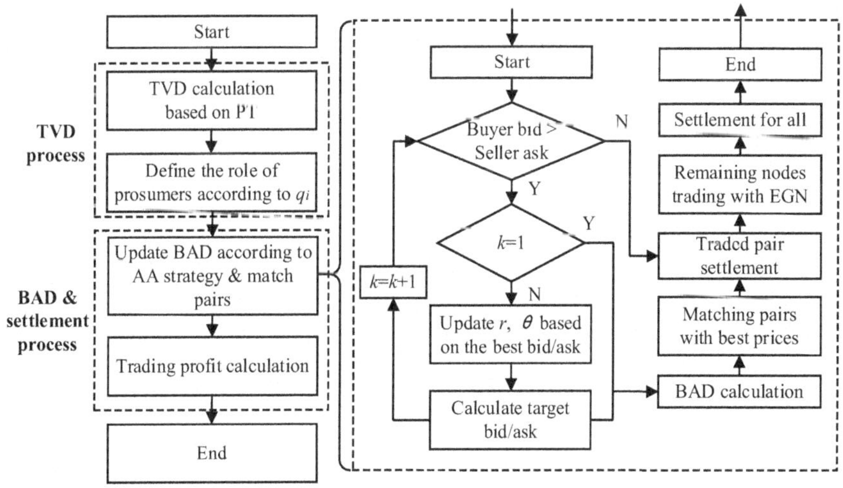

Fig. 3.3 Smart contract execution process

learning is conducted using the transaction information from round $k-1$. The bidding (asking) strategy is adjusted through short-term and long-term learning processes. After these adjustments, prosumers with optimal bids and asks are matched and traded via the smart contract. Prosumers who remain unmatched then trade with the external grid network (EGN).

(4) Transaction settlement. In each round, the settlement price for a trading pair is the average of the seller's ask price and the buyer's bid price. The settlement electricity volume is determined by the traded volume, as calculated by the smart contract. For prosumers trading with the EGN, sellers settle at the benchmark price, while buyers settle at the off-grid price.

(5) Trading profit calculation. The smart contract computes the market transaction profits for prosumers by substituting the actual transaction prices q_i and volumes ρ_i into Eq. (3.3–3.6). This involves using the realized values of the trading parameters to determine the financial outcomes for each prosumer based on their respective roles as buyers or sellers.

3.3 Solutions and Smart Contract Compilation

3.3.1 Solutions of TVD

To execute the smart contract process, the first step is to calculate the TVD of each prosumer by solving the model presented in Sect. 3.2.1. To determine the concavity or convexity of the model, the second-order partial derivative of Eq. (3.1) needs to be obtained:

$$\partial^2 u_i \partial^{-2} q_i = -\beta_i^2 (q_i + R_i + 1)^{-2} - 2\gamma_i < 0 \tag{3.22}$$

Therefore, it can be seen that the objective function is a concave function [36]. Define q_i^* to be the optimal TVD, i.e.,

$$q_i^* = \arg\max \Gamma_i^{\mathrm{PT}} \tag{3.23}$$

In the domain of definition, according to the first-order optimality condition, the optimal TVD can be solved as

$$\partial \Gamma_i^{\mathrm{PT}} \partial^{-1} q_i = 0 \Rightarrow \sum_{m \in \Omega_{\mathrm{P}}} \left[\beta_i (q_i + R_i + 1)^{-1} - \rho_i - 2\gamma_i q_i\right] \omega_{i,m} = 0 \tag{3.24}$$

Since $\omega_{i,m} \neq 0$, we can further get:

$$\frac{\beta_i}{q_i + R_i + 1} - \hat{\rho}_i - 2\gamma_i q_i = 0 \tag{3.25}$$

where $\hat{\rho}_i$ is written as:

$$\hat{\rho}_i = \sum_{m\in\Omega_P} \rho_i\omega_{i,m}\left(\sum_{m\in\Omega_P} \omega_{i,m}\right)^{-1} \tag{3.26}$$

By simplifying Eq. (3.26), a quadratic Eq. (3.27) can be obtained:

$$2\gamma_i q_i^2 + (\hat{\rho}_i + 2\gamma_i R_i)q_i + \left[(\hat{\rho}_i + 2\gamma_i)(R_i + 1) - \beta_i\right] = 0 \tag{3.27}$$

Solved as:

$$q_i^* = -2z_i\left(y_i - \sqrt{y_i^2 - 4x_i z_i}\right)^{-1} \tag{3.28}$$

where:

$$x_i = 2\gamma_i \tag{3.29}$$

$$y_i = \hat{\rho}_i + 2\gamma_i R_i \tag{3.30}$$

$$z_i = (\hat{\rho}_i + 2\gamma_i)(R_i + 1) - \beta_i \tag{3.31}$$

Once the role of each prosumer is defined based on the calculated TVD and the analytical solution of the TVD model, the smart contract utilizes this data for transaction matching and settlement. This approach enhances the execution speed of the smart contract by streamlining the process and ensuring efficient allocation of resources in the market. With clear roles assigned to buyers and sellers, the matching and settlement process can proceed smoothly, facilitating timely and accurate transactions.

3.3.2 Smart Contract Compilation

This section utilizes the Solidity language on the Remix platform to compile and deploy an Ethereum-based blockchain for achieving transaction matching and settlement through smart contracts [35]. Figure 3.4 illustrates the deployed Remix interface, showcasing the user interface for interacting with the smart contract functionalities.

First, the "javascriptVM" environment under Solidity is selected in the Remix compiler. Then, Remix provides virtual accounts through the Ethereum virtual machine, each initially containing 100 ether. Following Remix's compilation requirements, after successfully deploying the smart contract, initial information needs to

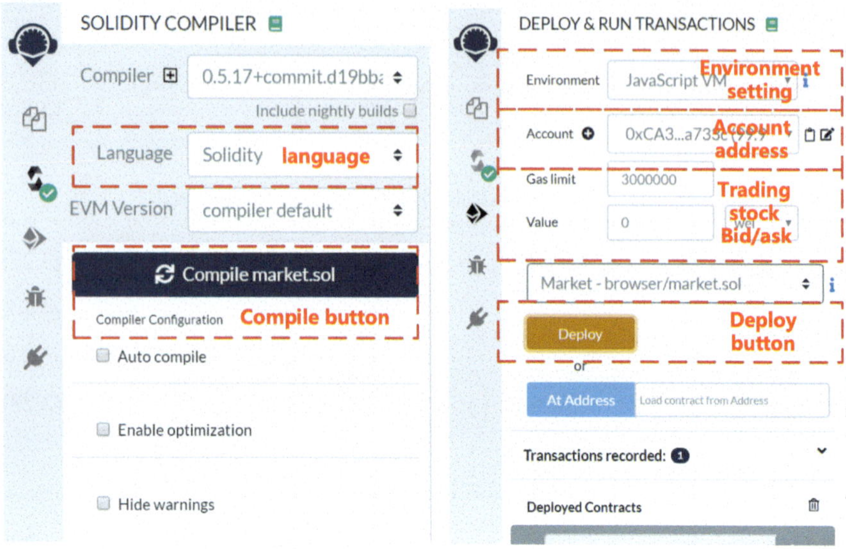

Fig. 3.4 Diagram of Remix interface

be submitted in the format of the constructor. This includes the addresses of the prosumers (Ethereum addresses), their TVDs, and BADs. Since the Remix platform's compilation requires integer types, the prosumers' TVDs, and BADs are multiplied by 100 to integrate the submission information.

The simulation environment is on the computer operating system used for Remix compilation and deployment is Windows 10 (×64), with 8.00 GB of RAM and an Intel Core i5-8250U CPU running at 1.6 GHz.

3.4 Case Studies

3.4.1 Case Description

To verify the validity and rationality of the model presented in this section, a modified IEEE-33 node distribution system, depicted in Fig. 3.5, is adopted for simulation and analysis. In this system, node 1 is the root node connected to the EGN. Each node, except the root node, corresponds to a prosumer characterized by a dual source-load characteristic. Certain nodes (e.g., nodes 14, 24, 25, 30, and 32) are equipped with higher capacity RE sources, while others have larger loads (e.g., nodes 6, 7, 24, 25, 30, 31, and 32). Figure 3.5 provides an overview of the load, RE installation, electric storage (ES) capacity, comfort factor β_i, and subjective degree α_i of each prosumer. Additionally, we calculate the distance sensitivity coefficient γ_i based on the average physical distance of each node's location from the node with high power generation capacity and the root node [37]. The resulting coefficients are illustrated in Fig. 3.6.

To analyze the influence of OMPs on transaction results, six scenarios have been established for comparative analysis, as outlined in Table 3.1a. Furthermore, Cases 4–6 are subdivided into three subcases each, to evaluate the impact of prosumers' subjectivity on their trading strategy and results. Subjective degrees α_H, α_M and α_L are assigned as α_0, $\alpha_0/2$ and $\alpha_0/4$ respectively, as shown in Table 3.1b. This ensures variation in prosumers' subjectivity across different cases. For instance, the subjectivity of prosumers increases gradually from Case 4-1 to Case 4-3, while their rationality decreases.

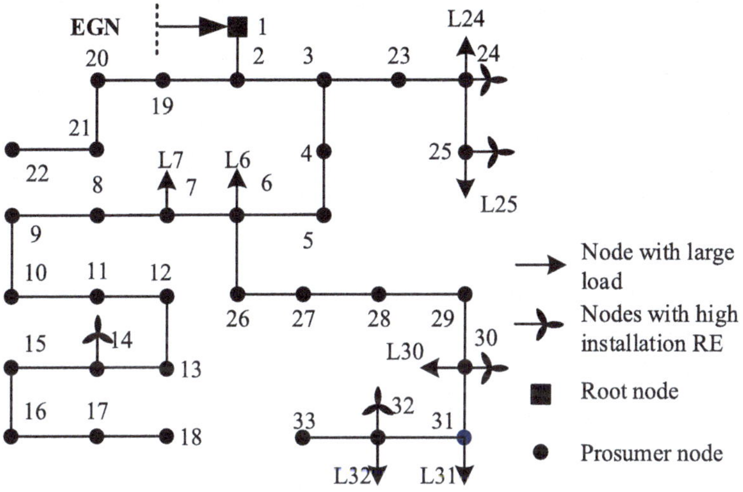

Fig. 3.5 33 node diagram of distribution network

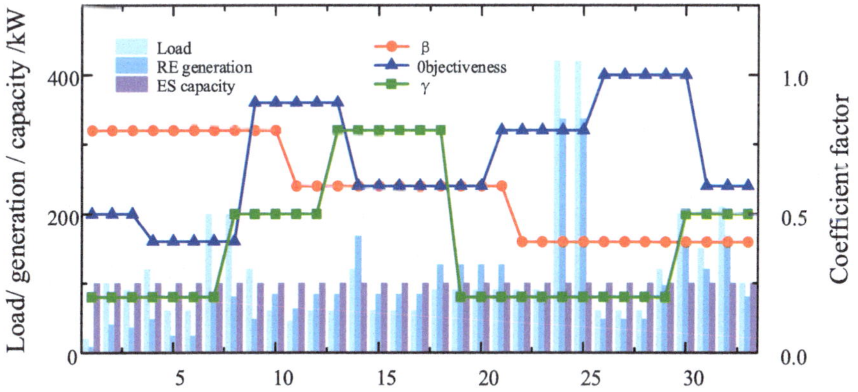

Fig. 3.6 Basic data

Table 3.1 Case setting

(a) OMP setting

Parameter	**Case 1**	**Case 2**	**Case 3**	**Case 4**	**Case 5**	**Case 6**
$P > P^*$	$P > P^*$	$P > P^*$	$P > P^*$	$P > P^*$	$P > P^*$	$P > P^*$
α_i	1	1	1	$\alpha_H, \alpha_M, \alpha_L$	$\alpha_H, \alpha_M, \alpha_L$	$\alpha_H, \alpha_M, \alpha_L$

(b) Subjectivity setting

Case	α_H	α_M	α_L
Case 4	Case 4–1	Case 4–2	Case 4–3
Case 5	Case 5–1	Case 5–2	Case 5–3
Case 6	Case 6–1	Case 6–2	Case 6–3

Table 3.2 Probability weighting of OMPs setting

$P > P^*$	$P > P^*$			$P > P^*$			$P > P^*$		
Price	0.2	0.5	0.8	0.3	0.6	0.8	0.4	0.6	0.8
P_m	64	22	14	62	24	14	14	24	62

[a] The unit of price is \$/kWh, and the unit of probability is %

The data settings for each Case and the objective price concerning the probability weighting of price events are presented in Table 3.2. Initially, prosumers are assumed to be entirely rational in Case 1, Case 2, and Case 3, meaning they make trading decisions based on objective assessments of expected utility. Additionally, ρ_m^L, ρ_m^M, and ρ_m^H represent the objective prices with low, medium, and high values, respectively. Furthermore, Case 4 to Case 6 is utilized to examine the decision-making behavior of prosumers with varying degrees of subjectivity in response to changes in the objective probability of market prices.

3.4.2 *Analysis of Objective TVD Changes*

This section primarily emphasizes objective decision-making as the basis for comparing subjective cases. It begins by analyzing changes in objective TVD.

Figure 3.7 illustrates the objective TVD from Case 1 to Case 3. It is evident from the figure that as the OMP increases, the negative value in prosumers' objective TVD also increases. This suggests a transition for most prosumers from being buyers to sellers. This shift occurs because higher OMPs elevate the cost of purchasing electricity, while selling electricity becomes more profitable for prosumers, thus enhancing their trading profit. For a given prosumer, as the OMP rises, the TVDs corresponding to nodes 13–18 exhibit minimal change compared to others. This phenomenon arises from these prosumers having a high distance sensitivity coefficient γ_i (as depicted in Fig. 3.6). Excessive transaction volume diminishes their

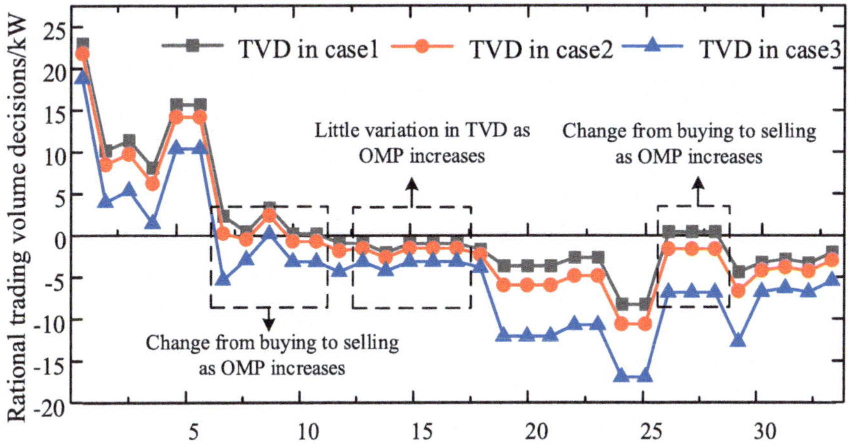

Fig. 3.7 TVD results of nodes in Case 1 to Case 3

trading profit, resulting in relatively small TVD values and reflecting minor fluctuations across the three cases. In summary, the devised TVD model demonstrates its adaptability to changes in OMP, thereby facilitating optimal decision-making for prosumers.

3.4.3 Comparative Analysis of Subjective and Objective TVD

Figure 3.8 illustrates the difference between objective and subjective TVD from Case 4 to Case 6, focusing on prosumers with varying degrees of subjectivity. The control variable method makes comparisons between cases with the same OMP. For example, Fig. 3.8a shows the discrepancy between TVDs in Case 1 and Case 4. From Fig. 3.8a and b, it is clear that when subjectivity increases in Case 4 and Case 5, the gap between objective and subjective TVD widens. Concurrently, the TVD of buyers exhibits a positive trend, gradually decreasing, while the TVD of sellers shows a negative trend, steadily increasing. This pattern occurs because in scenarios with low OMP and high OMP, the probability of low prices occurring is high, while the probability of high prices is low. We list the specific explanations as below for these findings.

The results in Case 4 and Case 5 can be explained by the probability weight function outlined in Eq. (3.2). The function's curve is depicted in Fig. 3.9 for α values of 0.3, 0.5, 0.7, and 1. The curve indicates that when $P > P^*$, the probability weight of the objective probability is less than that of the subjective probability. Conversely, if $P < P^*$, this relationship is reversed. As prosumers become more subjective (i.e., as α decreases), the discrepancy between subjective and objective probabilities increases. Referring to Table 3.2, it is evident that the objective probability of a low price event in

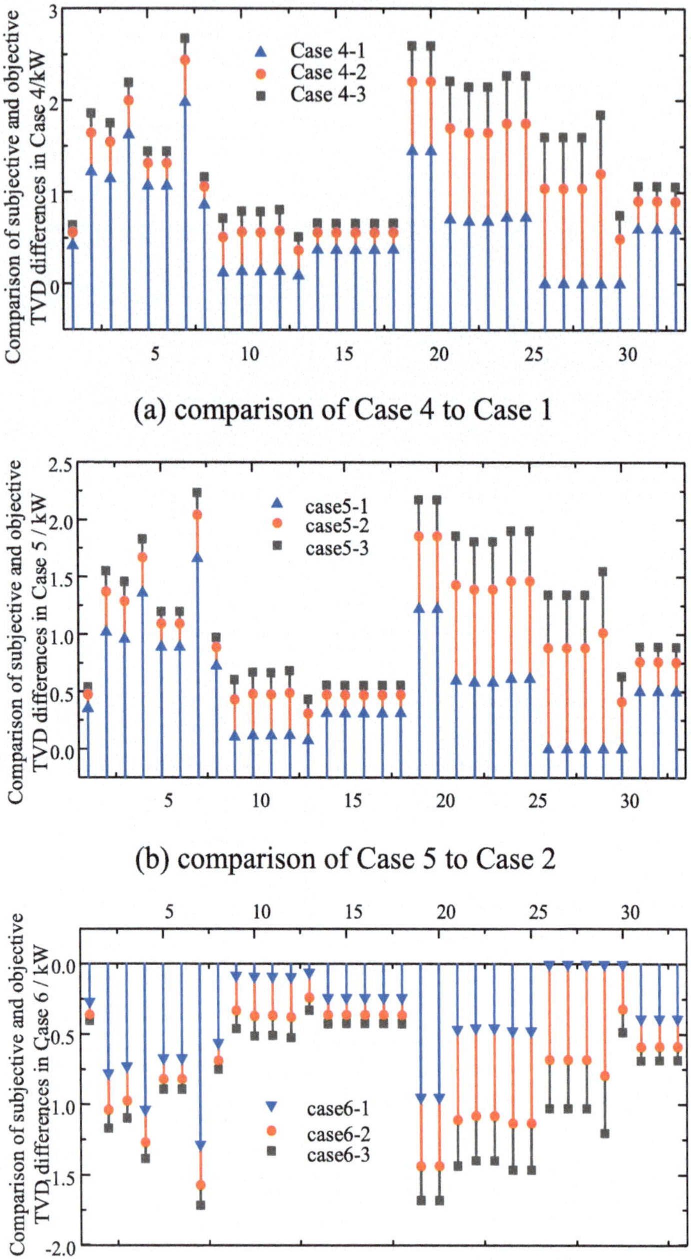

(a) comparison of Case 4 to Case 1

(b) comparison of Case 5 to Case 2

(c) comparison of Case 6 to Case 3

Fig. 3.8 Comparison of objective and subjective TVD

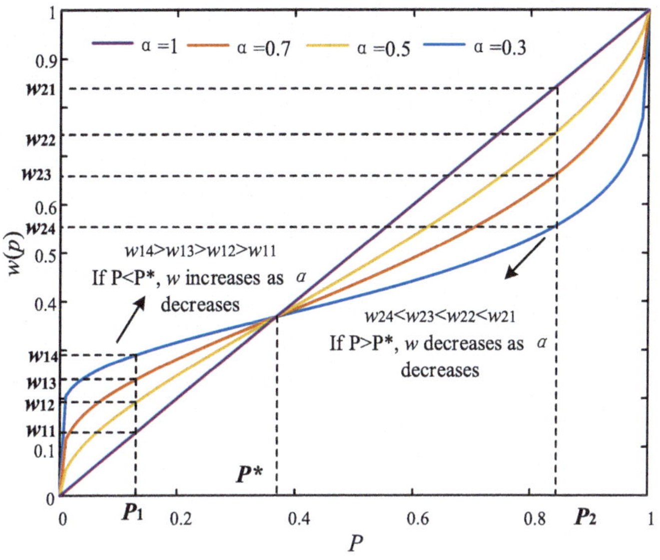

Fig. 3.9 Probability weight function graph

Case 4 and Case 5 is higher, but the subjective probability is lower, and this disparity grows with increased subjectivity. In summary, for prosumers, lower objective market prices (OMPs) are subjectively perceived as higher, and greater subjectivity leads to an even higher perceived market price. Consequently, as concluded in Sect. 3.4.2, the TVD for buyers decreases with increased subjectivity, while the opposite is true for sellers.

In contrast, Case 6 shows that as prosumer subjectivity increases, the TVD for buyers increases, and for sellers, it decreases. This is due to higher OMPs where the subjective projection of low prices probability is higher than the objective probability, increasing with greater subjectivity, opposite to higher price events. Thus, for prosumers, higher OMPs are subjectively perceived as lower, and stronger subjectivity results in a lower perceived market price. According to the conclusion in Sect. 3.4.2, the TVD for buyers rises with greater subjectivity, while the reverse is true for sellers.

3.4.4 Impacts of OMP on Transaction Profits

Figure 3.10 compares the objective transaction profits of prosumers in Cases 1–3. Transaction profits are influenced by changes in OMPs. This is because certain coefficient factors determining prosumers' trading strategies are sensitive to OMPs, leading to variations in trading profits as OMPs fluctuate. Firstly, consider nodes 13–18 and 29–33. In Case 3, their trading profit is nearly zero because their minimum

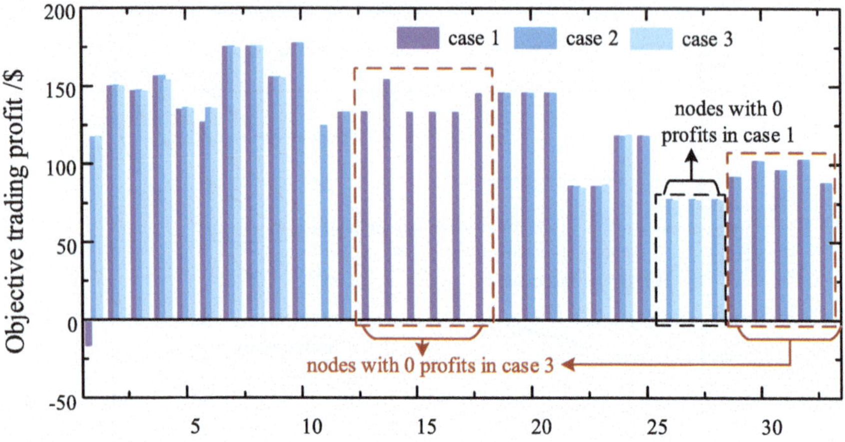

Fig. 3.10 Objective transaction profit of prosumers in Case 1, 2, 3

asking prices γ_i are higher, as shown in Fig. 3.6. Thus, they have limited ability to reduce their asking prices in the LEM, resulting in weak market competitiveness. Consequently, these nodes fail to secure transactions in the pair-matching process and must trade with EGN, leading to relatively low market transaction profits. When OMPs are high, the P2P trading market favors sellers. Sellers with lower asking prices can prioritize transactions and achieve higher market profits.

Secondly, nodes 26–28 have nearly no trading profit in Case 1 due to their relatively low unit load utility β_i. According to economic principles, buyers' unit purchase costs should be lower than the unit load utility [34], which means there is little scope to increase their bidding price in the LEM when β_i is the maximum bid. This results in weak market competitiveness and lower trading profits. When OMPs are low, the P2P trading market favors buyers. Buyers with high unit load utility can secure transactions more efficiently, increasing market profits.

3.4.5 Impacts of OMPs and Subjectivity on Market Transaction Profits

Figure 3.11 illustrates the contrast between subjective and objective market transaction returns for each prosumer in Cases 4 to 6 at various subjective degrees. The analysis begins by examining the impact of OMPs on trading profits. Comparing the three figures reveals that the differences in Case 6-1, 6-2, and 6-3 relative to Case 2 are more pronounced than in Case 4 or Case 5. This occurs because prosumers' subjective evaluations affect their TVD and CEP estimations, subsequently influencing their BAD. Consequently, as established in Sect. 3.4.3, subjective prosumers are more responsive to higher OMPs, leading to more significant variations in their market transaction profits than objective cases. In summary, when OMPs are high, prosumer subjectivity has a more significant impact on market trading profits than when OMPs are low.

Additionally, the effect of prosumer subjectivity on trading profits is further analyzed. For instance, considering node 19–20 in Fig. 3.11c, when rationality is high (Case 6-1), the trading profit exceeds that in cases with stronger subjectivity. This is because, in Case 6-1, the TVD for nodes 19–20 is higher than in Case 6-2 and Case 6-3 (refer to Fig. 3.7 and 3.8c). Using Eqs. (3.20) and (3.9), it is evident that the node's BAD increase rate is relatively higher and the CEP estimation is more significant than in Case 6-2 and Case 6-3, enhancing their competitiveness in market trading. As a result, their BAD strategy is more advantageous, yielding higher market trading profits. In conclusion, the prosumers' subjectivity influences trading profits, with more rational attitudes leading to greater trading profits in high OMP scenarios.

3.4.6 Analysis of Adaptive Learning Process

(1) Short-term learning

This section delves into the model's short-term learning process, focusing on the change in prosumers' aggressiveness during trading. It illustrates this using node 6 in cases 1 and 4.

Figure 3.12 shows that for node 6, aggressiveness r initially increases and then decreases across different scenarios. This pattern emerges because node 6 fails to strike a transaction in the first round, prompting an adjustment in the second round. When the estimated CEP falls below the traded price from the previous round, indicating r_{op} is positive, node 6 must increase its aggressiveness r to improve its chances of the transaction according to Eq. (3.14). Consequently, r increases until the fifth round. However, despite bidding optimally in the sixth round, node 6 still fails to transact due to limited volume. Subsequently, it chooses to safeguard its profit by reducing r. In summary, short-term learning enables quick adjustment of prosumers' BAD to an optimal level within constraints, but when transactions fail, the learning process adapts to protect the trader's profit.

In addition, Fig. 3.12 demonstrates that prosumers' subjectivity influences aggressiveness r in short-term learning. As subjectivity increases, r initially rises and then declines ($r4 < r1 < r2 < r3$). This trend emerges because heightened subjectivity leads to higher subjective market CEPs compared to rational cases, increasing the competitiveness of buyers' bids. Consequently, prosumers must increase r to enhance their chances of successful transactions. However, in scenarios of strong subjectivity, subjective market prices elevate, shifting the market from a buyer's to a seller's market. Consequently, optimal bidding prices decrease, prompting prosumers to

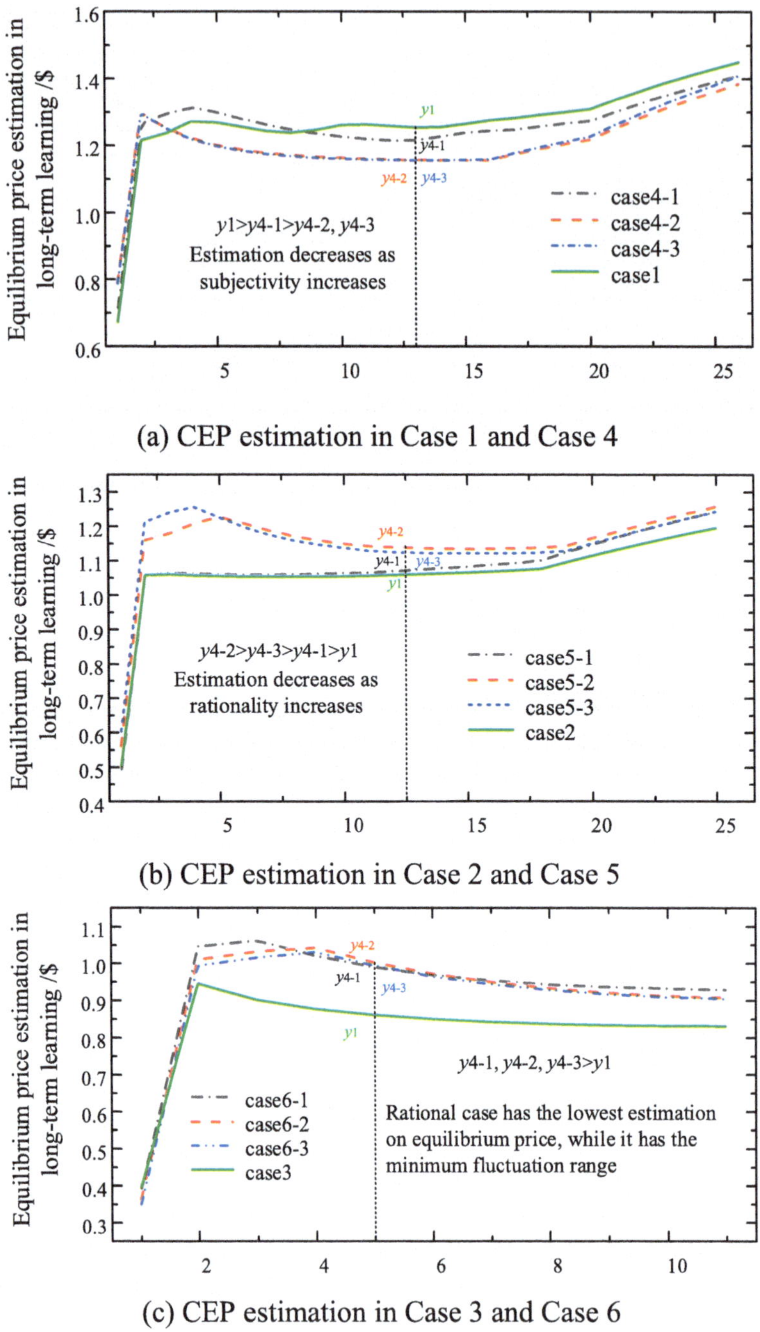

Fig. 3.13 Analysis of CEP estimation in long-term learning

value of first ε as 1, and the theoretical curve of $\theta(p_v)$ is plotted in Fig. 3.14a. The fluctuation in CEP in the first round appears greater than that in the second round, where a greater CEP fluctuation necessitates a quicker adjustment of the target BAD, leading to simultaneous changes in the two curves.

Additionally, the relationship curve between θ and CEP is depicted based on Eq. (3.17) in Fig. 3.14b. When $\varepsilon = 1$, as CEP increases, θ gradually decreases. Therefore, the two curves in Fig. 3.14a exhibit opposite trends.

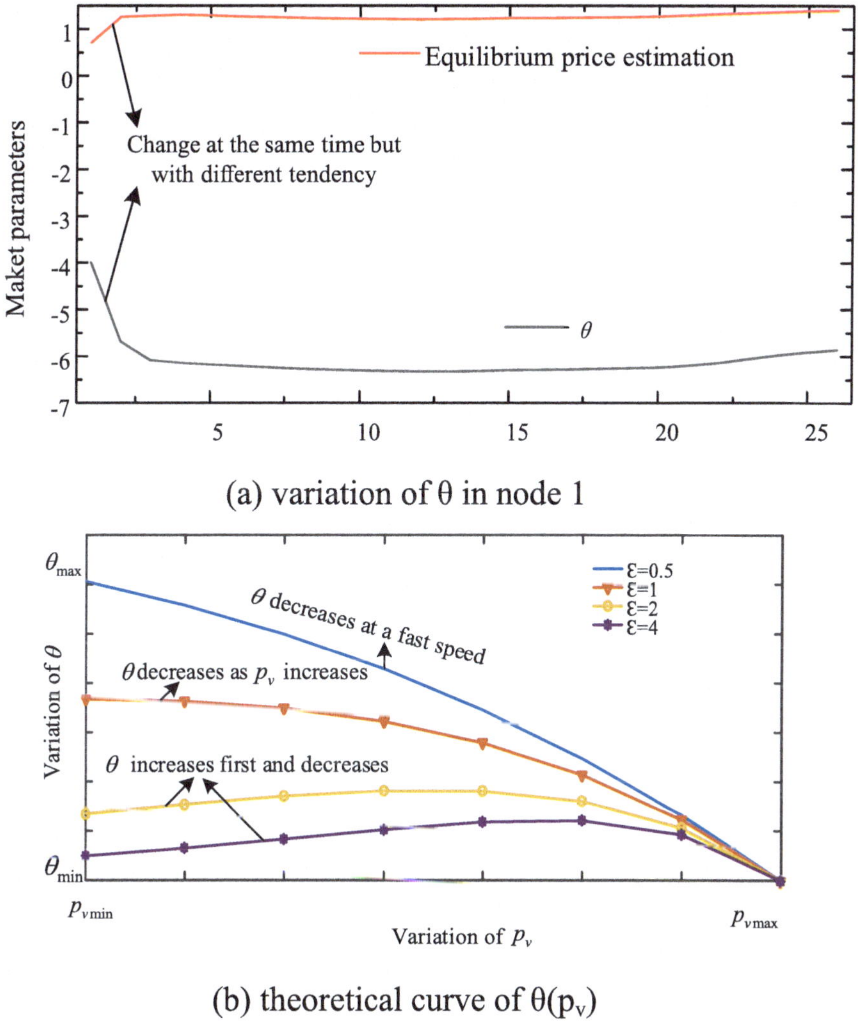

Fig. 3.14 Analysis of long-term learning parameters

Table 3.3 Traded nodes in each round in Case 6-1

Round	Buyer node: seller node	Round	Buyer node: seller node
1	9:7	8	6:26
2	4:7	9	1:26
3	2:11	10	1:27
4	2:24	11	1:28
5	3:24	12	1:19
6	5:19	13	1:20
7	6:7	14	External grid network: Untraded nodes

3.4.7 *Transaction Matching Analysis*

The section discusses the change curve of traded price and volume in LEMs, mainly focusing on prosumers with relatively higher rationality. As the OMP (Offered Market Price) increases, the number of transaction rounds in the LEM decreases, indicating a shift from a buyer's market to a seller's market. In Case 6, the limited buyer's power volume is traded, leaving a large amount of untraded seller's electricity to be settled with the EGN, leading to a decrease in transaction rounds.

In Case 6-1, Table 3.3 displays the nodes that successfully matched in the transaction process, with untraded nodes listed in the last round. The "BAD" strategy, which involves matching the highest bid of a buyer with the lowest ask of a seller, provides an advantage to nodes that can be traded first. As the CEP and prosumers' aggressiveness varies, the remaining nodes adjust their BAD strategies round by round to strike transactions. The last-traded node, constrained by the upper/lower limit of BAD, adapts slowly in the market, reaching a transaction at the end of the matching period.

3.4.8 *Social Welfare and Market Efficiency Analysis*

To assess the effectiveness of the proposed trading model, we conduct a comparative analysis with the ZI Strategy [28] concerning both social welfare [38] and market efficiency [39]. Utilizing data from Case 4-1, we generate 40 sets of unit load utility β_i and distance sensitivity coefficient γ_i through uniform distribution. These parameters represent the maximum and minimum price of the prosumers' BAD, respectively, facilitating the calculation of social welfare and market efficiency within the P2P trading model framework.

It's crucial to note that social welfare is determined as the aggregate profits acquired by all participators, as per Eq. (3.3), within the LEM, Market efficiency, on the other hand, is quantified as the ratio of social welfare to the total profits of all prosumers under centralized clearing mechanisms [39]. The outcomes are presented in Fig. 3.16a, b.

Upon examining Fig. 3.16a, b, it becomes evident that the trading strategies proposed in this study outperform those in [28]. Aligned with economic principles [34], centralized market clearing typically maximizes social welfare. Furthermore, our trading strategy incorporates CEP estimation based on market transaction prices and adjusts BADs by fine-tuning aggressiveness. In contrast, market participants in [28] rely on random bidding decisions within preset bounds, lacking adaptation to market information. Consequently, social welfare in [28] falls short compared to our proposed strategy.

Moreover, market efficiency, defined as the value ratio calculated in Fig. 3.16a to the social welfare under centralized clearing, also favors our trading model. Given that the test data samples are uniformly distributed random numbers in (0, 1), the disparities in social welfare under centralized clearing are minimal. Consequently, the market efficiency of our trading model surpasses that of [28], with a similar trend observed in Fig. 3.16b.

Regarding subjectivity or OMP adjustments, our trading strategies uphold optimal TVD and incorporate adaptive learning-based BADs. Thus, the advantages in social welfare and market efficiency remain consistent across all scenarios. In summary, our proposed trading strategy demonstrates superiority over traditional approaches concerning both social welfare and market efficiency.

3.4.9 Smart Contract Compilation Results

The Remix platform is utilized to create 33 virtual account addresses, with nodes 1–6 presented as examples to display their corresponding virtual account addresses, as outlined in Table 3.4. Additionally, Table 3.5 provides the trading outcomes of the LEM within Case 6-1.

Table 3.4 Smart contract compilation initialization information

Node	Ethereum address
1	0x75f44e063Df873eE5Df4c99f9797876136D65c00
2	0x4B718b6094a2363097FfDA0D9C34eF2eD25C19B6
3	0xc21751644191b13F54cF49cB7C543EA948b053b5
4	0x102b641c3B750A98f8AA4ea0393ed9025dE98ce5
5	0x3d702Abb841795FF8364BdaD9166A7D2Bf842cC2
6	0xc728E117DaEEbf44c53b6f904D29849D5a593073
…	…

Table 3.5 Trading results in Case 6-1

Iteration	Transaction address information	Traded quantity	Traded price
1	0x5B...3AE7 → 0x11...a134	52	18
2	0x10...8ce5 → 0x11...a134	52	147
3	0x4B...9B6 → 0x53...3A64	52	313
4	0x4B...9B6 → 0xa1...6923	46	87
5	0xc2...53b5 → 0xa1...6923	46	543
6	0x3d...2cC2 → 0xCA...733c	46	1042
7	0xc7...3073 → 0x11...a134	46	367
8	0xc7...3073 → 0x14...160C	46	675
9	0x5B...3AE7 → 0x14...160C	45	11
10	0x5B...3AE7 → 0x4B...D2dB	45	686
11	0x5B...3AE7 → 0x58...0225	45	686
12	0x5B...3AE7 → 0xCA...733c	45	166
13	0x5B...3AE7 → 0xdD...2148	45	336

The transaction results obtained through the trading platform based on blockchain align closely with the traded quantity and price data presented in Table 3.3 and Fig. 3.15c. This congruence underscores the effectiveness of the P2P transaction mechanism proposed in this section, which leverages smart contracts to orchestrate transactions.

3.5 Conclusions

The proliferation of prosumers with distributed energy generation ability underscores the necessity for an effective P2P trading mechanism to facilitate transactions and settlements among users on the distribution network side. This chapter proposes a P2P trading strategy based on PT and an adaptive learning model, which accounts for various features of prosumers. The model comprises optimal trading quantity and bidding decision models for prosumers, alongside a transaction matching and market settlement process facilitated by smart contracts. By utilizing the Remix Ethereum platform, we can verify the feasibility of the proposed P2P trading model.

Simulation results demonstrate several key findings: (1) The P2P trading model offers optimal trading strategies for prosumers while promptly adapting to changes in market information and prosumers' subjectivity. (2) The smart contract matching process facilitates successful transactions among competitive market members, particularly benefiting prosumers employing strategic bidding and swift adaptation in the LEM. (3) The adaptive learning process within the P2P trading model effectively stabilizes market trading prices and accelerates the convergence of prosumers' bids toward optimal ones. (4) Compared to traditional trading methods, the proposed P2P

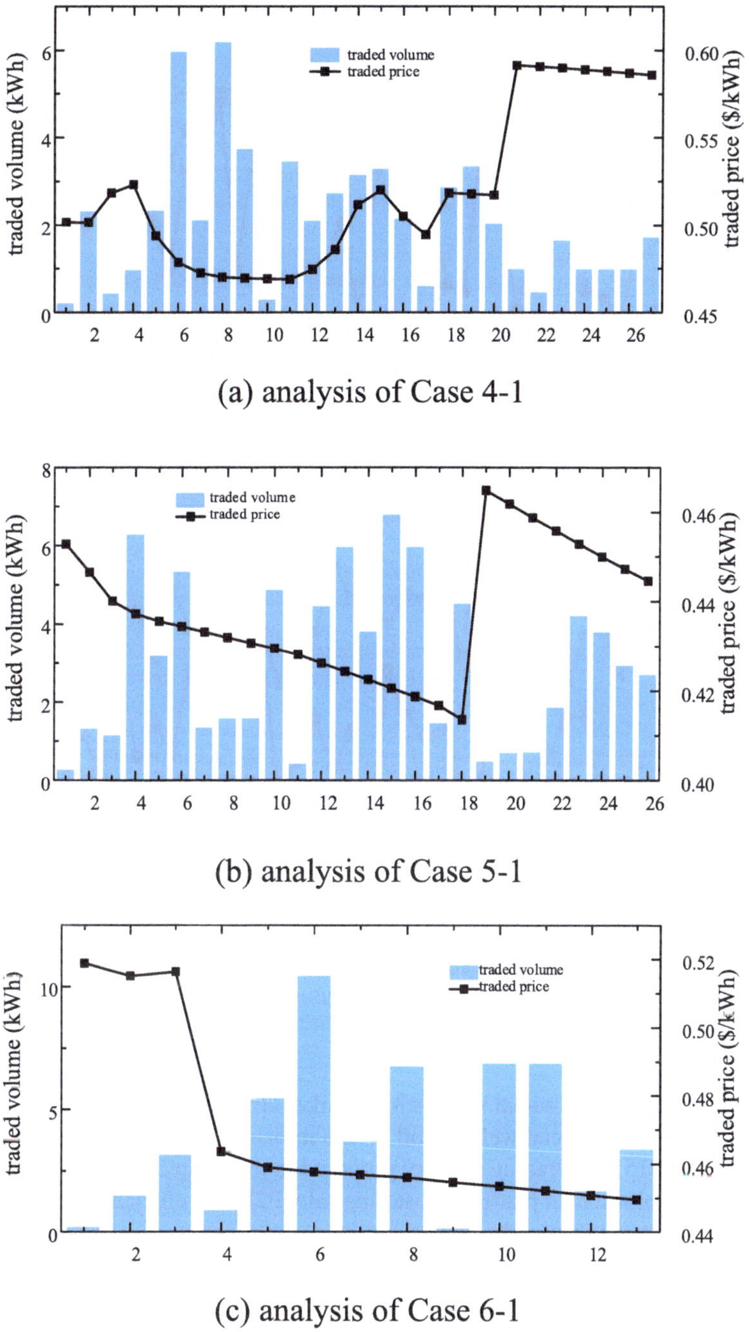

Fig. 3.15 Change in traded price and volume in the LEM

20. Krishna, V. (2002). *Auction theory*. Academic Press.
21. Minghua, H., Hofung, L., & Jennings, N. R. (2003). A fuzzy-logic based bidding strategy for autonomous agents in continuous double auctions. *IEEE Transactions on Knowledge and Data Engineering, 15*(6), 1345–1363.
22. Yinxin, Y., Bin, D., Xigao, W., & Ying, Z. (2018). A novel generation right trade in blockchain-enabled continuous double auction system. In *The 11th IET International Conference on Advances in Power System Control, Operation and Management (APSCOM 2018)* (pp. 1–6).
23. Chen, K., Lin, J., & Song, Y. (2019). Trading strategy optimization for a prosumer in continuous double auction-based peer-to-peer market: A prediction-integration model. *Applied Energy, 242*, 1121–1133.
24. Sorin, E., Bobo, L., & Pinson, P. (2019). Consensus-based approach to peer-to-peer electricity markets with product differentiation. *IEEE Transactions on Power Systems, 34*(2), 994–1004.
25. Khorasany, M., Mishra, Y., & Ledwich, G. (2020). A decentralized bilateral energy trading system for peer-to-peer electricity markets. *IEEE Transactions on Industrial Electronics, 67*(6), 4646–4657.
26. Vytelingum, P., Ramchurn, S. D., Voice, T. D., Rogers, A., & Jennings, N. R. (2010). Trading agents for the smart electricity grid. In *Proceedings of the Ninth International Conference on Autonomous Agents and Multiagent Systems (AAMAS 2010).*
27. Yan, X., Lin, J., Hu, Z., & Song, Y. (2017). P2P trading strategies in an industrial park distribution network market under regulated electricity tariff. In *2017 IEEE Conference on Energy Internet and Energy System Integration* (pp. 1–5).
28. Gode, D. K., & Sunder, S. (1993). Allocative efficiency of markets with zero-intelligence traders: Market as a partial substitute for individual rationality. *Journal of Political Economy, 101*(1), 119–137.
29. Goncalves Da Silva, P., Ilic, D., & Karnouskos, S. (2014). The impact of smart grid prosumer grouping on forecasting accuracy and its benefits for local electricity market trading. *IEEE Transactions on Smart Grid, 5*(1), 402–410.
30. Jhala, K., Natarajan, B., & Pahwa, (2019). A prospect theory-based active consumer behavior under variable electricity pricing. *IEEE Transactions on Smart Grid, 10*(3), 2809–2819.
31. Drazen, P. (1998). The probability weighting function. *Econometrica, 66*, 497–528.
32. Faqiry, M. N., & Das, S. (2016). A budget balanced energy distribution mechanism among prosumers and prosumers in microgrid. In *2016 IEEE International Conference on Internet of Things (iThings) and IEEE Green Computing and Communications (GreenCom) and IEEE Cyber, Physical and Social Computing (CPSCom) and IEEE Smart Data (SmartData)* (pp. 516–520).
33. Jiang, H., & Jian, C. P. (2006). A methodology for allocating transmission losses to pool and bilateral transactions and its features. *Proceeding of the CSEE, 26*(8), 49–54.
34. Daniel, K., & Goran, S. (2005). *Fundamentals of power system economics*. Wiley.
35. Bistarelli, S., et al. (2020). Ethereum smart contracts: Analysis and statistics of their source code and opcodes. *Internet of Things, 11*, 100198.
36. Dimitri, B. (2009). *Convex optimization theory*. Athena Scientific.
37. Huang, H., Zhang, L., & Su, J. (2009). Marginal transmission loss allocation to bilateral exchanges based on EBE principal. In *2009 4th IEEE Conference on Industrial Electronics and Applications* (pp. 1305–1309).
38. Peng, T., et al. (2004). Analysis of economic mechanism of two regional electricity market models. *Automation of Electric Power Systems, 28*, 20–23.
39. Nicolaisen, J., Petrov, V., & Tesfatsion, L. (2001). Market power and efficiency in a computational electricity market with discriminatory double-auction pricing. *IEEE Transactions on Evolutionary Computation, 5*(5), 504–523.

Chapter 4
Optimal Trading Strategy of Data-Center Prosumer in Multiple Local Energy Markets

Abstract The local energy market (LEM) has emerged as an effective way to automatically regulate the power balance and promote distributed energy utilization of the distribution system. With the increasing integration of data centers (DCs) with high load-shifting flexibility into the distribution network, they have gradually become significant market players in the LEMs. However, limited research has been done on the optimal trading strategy of DCs within LEMs. A bi-level optimization model for networked DCs across multiple LEMs is proposed to bridge the research gap. Firstly, since DCs with substantial power consumption can directly impact the market clearing prices, LEM participants are classified into DC-type and conventional-type prosumers, both analytically modeled. Secondly, a LEM model incorporating DCs is created, employing the alternating direction method of multipliers (ADMM) to clear the market in a distributed method, thereby ensuring end-user privacy. Thirdly, a bi-level optimal trading strategy for interconnected DC prosumers across various LEMs is proposed to distribute workloads among DCs optimally. Fourth, the differential evolution and meta-model address the bi-level optimization problem efficiently. The simulation results indicate that this strategy enhances networked DCs' benefits, boosts renewable energy utilization, and improves overall social welfare.

4.1 Introduction

With the rapid development of renewable energy in distribution systems, power system operators face significant challenges in balancing supply and demand, which requires more flexibility for reliable and economical operation. Traditional controllable generators can address these issues by adjusting power generation atomically. However, their generation proportion is limited by the high penetration of renewable energy, operational costs, and pollution. Alternatively, numerous flexible resources on the demand side, such as electric vehicles [1, 2], air conditioning [3], and data centers (DCs) [4], can adjust power consumption spatially to balance the fluctuation of renewable energy.

M. Song et al., *Local Energy Markets*, https://doi.org/10.1007/978-981-97-9750-9_4

Numerous demand-side resources exist, but they cannot directly participate in the wholesale market because of their small individual capacities. Meanwhile, end-users cannot perceive electricity market prices, which contain power system operation information, leading to a lack of motivation for flexible resource operators to aid in the supply–demand balance. To address this, the local energy market (LEM) has been proposed, as '*a set of economic and control mechanisms that allow the dynamic balance of supply and demand across the entire electrical infrastructure using value as a key operational parameter*.' [5]. In LEM, end users shift from traditional customers who only passively receive price information to prosumers who can actively trade with others [6]. This motivates end users to regulate their flexible resources in response to LEM's time-varying supply and demand information. Regarding theory, current studies mainly focus on LEM's mechanism design [7], optimal clearing methods [8, 9] and trading platforms in distribution systems [10, 11]. Regarding pilot projects, countries like the Netherlands [12], the U.S. [13], Germany [14], the Philippines [15], and China [16] have shown initiative in pilot projects to validate LEM's effectiveness. Regarding policy, the National Energy Administration of China issued a document in 2022 to incentivize distributed prosumers' participation in LEM, promoting renewable energy consumption [17]. In summary, the LEM has gained global attention and is beginning to be phased into application.

Meanwhile, digital industries such as big data, cloud computing, and 5G communication are rapidly advancing. As the primary storage and computing entities for massive amounts of data, DCs are expanding significantly. In China, DCs, the 5G network, and the industrial internet have become crucial 'new infrastructures' in the national development plan [18]. The rapid development of DCs has led to increasing power consumption. In 2021, China's DCs consumed 216.1 billion kWh of electricity, about 2.6% of the total electricity consumption [19]. Unlike flexible loads confined in one place, multiple DCs operated by a single company across various distribution systems can dynamically adjust and distribute workloads through the fiber optical network [4]. In other words, DCs are fiber-connected and can flexibly adjust in both time and space dimensions. Recently, some studies have focused on enhancing the energy efficiency of DCs for lower costs. Reference [20] minimizes the DC energy costs using a deep reinforcement learning algorithm. In Reference [21], an AI-driven method is proposed to manage the dynamic thermal conditions of DCs, resulting in an 8.84% power saving compared to traditional methods. To reduce the data bandwidth and storage space of DCs, Reference [22] developed a real-time operating behavior model of DCs based on the Markov chain. However, these studies focus solely on the energy utilization of DCs from a device perspective. In reality, the energy cost of DCs is also tied to electricity prices. Therefore, more attention should be given to the optimal operation of networked DCs within the electricity market.

To minimize the cost of DCs, some papers have studied the optimal trading strategy of networked DCs in electricity markets by workload scheduling. References [23, 24] propose a geographical load-balancing strategy to minimize the electricity cost of DCs, assuming that DCs are price takers. Reference [25] suggests that due to the large volume of DCs, transferring their workloads can impact the real-time market price. In addition, some studies focus on how electricity retail companies can adjust the workloads of DCs using reasonable pricing mechanisms across different distribution systems. Reference [26] establishes a bi-level trading model between electricity retail companies and DCs to minimize electricity costs. In [27], a non-cooperative game pricing model among electricity retail companies is established to reduce the cost of networked DCs. However, most papers focus on the trading strategies of DCs in the wholesale market, few papers address how to dispatch the workloads of networked DCs in LEMs.

The rise of LEM requires DCs in various distribution systems to engage in multiple LEMs. Networked DCs are connected with several markets, encountering a complex market environment. Additionally, a single DC consumes significant power, exceeding that of other local loads. DCs are not passive price takers and can influence the market clearing price. Thus, it is crucial to study the trading strategy of DCs in multiple LEMs to optimize the scheduling of networked DCs. The main contributions of this chapter are summarized as follows.

(1) Considering DCs' considerable power consumption, they are treated as price-makers, and the market players are divided into DC and conventional prosumers in LEMs. Then, a LEM model integrating DCs is proposed and the direction method of multipliers (ADMM) is employed to clear the market for users' privacy.
(2) A bi-level optimization model for networked DCs in multiple LEMs is proposed to optimize the workload allocation efficiency. At the upper level, the objective is to maximize networked DCs' benefits. At the lower level, LEMs are simultaneously cleared in different distribution systems considering various prosumers' behaviors. This is the first attempt to optimize the workload allocation of networked DCs while participating in multiple LEMs.
(3) Differential evolution and meta-model techniques are utilized to solve the bi-level optimization model of networked DCs efficiently.

4.2 Operation Framework of Networked DCs in Multiple LEMs

Multiple DCs across different distribution systems are geographically interconnected and managed by a single operator. When DCs participate in multiple LEMs, the operator must decide how to distribute workloads among the DCs to maximize the revenue. Figure 4.1 illustrates the optimal operation framework of networked DCs in multiple LEMs. Users' data service requests are first aggregated at the local front-end agent of DCs and then efficiently distributed to the networked DCs by the operator, considering the operating information of LEMs. Since the significant power consumption of DCs and their role as price makers in LEMs, the optimal trading strategy of networked DCs is closely coupled with the clearing process of LEMs, which is the point of this chapter. The detailed modeling of LEM and DCs is given in Sect. 4.3. The optimal trading strategy of networked DCs in different LEMs is provided in Sect. 4.4.

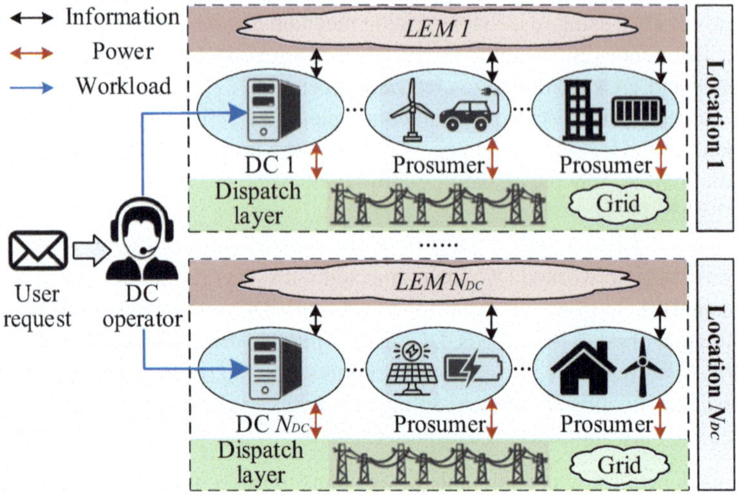

Fig. 4.1 Operation framework of networked DCs in multiple LEMs

4.3 LEM Modeling and Clearing Algorithm

It is assumed that there are N_{DC} DCs operated by an operator in the LEM. Due to DCs' higher power consumption, market participants are categorized into DC-type and conventional-type prosumers. The total number of prosumers in the ξ-th LEM is denoted by K_{ξ} interaction between prosumers and the platform is shown in Fig. 4.2.

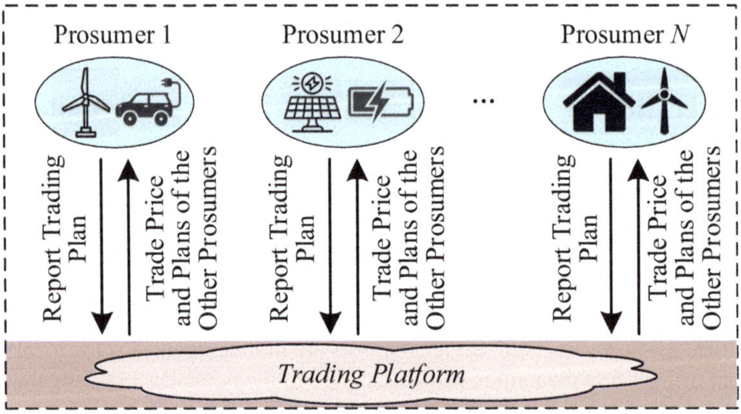

Fig. 4.2 Information interaction between prosumers and the platform

4.3.1 Objective Function of the Market Clearing Model

The clearing objective of the ξ-th LEM is to maximize social welfare, with the objective function as follows:

$$\max\ W_{\xi} = \sum_{k=1}^{K_{\xi}} S_{\xi,k} \tag{4.1}$$

where W_{ξ} is the total social welfare; $S_{\xi,k}$ is the revenue of the k-th prosumer. The detailed analysis for DC- and conventional-type prosumers is given as follows.

1. *Revenue of DC-type prosumer* ($k = 1$).

The revenue of DC in the ξ-th market is given as:

$$S_{\xi,1} = \sum_{t=1}^{T} (B_{\xi,1,t}^{\text{DC}} + B_{\xi,1,t}^{\text{grid}} + B_{\xi,1,t}^{\text{LEM}} - C_{\xi,1,t}^{\text{net}}) \tag{4.2}$$

where $B_{\xi,1,t}^{\text{DC}}$ is the revenue of DC by serving users; $B_{\xi,1,t}^{\text{grid}}$ is the revenue when DC buys or sells electricity from the grid; $B_{\xi,1,t}^{\text{LEM}}$ is the revenue of DC when trading with other prosumers in the LEM; $C_{\xi,1,t}^{\text{net}}$ is the network transmission fee paid by DC; T is the total number of time slots; t denotes a specific time slot.

The main function of a DC is to process users' data, and the revenue from this service can be expressed as a linear function of the processed workload:

$$B_{\xi,1,t}^{\text{DC}} = \nu^{\text{base}} \sum_{\delta=1}^{\Delta} \lambda_{\xi,1,\delta,t} \tag{4.3}$$

where Δ is the number of front-end agents; $\lambda_{\xi,1,\delta,t}$ is the average arrival rate of workloads from front-end agent δ of DC; ν^{base} is the basic income of DC for processing a unit workload.

The revenue of a DC when purchasing or selling electricity with distribution systems is calculated as:

$$B_{\xi,1,t}^{\text{grid}} = (\pi_{\xi,t}^{\text{grid},s} p_{\xi,1,t}^{\text{grid},s} - \pi_{\xi,t}^{\text{grid},b} p_{\xi,1,t}^{\text{grid},b})\Delta t \tag{4.4}$$

where $p_{\xi,1,t}^{\text{grid},s}$ and $p_{\xi,1,t}^{\text{grid},b}$ are the selling and purchasing power amount from the distribution system; Δt is the duration of one slot; $\pi_{\xi,t}^{\text{grid},s}$ and $\pi_{\xi,t}^{\text{grid},b}$ are the corresponding time-of-use price and feed-in tariff, respectively.

The revenue of DC in LEM is given as:

$$B_{\xi,1,t}^{\mathrm{LEM}} = \sum_{m=2}^{K_\xi} \pi_{\xi,1m,t} p_{\xi,1m,t}^{\mathrm{LEM}} \Delta t \tag{4.5}$$

where $p_{\xi,1m,t}^{\mathrm{LEM}}$ is the power sold by DC to the m-th prosumer in time t, and $\pi_{\xi,1m,t}$ is the trading price between them.

When DC trades with other prosumers in the LEM, it must pay a network transmission fee to the distribution system operator. This fee depends on the electrical distance between the two parties [7], which is defined as:

$$C_{\xi,1,t}^{\mathrm{net}} = \frac{\pi_\xi^{\mathrm{unit}}}{2} \sum_{m=2}^{K_\xi} d_{\xi,1m}^{\mathrm{elec}} |p_{\xi,1m,t}^{\mathrm{LEM}}| \Delta t \tag{4.6}$$

where $d_{\xi,1m}^{\mathrm{elec}}$ is the electrical distance between DC and the m-th prosumer; π_ξ^{unit} is the unit price of the transmission fee; $p_{\xi,1m,t}^{\mathrm{LEM}}$ is the power sold by DC to the m-th prosumer in time t.

This chapter uses Thevenin impedance distance [28] to describe the electrical distance of trading, which is given as follows:

$$d_{km}^{\mathrm{elec}} = |Z_{kk} + Z_{mm} - Z_{km} - Z_{mk}| \tag{4.7}$$

where Z_{km} is the corresponding impedance between nodes k and m, found in the node impedance matrix Z.

2. *Revenue of conventional-type prosumers* ($k = 2, \ldots, K_x$).

A conventional-type prosumer consists of a distributed generator (DG), energy storage system (ESS), and flexible load (FL). The total revenue of conventional-type prosumers can be calculated as follows:

$$S_{\xi,k} = \sum_{t=1}^{T} \left(-C_{\xi,k,t}^{\mathrm{DG}} - C_{\xi,k,t}^{\mathrm{ESS}} + U_{\xi,k,t}^{\mathrm{FL}} + B_{\xi,k,t}^{\mathrm{grid}} + B_{\xi,k,t}^{\mathrm{LEM}} - C_{\xi,k,t}^{\mathrm{net}} \right) \tag{4.8}$$

where $C_{\xi,k,t}^{\mathrm{DG}}$ is the DG power generation cost of the k-th prosumer; $C_{\xi,k,t}^{\mathrm{ESS}}$ is the depreciation cost of the ESS; $U_{\xi,k,t}^{\mathrm{FL}}$ is the FL utility; $B_{\xi,k,t}^{\mathrm{grid}}$ is the revenue when the prosumer buys or sells electricity from the grid; $B_{\xi,k,t}^{\mathrm{LEM}}$ is the revenue when the prosumer trades with others in LEM; $C_{\xi,k,t}^{\mathrm{net}}$ is the network transmission fee.

The DG includes photovoltaics (PV), wind turbines (WT), and micro gas turbines. The power generation cost is given as:

$$C_{\xi,k,t}^{\mathrm{DG}} = \alpha_{\xi,k} (p_{\xi,k,t}^{\mathrm{DG}} \Delta t)^2 + \beta_{\xi,k} p_{\xi,k,t}^{\mathrm{DG}} \Delta t + \gamma_{\xi,k} \tag{4.9}$$

where $p_{\xi,k,t}^{\mathrm{DG}}$ is the output of the DG; $\alpha_{\xi,k}$, $\beta_{\xi,k}$ and $\gamma_{\xi,k}$ are the cost coefficients, respectively.

The ESS depreciation cost is described as:

$$C_{\xi,k,t}^{\mathrm{ESS}} = c_{\xi,k}^{\mathrm{dep}} p_{\xi,k,t}^{\mathrm{ESS,ch}} \eta_{\xi,k}^{\mathrm{ch}} \Delta t \tag{4.10}$$

where $p_{\xi,k,t}^{\mathrm{ESS,ch}}$ is the ESS charging power; $\eta_{\xi,k}^{\mathrm{ch}}$ is the charging efficiency, and $c_{\xi,k}^{\mathrm{dep}}$ is the depreciation cost factor of ESS.

The utility of FL is stated as:

$$U_{\xi,k,t}^{\mathrm{FL}} = \begin{cases} \omega_{\xi,k} p_{\xi,k,t}^{\mathrm{FL}} \Delta t - \frac{\sigma_{\xi,k}}{2}\left(p_{\xi,k,t}^{\mathrm{FL}} \Delta t\right)^2, & p_{\xi,k,t}^{\mathrm{FL}} \Delta t < \frac{\omega_{\xi,k}}{\sigma_{\xi,k}} \\ \frac{\omega_{\xi,k}^2}{2\sigma_{\xi,k}}, & p_{\xi,k,t}^{\mathrm{FL}} \Delta t \geq \frac{\omega_{\xi,k}}{\sigma_{\xi,k}} \end{cases} \tag{4.11}$$

where $p_{\xi,k,t}^{\mathrm{FL}}$ is the electric power of FL; $\omega_{\xi,k}$ and $\sigma_{\xi,k}$ are the coefficients of the utility function.

The revenue of the prosumer trading with the distribution system is given as:

$$B_{\xi,k,t}^{\mathrm{grid}} = (\pi_{\xi,t}^{\mathrm{grid},s} p_{\xi,k,t}^{\mathrm{grid},s} - \pi_{\xi,t}^{\mathrm{grid},b} p_{\xi,k,t}^{\mathrm{grid},b}) \Delta t \tag{4.12}$$

where $p_{\xi,k,t}^{\mathrm{grid},s}$ and $p_{\xi,k,t}^{\mathrm{grid},b}$ are the selling and purchasing power from the distribution system, respectively.

The revenue of the prosumer in LEM is calculated as:

$$B_{\xi,k,t}^{\mathrm{LEM}} = \sum_{m=1,m\neq k}^{K_\xi} \pi_{\xi,km,t} p_{\xi,km,t}^{\mathrm{LEM}} \Delta t \tag{4.13}$$

where $p_{\xi,km,t}^{\mathrm{LEM}}$ is the power sold by the k-th prosumer to the m-th prosumer; $\pi_{\xi,km,t}$ is the trading price between them.

The network transmission fee is given as:

$$C_{\xi,k,t}^{\mathrm{net}} = \frac{\pi_\xi^{\mathrm{unit}}}{2} \sum_{m=1,m\neq k}^{K_\xi} d_{\xi,km}^{\mathrm{elec}} |p_{\xi,km,t}^{\mathrm{LEM}}| \Delta t \tag{4.14}$$

where $p_{\xi,km,t}^{\mathrm{LEM}}$ is the power sold by the k-th prosumer to the m-th prosumer in time t.

4.3.2 Constraints of the Market Clearing Model

1. *Constraints of DC-type prosumer* ($k = 1$).

(a) *Constraints with DC Power*

It is assumed that all servers in the DC have the same performance and rated power. Thus, the average utilization of the server's CPU, $u_{\xi,1,t}^{\mathrm{IW}}$ is calculated as:

$$u_{\xi,1,t}^{\mathrm{IW}} = \frac{\sum_{\delta=1}^{\Delta} \lambda_{\xi,1,\delta,t}}{\mu_{\xi,1}^{\mathrm{IW}} m_{\xi,1,t}^{\mathrm{IW}}} \tag{4.15}$$

where $\lambda_{\xi,1,\delta,t}$ is the average arrival rate of the workload from the front-end agent δ; $\mu_{\xi,1}^{\mathrm{IW}}$ is the average rate at which servers process workloads; $m_{\xi,1,t}^{\mathrm{IW}}$ is the number of the active servers.

The power of a single server can be expressed as a linear function of the CPU utilization rate [29]. Therefore, the server's power, $p_{\xi,1,t}^{\mathrm{server}}$ can be calculated as:

$$p_{\xi,1,t}^{\mathrm{server}} = m_{\xi,1,t}^{\mathrm{IW}}(p_{\xi,1}^{\mathrm{idle}} + u_{\xi,1,t}^{\mathrm{IW}}(p_{\xi,1}^{\mathrm{peak}} - p_{\xi,1}^{\mathrm{idle}})) = m_{\xi,1,t}^{\mathrm{IW}} p_{\xi,1}^{\mathrm{idle}} + \frac{\sum_{\delta=1}^{\Delta} \lambda_{\xi,1,\delta,t}}{\mu_{\xi,1}^{\mathrm{IW}}}(p_{\xi,1}^{\mathrm{peak}} - p_{\xi,1}^{\mathrm{idll}}) \tag{4.16}$$

where $p_{\xi,1}^{\mathrm{idle}}$ and $p_{\xi,1}^{\mathrm{peak}}$ are the idle and peak power of the active servers, respectively. Therefore, the total power of DC, $p_{\xi,1,t}^{\mathrm{DC}}$ can be calculated as:

$$p_{\xi,1,t}^{\mathrm{DC}} = a_{\xi,1} p_{\xi,1,t}^{\mathrm{server}} + b_{\xi,1} \tag{4.17}$$

where $a_{\xi,1}$ and $b_{\xi,1}$ are the power coefficients of DC.

(b) *Constraints on workload processing latency.*

The M/M/1 queuing model [30] is used to calculate the processing delay time of the workload, which must satisfy:

$$0 < \frac{1}{\mu_{\xi,1}^{\mathrm{IW}} - \frac{\sum_{\delta=1}^{\Delta} \lambda_{\xi,1,\delta,t}}{m_{\xi,1,t}^{\mathrm{IW}}}} \le D \tag{4.18}$$

D is the maximum latency of the workload. Equation (4.18) can be re-written as:

$$m_{\xi,1,t}^{\mathrm{IW}} \ge \frac{\sum_{\delta=1}^{\Delta} \lambda_{\xi,1,\delta,t}}{\mu_{\xi,1}^{\mathrm{IW}} - \frac{1}{D}} \tag{4.19}$$

(c) *Constraints on the number of active servers:*

$$0 \le m_{\xi,1,t}^{\mathrm{IW}} \le M_{\xi,1} \tag{4.20}$$

where $M_{\xi,1}$ is the total number of servers.

(d) *Constraints on DC' electricity trading with distribution systems*:

$$\begin{cases} 0 \le p_{\xi,1,t}^{\text{grid},s} \le p_{\xi,1}^{\text{grid,max}} \\ 0 \le p_{\xi,1,t}^{\text{grid},b} \le p_{\xi,1}^{\text{grid,max}} \end{cases} \tag{4.21}$$

where $p_{\xi,1}^{\text{grid,max}}$ is the maximum power that is bought from/sold to distribution systems by DC.

(e) *Constraints on power balance:*

$$p_{\xi,1,t}^{\text{grid},b} = p_{\xi,1,t}^{\text{grid},s} + \sum_{m=2}^{K_\xi} p_{\xi,1m,t}^{\text{LEM}} + p_{\xi,1,t}^{\text{DC}} \tag{4.22}$$

2. *Constraints of conventional-type prosumers* ($k = 2, \ldots, Kx$).

(a) *Constraints on the power generation of DG.*

$$0 \le p_{\xi,k,t}^{\text{DG}} \le p_{\xi,k,t}^{\text{DG,max}} \tag{4.23}$$

where $p_{\xi,k,t}^{\text{DG,max}}$ is the maximum output of DG.

(b) *Constraints on ESS:*

$$\begin{cases} E_{\xi,k,t}^{\text{ESS}} = E_{\xi,k,t-1}^{\text{ESS}} + \left(p_{\xi,k,t}^{\text{ESS,ch}} \eta_{\xi,k}^{\text{ch}} - \frac{p_{\xi,k,t}^{\text{ESS,dis}}}{\eta_{\xi,k}^{\text{dis}}} \right) \Delta t \\ E_{\xi,k}^{\text{ESS,min}} \le E_{\xi,k,t}^{\text{ESS}} \le E_{\xi,k}^{\text{ESS,max}} \\ 0 \le p_{\xi,k,t}^{\text{ESS,ch}} \le p_{\xi,k}^{\text{ESS,max}} \\ 0 \le p_{\xi,k,t}^{\text{ESS,dis}} \le p_{\xi,k}^{\text{ESS,max}} \end{cases} \tag{4.24}$$

where $p_{\xi,k,t}^{\text{ESS,dis}}$ is the discharge power of ESS; $\eta_{\xi,k}^{\text{dis}}$ is the discharge efficiency of ESS; $E_{\xi,k,t}^{\text{ESS}}$ is the stored energy in ESS; $E_{\xi,k}^{\text{ESS,min}}$ and $E_{\xi,k}^{\text{ESS,max}}$ are the minimum and maximum stored energy in ESS; $p_{\xi,k}^{\text{ESS,max}}$ is the maximum charging and discharging power, respectively.

(c) *Constraints on FL*:

$$p_{\xi,k,t}^{\text{FL,min}} \le p_{\xi,k,t}^{\text{FL}} \le p_{\xi,k,t}^{\text{FL,max}} \tag{4.25}$$

where $p_{\xi,k,t}^{\text{FL,min}}$ and $p_{\xi,k,t}^{\text{FL,max}}$ are the minimum and maximum power of FL, respectively.

(d) *Constraints on prosumer' trading with distribution systems*:

$$\begin{cases} 0 \le p_{\xi,k,t}^{\text{grid},s} \le p_{\xi,k}^{\text{grid,max}} \\ 0 \le p_{\xi,k,t}^{\text{grid},b} \le p_{\xi,k}^{\text{grid,max}} \end{cases} \tag{4.26}$$

where $p_{\xi,k}^{\text{grid,max}}$ is the maximum power that is bought from/sold to distribution systems by the prosumer.

3. *Constraints of LEM*

(a) *Constraints on the power balance of prosumers*:

$$p_{\xi,k,t}^{\text{grid},s} + \sum_{m=1,m\neq k}^{K_\xi} p_{\xi,km,t}^{\text{LEM}} + p_{\xi,k,t}^{\text{ESS,ch}} + p_{\xi,k,t}^{\text{FL}} = p_{\xi,k,t}^{\text{grid},b} + p_{\xi,k,t}^{\text{DG}} + p_{\xi,k,t}^{\text{ESS,dis}} \tag{4.27}$$

(b) *Constraints on power balance in LEM*

The transaction power between two prosumers should be consistent, which is stated as:

$$p_{\xi,km,t}^{\text{LEM}} + p_{\xi,mk,t}^{\text{LEM}} = 0 \tag{4.28}$$

4.3.3 LEM Clearing Algorithm

The ADMM algorithm is employed to clear the LEM in a distributed way. Each prosumer only needs to exchange limited information with the other prosumers. Transaction consensus is then achieved through the continuous optimization of prosumers for decision-making. The iterative method for prosumers' decision-making based on the ADMM algorithm is described in Eq. (4.29–4.32). The detailed derivation process is provided in Appendix 4.1.

$$\mathrm{x}_{\xi,k}^{l+1} = \arg\min_{\mathrm{x}_{\xi,k}} \left(\sum_{t=1}^{T} \sum_{m=1}^{M} \left(\begin{array}{l} \tilde{f}_{\xi,k}(\mathrm{x}_{\xi,k}) \\ +\pi_{\xi,km,t}^{l} \left(\frac{p_{\xi,km,t}^{\text{LEM},l} - p_{\xi,mk,t}^{\text{LEM},l}}{2} - p_{\xi,mk,t}^{\text{LEM},l} \right) \Delta t \\ +\frac{\rho_\xi}{2} \left(\frac{p_{\xi,km,t}^{\text{LEM},l} - p_{\xi,mk,t}^{\text{LEM},l}}{2} - p_{\xi,km,t}^{\text{LEM}} \right)^2 \Delta t^2 \end{array} \right) \right) \tag{4.29}$$

$$\pi_{\xi,km,t}^{l+1} = \pi_{\xi,km,t}^{l} - \rho_\xi \frac{p_{\xi,km,t}^{\text{LEM},l+1} + p_{\xi,mk,t}^{\text{LEM},l+1}}{2} \Delta t \tag{4.30}$$

where l is the number of iterations; $\mathrm{x}_{\xi,k}^{l+1}$ is the decision variable of the prosumer; The definition of the function $\tilde{f}_{\xi,k}$ is given in Appendix 4.1; $\pi_{\xi,km,t}^{l}$ is the transaction price between the k-th and m-th prosumers; ρ_ξ is the penalty coefficient of ADMM algorithm.

The primal residual, $\boldsymbol{r}^{l+1}$, and the dual residual, $\boldsymbol{s}^{l+1}$, in the $l+1$ iteration, are calculated as:

$$\begin{cases} \left\| \mathrm{r}^{l+1} \right\|_2 = \frac{\Delta t}{2} \left(\sum_{k=1}^{K_\xi} \sum_{t=1}^{T} \sum_{m=1}^{K_\xi} \left(p_{\xi,km,t}^{\mathrm{LEM},l+1} + p_{\xi,mk,t}^{\mathrm{LEM},l+1} \right)^2 \right)^{\frac{1}{2}} \\ \left\| \mathrm{s}^{l+1} \right\|_2 = \frac{\rho_\xi \Delta t}{2} \left(\sum_{k=1}^{K_\xi} \sum_{t=1}^{T} \sum_{m=1}^{K_\xi} \left(p_{\xi,km,t}^{\mathrm{LEM},l+1} - p_{\xi,mk,t}^{\mathrm{LEM},l+1} \right)^2 \right)^{\frac{1}{2}} \end{cases} \tag{4.31}$$

The iteration termination condition is as follows.

$$\begin{cases} \left\| \mathrm{r}^{l+1} \right\|_2 \le \varepsilon^{\mathrm{pri}} \\ \left\| \mathrm{s}^{l+1} \right\|_2 \le \varepsilon^{\mathrm{dual}} \end{cases} \tag{4.32}$$

where $\varepsilon^{\mathrm{pri}}$ and $\varepsilon^{\mathrm{dual}}$ are the convergence bounds of the primal residual and the dual residual, respectively.

To sum up, LEM clearing process is summarized as follows:

Step 1: Each prosumer collects information, initializes the transaction plan $p_{\xi,km,t}^{\mathrm{LEM},0}$, and then reports it to the trading platform. The platform initializes the price $\pi_{\xi,km,t}^{0}$ of each transaction and the iteration time l is set to 0;
Step 2: Each prosumer obtains the transaction plan $p_{\xi,mk,t}^{P2P,l}$ and transaction price $\pi_{\xi,km,t}^{l}$ of the other prosumers from the trading platform. Then, it optimizes its operating behaviors $\boldsymbol{x}_{\xi,k}^{l+1}$ according to Eq. (4.29) and reports the updated trading plan $p_{\xi,km,t}^{\mathrm{LEM},l+1}$ to the trading platform;
Step 3: The trading platform calculates the price $\pi_{\xi,km,t}^{l+1}$ according to Eq. (4.30);
Step 4: The trading platform calculates the primal residual and dual residual in the $l+1$ iteration according to Eq. (4.31). If the iteration termination condition Eq. (4.32) is satisfied, the market is cleared. Otherwise, $l \leftarrow l+1$, and skip to *Step* 2.

4.4 Optimal Trading Strategy of Networked DCs in Multiple LEMs

Given the close relationship between the optimal scheduling of networked DCs and the operation of LEMs, a bi-level optimization model is proposed to address this issue within multiple LEMs (refer to Fig. 4.3). The upper-level problem focuses on maximizing the profit of networked DCs while determining workload allocation among them. Meanwhile, the lower-level problems involve clearing LEMs in various distribution systems. Detailed formulations are provided in the following sections.

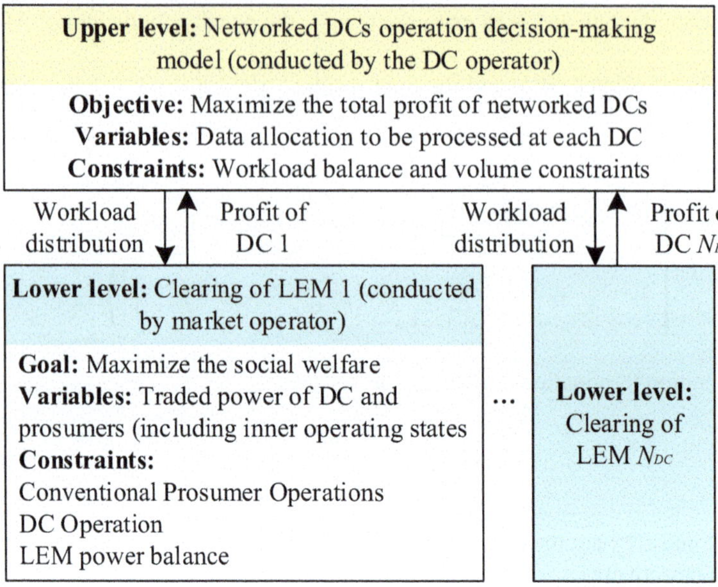

Fig. 4.3 Bi-level optimization structure of networked DCs in multiple LEMs

4.4.1 Optimization Model at the Upper Level

1. *Objective function*

The objective function is to maximize the total profit of networked DCs, which is given as:

$$\max S = \sum_{\xi=1}^{N_{DC}} S_{\xi,1} \tag{4.33}$$

where $S_{\xi,1}$ is decided at the lower level.

2. *Constraints*

All workloads should be distributed to DCs for processing, stated as:

$$L_{\delta,t} = \sum_{\xi=1}^{N_{DC}} \lambda_{\xi,1,\delta,t} \tag{4.34}$$

where $L_{\delta,t}$ is the average arrival rate of the workloads.

Workload quantity should satisfy:

$$\lambda_{\xi,1,\delta,t} \geq 0 \tag{4.35}$$

4.4.2 Optimization Model at the Lower Level

Once a DC receives the workload allocation result $\lambda_{\xi,1,\delta,t}$ from the upper level, it will then optimize its trading behaviors in the LEM. As detailed in Sect. 4.3, the LEMs in different distribution systems are cleared simultaneously.

4.5 Solutions for the Optimal Trading Strategy

Generally, the Karush–Kuhn–Tucker (KKT) optimality criterion is employed to transform the bi-level problem into a single-level problem for resolution. However, the operator of the DC at the upper level lacks operational information of the prosumers and LEM at the lower level. Consequently, the bi-level problem cannot be resolved using KKT conditions. Therefore, the differential evolution approach addresses the bi-level optimization model of networked DCs in multiple LEMs [31]. Thus, the differential evolution approach is utilized to solve the bi-level optimization model of networked DCs in multiple LEMs [31]. During interactions with LEMs, the DC operator optimizes workload allocation among DCs. Meanwhile, the meta-model [32] approximates the LEM clearing process and minimizes iteration times.

4.5.1 Differential Evolution Algorithms

The differential evolution method is a heuristic random search algorithm based on population evolution, designed to address global optimization problems. Its operations include differential mutation, crossover, and selection [33, 34]. Algorithm 1 depicts the solution steps.

1. *Differential mutation*

The mutant individual is divided into the base vector and the difference vector, which is shown in Eq. (4.36).

$$\upsilon_i = \lambda_{r_1} + F \cdot (\lambda_{r_2} - \lambda_{r_3}) \tag{4.36}$$

where λ_{r_1}, λ_{r_2}, and λ_{r_3} are random individuals in the population Λ; λ_{r_1} is the base vector, and $(\lambda_{r_2} - \lambda_{r_3})$ is the difference vector. Since the base vector is randomly selected, there is only one difference vector, and this difference variation is recorded as DE/rand/1.

2. *Crossover*

Crossover, the exchange between elements of current and mutant individuals, can be categorized as binomial or exponential. Assuming that the dimension of the individual is n, $\lambda_i = [\lambda_{i,1}, \ldots, \lambda_{i,j}, \ldots, \lambda_{i,n}]^{\mathrm{T}}$, $\upsilon_i = [\upsilon_{i,1}, \ldots, \upsilon_{i,j}, \ldots, \upsilon_{i,n}]^{\mathrm{T}}$ and $o_i = [o_{i,1}, \ldots, o_{i,j}, \ldots, o_{i,n}]^{\mathrm{T}}$ are satisfied. Taking the binomial crossover as an example, the j-th element of the test individual is determined as follows.

$$o_{i,j} = \begin{cases} \upsilon_{i,j} & \text{rand}(0, 1) \leq C_r \\ \lambda_{i,j} & \text{others} \end{cases} \tag{4.37}$$

where rand (0, 1) is a random number between [0, 1].

3. *Selection*

Comparing $\boldsymbol{\lambda}_i$ and $\boldsymbol{o}_i$ the greedy criterion is applied to choose the better individual between the present and test individuals entering into the new generation population.

Algorithm 1: Differential Evolution (max S as an example)

Initialize the population size N_p, scaling factor F, crossover probability C_r, and the iteration time l is set to 0

Initialize the population $\Lambda = \{\lambda_1^l, \lambda_2^l, \ldots, \lambda_{N_p}^l\}$

while the iteration termination condition is not met **do**

  Λ differential mutation to get the mutant population $Y = \{\upsilon_1^l, \upsilon_2^l, \ldots, \upsilon_{N_p}^l\}$

  Λ and Y are crossed to obtain the test population $O = \{o_1^l, o_2^l, \ldots, o_{N_p}^l\}$

  for all $i = \{1, 2, \ldots, N_p\}$ **do**

    if $S(o_i^l)\ S(o_i^l)$ **then**

      $\lambda_i^{l+1} \leftarrow o_i^l$

    else

      $\lambda_i^{l+1} \leftarrow \lambda_i^l$

    end if

  end for

  $l \leftarrow l+1$

end while

4.5.2 *Meta Model*

The meta-model, known as the "model of the model", offers an approximate mathematical portrayal of specific numerical computations. By incorporating the meta-model into the optimization analysis, the calculation efficiency can be greatly improved while the solving difficulty is also reduced [35, 36]. This chapter uses the Kriging meta-model [37], celebrated for its robust nonlinear function approximation and unique error estimation. Kriging [35, 38] acts as an interpolation scheme for $y(\boldsymbol{x})$, depicted as a linear amalgamation of diverse functions $f_i(\boldsymbol{x})$ coupled with a particular instantiation of the stochastic process $Z(\boldsymbol{x})$, which can be stated as:

$$y(\boldsymbol{x}) = \sum_{i=1}^{N} \kappa_i f_i(\mathrm{x}) + Z(\boldsymbol{x}) \tag{4.38}$$

where N is the number of functions $f_i(\boldsymbol{x})$; κ_i is the i-th linear regression coefficient. The mean of $Z(\boldsymbol{x})$ is 0, and its covariance function is given as:

$$\mathrm{cov}[Z(\boldsymbol{x}^{(i)}), Z(\boldsymbol{x}^{(j)})] = \sigma^2 R(\boldsymbol{x}^{(i)}, \boldsymbol{x}^{(j)}) \tag{4.39}$$

where σ^2 is the variance of the random process; R is a correlation function.

The indeterminate parameters of Kriging are trained using input and output data as sample points to create the optimal Kriging meta-model. High-dimensional input variables necessitate a significant number of calculations for model training. In such cases, the partial least squares (PLS) method can pinpoint influential model parameters, reducing their number and shortening training time. The meta-model built using this method is the KPLS meta-model (Kriging meta-model combined with partial least squares) [39].

4.5.3 Solution Process

A meta-model is developed for each LEM to reduce LEM clearing times, with the workload of the DC as input and its profit as output. In this way, the model training time can be reduced by utilizing the KPLS meta-model. However, without correction during optimization, the solution heavily relies on the initial meta-model's accuracy, complicating the search for the best solution. Hence, a dynamic update mechanism is proposed, calculating the meta-model error for the optimal individual in each generation [40]. In summary, differential evolution and dynamic KPLS meta-model are combined to form the hybrid solution algorithm HA-DEDKM (hybrid algorithm of differential evolution and dynamic KPLS meta-model). The detailed solution process is shown in Fig. 4.4.

Step 1: Initialize the parameters, including the population size N_p of the differential evolution algorithm, the scaling factor F, the crossover probability C_r, the number of initial samples number N_s, and the maximum tolerable prediction error $\varepsilon^{\max}$.

Step 2: Build the initial KPLS meta-model. Latin hypercube sampling is used to generate N_s initial sample points. Each of sample represents a DC workload distribution and is recorded as $\lambda_i^{\mathrm{T}} = [\lambda_{i,\xi,1,\delta,t}](\xi \in N_{\mathrm{DC}}, \delta \in \Delta, t \in T)$ and satisfies the constraints Eqs. (4.34) and (4.35). Each distribution plan is substituted into the LEM clearing model in Sect. 4.3.2 to obtain the profits of DC in each

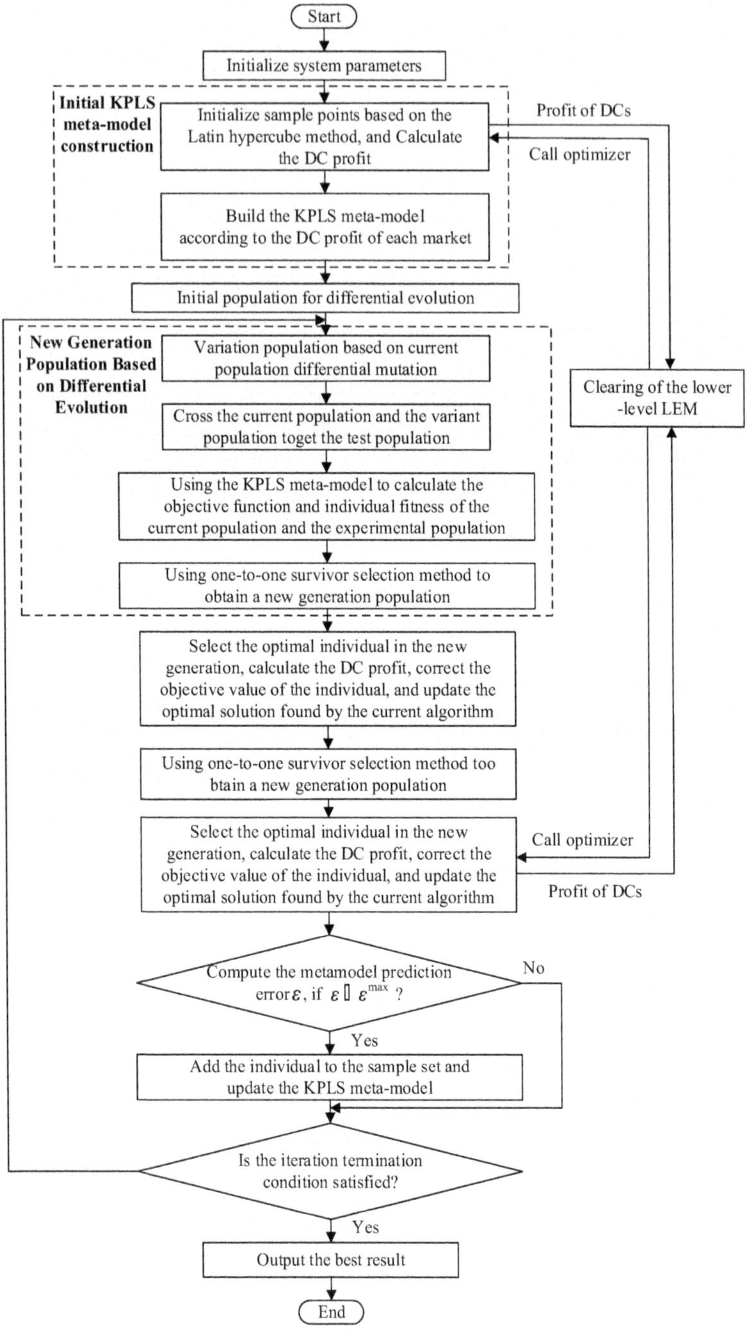

Fig. 4.4 Bi-level optimization of networked DCs in multiple LEMs

market. Construct the KPLS meta-model for the ξ-th market: workload distribution $\lambda_{i,\xi}^{\mathrm{T}} = [\lambda_{i,\xi,1,\delta,t}](\delta \in \Delta, t \in T)$ as input, DC profit $S_{i,\xi,1}$ as output and sample data set $\{(\lambda_{i,\xi}, S_{i,\xi,1})|i = 1, 2, ..., N_s\}$.

Step 3: The number of iterations l is set to 0. Initialize the population of differential evolution Λ^l.

Step 4: To create the new generation population Λ^{l+1}, the present population Λ^l is subject to differential mutation, crossover, and selection. The KPLS meta-model determines the objective function value of population members, and the LEM clearing model is no longer called.

Step 5: Select the optimal individual $\lambda_{\text{best}}^{l+1}$ in the new generation population Λ^{l+1}. Call the LEM clearing model to calculate the individual's objective function value S_{best}^{l+1}. Suppose the optimal solution obtained by the algorithm is S_{opt}^{l}, if $S_{\text{best}}^{l+1} \geq S_{\text{opt}}^{l}$ and constraints Eq. (4.34) and Eq. (4.35) are satisfied, then $\lambda_{\text{opt}}^{l+1} \leftarrow \lambda_{\text{best}}^{l+1}, S_{\text{opt}}^{l+1} \leftarrow S_{\text{best}}^{l+1}$. Otherwise, $\lambda_{\text{opt}}^{l+1} \leftarrow \lambda_{\text{opt}}^{l}, S_{\text{opt}}^{l+1} \leftarrow S_{\text{opt}}^{l}$.

Step 6: Calculate the prediction error ε of the meta-model λ_{best}^{l+1}. If $\varepsilon \geq \varepsilon^{\max}$, add the point to the sample set and update the KPLS meta-model.

Step 7: If the iteration termination condition of differential evolution is satisfied, the optimal result is S_{opt}^{l+1}. Otherwise, $l \leftarrow l + 1$, and go to *Step 4*.

4.6 Case Studies

4.6.1 Parameter Settings

Three geographically networked DCs are operated by one Internet company, with the total number of servers per DC is 65,000. Each server has a processing rate of 20 requests per second. The active server consumes 0.1 kW when idle and 0.2 kW at full load. Assuming the company employs a single front-end agent, the workload arrival rate for a given day can be referenced from Fig. 4.11 in Appendix 4.2, sourced from [41]. The maximum delay time of the workload is 0.1 s, with each DC earning a basic income of 0.025 CNY per unit workload processed. To simulate the operations of multiple distribution systems with LEMs, three 10-node distribution systems, each incorporating three DCs, are utilized with parameters detailed in [42]. Prosumers within the same distribution system can trade with both the system and other prosumers, with a time interval Δt set at 1 h and the total duration T at 24 h. The diagram of networked DCs in distribution systems is shown in Fig. 4.5.

The time-of-use (ToU) electricity prices are detailed in Table 4.5 of Appendix 4.2, following the peak and valley periods as per the TOU electricity price regulations of Tianjin, Gansu, and Jiangsu provinces in *China*. PV and load data are from [43], while wind power data is sourced from [44]. The marginal cost of PV and wind energy is zero, and the load has a $\pm$ 10% margin of flexibility. Parameters α and β are randomly selected from 0.0001 to 0.0002 CNY/kWh and 0.3–0.4 CNY/kWh, respectively. Generation capacity ranges randomly between 1000 and 2000 kW. The depreciation

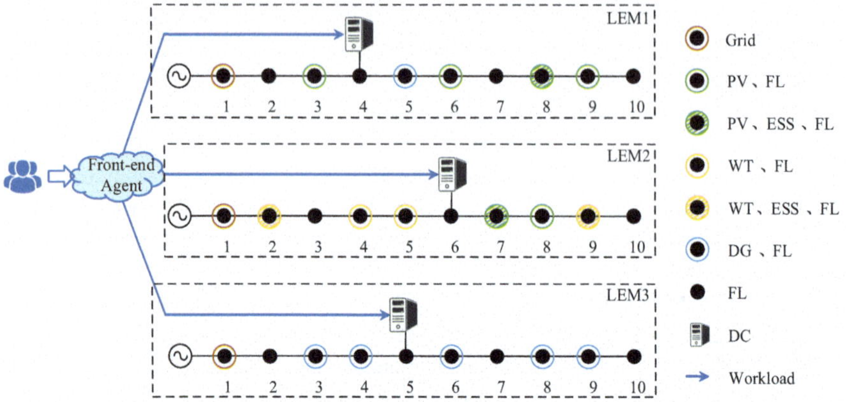

Fig. 4.5 Diagram of networked DCs in distribution systems

cost coefficient of ESS is 0.1 CNY/kWh. The charge and discharge efficiency of ESS is 95%, with a discharge depth of 90% and a discharge duration of 4 h. The unit network fee for LEM transactions is 0.005 CNY/(kWh Ω). Model simulation programming is based on the Python 3.7 environment. The convex optimization problems are solved using the commercial solver GUROBI.

4.6.2 *Trading Analysis of LEM*

1. *Economic Analysis of LEM*

Table 4.1 compares the total welfare before and after implementing LEM in distribution systems. Without LEM, prosumers can only trade electricity with the distribution system, resulting in a total market welfare of 311,639.02 CNY. When there is an LEM, electricity can be traded among prosumers, with the LEM clearing price falling between the TOU price and the feed-in tariff. This allows consumers to trade energy at better prices, earning more profits. Consequently, the total market welfare rises by 10.48%.

When the distribution system adopts LEM, the convergence process of the ADMM algorithm is shown in Fig. 4.12 of Appendix 4.3. Both the primal and dual residuals fall below the preset maximum limit after 507 iterations. At this point, the total welfare is 344,304.32 CNY, with an error of 0.00019% compared to the centralized clearing method. Additionally, the overall welfare has converged to within the error

Table 4.1 Welfare comparisons between the distribution systems with and without LEM

Profit/ CNY	No LEM	With LEM	
		Centralized	ADMM
Node 2	35,846.50	39,054.32	39,054.04
Node 3	35,207.43	37,909.15	37,909.72
Node 4	45,157.42	57,775.73	57,774.88
Node 5	32,434.38	34,610.64	34,610.51
Node 6	31,265.46	31,774.76	31,775.22
Node 7	28,890.90	31,305.97	31,305.99
Node 8	41,089.13	45,350.94	45,350.90
Node 9	38,065.77	41,390.34	41,389.90
Node 10	23,682.03	25,133.14	25,133.16
Market welfare	311,639.02	344,304.98	344,304.32

range of 0.1% after 27 times, which demonstrates that the algorithm can bring the objective function error into an acceptable range after a few repetitions.

2. *Operation Behavior Analysis of Prosumers*

To analyze their behaviors in LEM, we take DC at node 4, PV + FL at node 6, and PV + ESS + FL at node 8 as examples. Figure 4.6 shows the optimization plan for the DC at node 4. The DC's workload delay time is 0.1 s each period, reducing electricity costs while maintaining service quality. DC buys cheaper electricity from distribution systems during the price valley from 23:00 to 7:00 to meet its demand. DC prioritizes purchasing lower-cost power from other prosumers during power shortages before turning to distribution systems. Meanwhile, DC absorbs surplus PV energy from 9:00 to 17:00 when output peaks, enhancing the economic benefits for prosumers.

The operation plan for node 6's PV + FL is shown in Fig. 4.13 of Appendix 4.3. The prosumer buys electricity from the grid between 23:00 and 7:00. From 7:00 to 17:00, the prosumer uses its own PV power to meet the load. If the PV power is insufficient, the prosumer buys power from other prosumers due to LEM's price advantage. Meanwhile, excess PV power is sold to other prosumers. Between 17:00 and 23:00, when PV output drops to zero and there is no surplus power from other prosumers, the prosumer purchases electricity from the distribution system.

Figure 4.14 in Appendix 4.3 shows the operation plan of PV + ESS + FL at node 8. ESS enhances the flexibility of prosumer operations. ESS charges to store cheap electricity during low electricity prices and excess PV generation. At other times, ESS discharges to meet the prosumer's electricity demand and can sell stored electricity to other prosumers. Thus, ESS provides lower-cost electricity to all prosumers in the distribution system, improving the market's overall welfare.

3. *Clearing Price Analysis of LEM*

When the LEM is adopted, the clearing price falls between the time-of-use price and feed-in tariff. Taking the DC at node 4 as an example, Fig. 4.7 shows the transaction

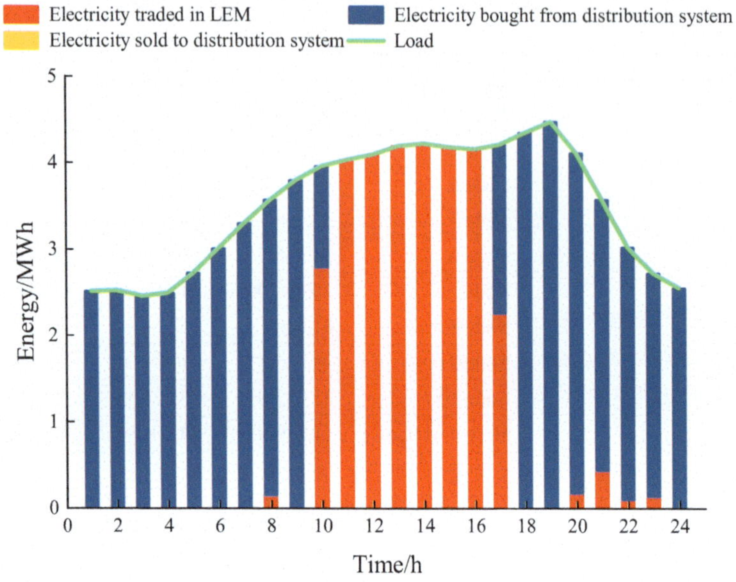

Fig. 4.6 Operation scheme of DC at node 4

electricity and corresponding clearing price in LEM from 9:00 to 10:00. The clearing prices between DC and prosumers vary. Each transaction between prosumers includes a network usage fee proportional to the electrical distance. For DC, the actual price paid per transaction (power purchase price) is the clearing price plus the network fee per unit of electricity. Additionally, the power purchase price for all DC transactions during the same period is the same.

Figure 4.15 in Appendix 4.3 shows the transaction power and price of DC from 10:00 to 11:00. During this period, the power purchase price for DC and other prosumers is the same. From 10:00 to 11:00, DC does not buy electricity from the grid, so the prosumer with the highest marginal generation cost determines the purchase price. The marginal generation cost for the prosumer DG at node 5 is 0.6237 CNY/kWh. The network fee for unit power consumption in LEM is 0.0023 CNY/kWh. To protect the prosumer's interests, the LEM clearing price between DC and prosumers should be at least 0.6260 CNY/kWh. Thus, the DC power purchase price is 0.6283 CNY/kWh. In conclusion, the purchase/sale price for prosumers in LEM during one period is the same, ensuring maximum profits for all prosumers.

4.6.3 Bi-level Optimization Results of Networked DCs

The bi-level optimization of networked DCs in multiple LEMs is analyzed. Differential evolution adopts the classic DE/target-to-best/1/L algorithm [45]. The population

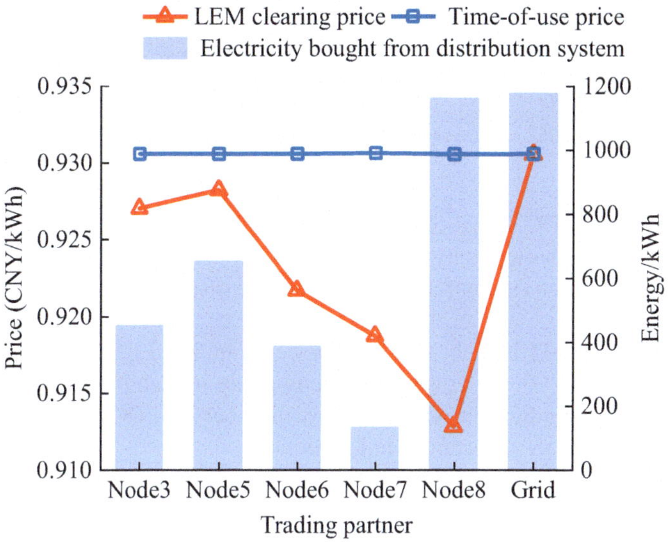

Fig. 4.7 LEM trading power and price of the DC at 9:00–10:00

size is 500, the scaling factor F is 0.7, the crossover probability C_r is 0.3, and the maximum evolutionary generation is 8000. The number of initial sample points in the meta-model is 500. The maximum prediction error $\varepsilon^{\max}$ is 0.001.

The following two scenarios are compared to demonstrate the proposed model's superiority.

Scenario 1: The workload is evenly allocated among three DCs.

Scenario 2: The workload is distributed among DCs to optimize total profit while taking into account the price information of multiple LEMs.

1. *Economic Analysis of Networked DCs*

Simulation results are displayed in Table 4.2. The total profit of DCs in Scenario 2 exceeds that in Scenario 1 by 41,821.93 CNY, a 23.54% increase. The supply–demand balance varies across distribution systems where DCs are located, leading to fluctuating LEM prices. Therefore, networked DC operators can optimize workload allocation among DCs based on price information to reduce costs and enhance profitability.

2. *Trading Results of Networked DCs*

Figure 4.8 illustrates the workload allocation strategy in Scenario 2, while Fig. 4.9 shows the transaction results of the DC in LEM. When there is surplus wind power in LEM 2, DC2 primarily handles the workload between 0:00 and 8:00, buying electricity at a meager cost from other prosumers. When wind power in LEM 2 is

Table 4.2 Profits of DCs in LEMs

	LEM 1 (CNY)	LEM 2 (CNY)	LEM 3 (CNY)	Total profit (CNY)
Scenario 1	57,775.73	75,415.52	44,478.43	177,669.67
Scenario 2	43,262.05	142,003.90	34,225.64	219,491.60

insufficient to meet DC's total electricity demand, some workloads are assigned to DC3, which purchases electricity from distribution system 3 to meet the load demand. At this point, the electricity purchase price from the distribution system in LEM3 is the lowest.

PV output in LEM1 is high, and wind power output in LEM2 is also very high between 8:00 and 16:00. As a result, electricity prices in both markets are relatively low. Workload during this period is primarily distributed to DC1 and DC2, and the demand is met through transactions in LEMs.

From 16:00 to 24:00, the PV output is zero, and wind power is insignificant. Therefore, DC mainly purchases power from the distribution system to meet the load demand and the workload is distributed to the LEM with the lowest electricity purchase price. Specifically, the workload is assigned to DC2 during 16:00–21:00, DC1 during 21:00–22:00, DC3 during 22:00–23:00, and DC1 during 23:00–24:00. It is worth noting that a small amount of workload is assigned to DC2 during 21:00–22:00 and 23:00–24:00. This is because the algorithm converges to a locally optimal solution. If all the workload is distributed to DC1 during these times, the objective function's value can be slightly improved, although the improvement would be minimal.

3. *Impacts of Networked DCs on the Other Prosumers*

The proposed bi-level optimization not only boosts networked DCs' profits but also influences other prosumers' earnings Fig. 4.16 in Appendix 4.4 illustrates how other prosumers' profits change under various scenarios. It shows that more workloads are

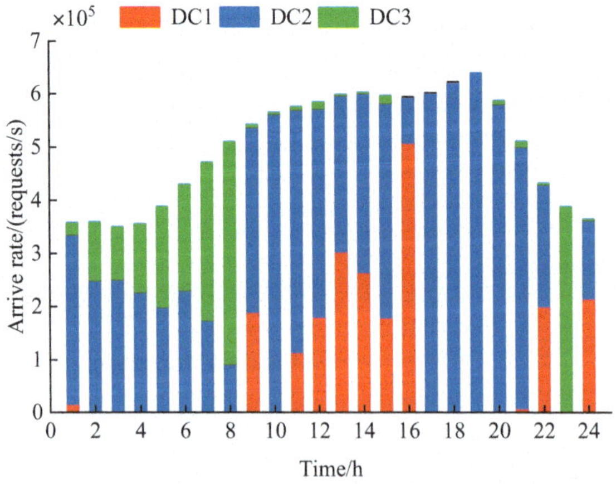

Fig. 4.8 Workload allocation result among DCs

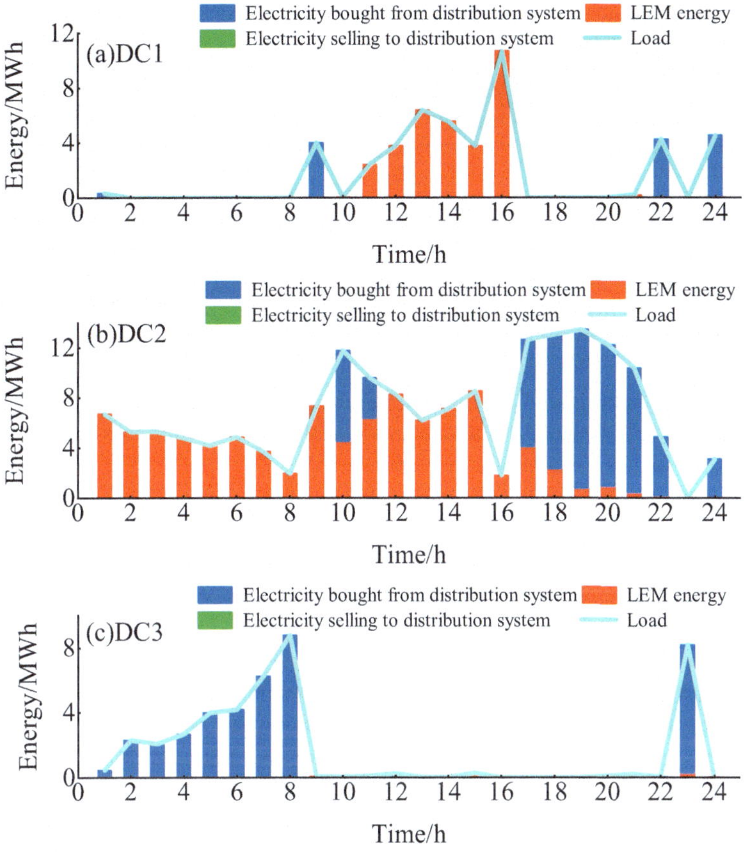

Fig. 4.9 Trading results of networked DCs

assigned to DC2 to utilize excess wind power in LEM 2. Generally, the three DCs' purchased power from the distribution system decreases while their purchases from other prosumers rise. Allocating fewer workloads to DCs in LEM 1 and 3 reduces the trading price, whereas assigning more to the DC in LEM2 raises it. The altered LEM price can impact other prosumers' profits in LEM.

Taking LEM 1 as an example, let's illustrate the impacts of the proposed method on other prosumers. Prosumers at nodes 2, 5, 7, and 10 primarily purchase electricity through LEM transactions. As the transaction price decreases, their profits will increase. Prosumers with PVs at nodes 3, 8, and 9 mainly participate in LEM as sellers. Due to the lower transaction price, their benefits decrease. The prosumer at node 6 sells electricity to the distribution system when its PV output is surplus and buys electricity from the prosumer at node 8 when its PV output is 0.

Table 4.3 compares the self-consumption rate of renewable energy, the charging and discharging energy of ESS, and the total social welfare of LEMs in Scenarios 1 and 2.

Table 4.3 Resource utilization and total welfare of all markets

	Renewable energy self-consumption rate (%)	Charging and discharging energy of ESS (kWh)	Total social welfare (10^4 CNY)
Scenario1	82.13	29,021.70	112.8802
Scenario2	93.39	31,784.71	116.9352

Table 4.3 shows that DCs increase their consumption of renewable energy due to its lower price, reducing the electricity sold to the grid by prosumers in LEMs. This also results in an increase in the renewable energy self-consumption rate from 82.13 to 93.39%. Additionally, the charging and discharging energy of the ESS has increased by 9.52%. The total social welfare of LEMs has increased by 40,550 CNY. These findings suggest that the proposed bi-level optimization method for networked DCs can enhance the utilization efficiency of distributed resources and improve social welfare.

4. *Impacts of Networked DCs on the Other Prosumers*

The bi-level optimization model of networked DCs in multiple LEMs is solved using the HA-DEDKM algorithm, and the convergence process is illustrated in Fig. 4.10. It is evident that the algorithm converges to a local optimum after 8000 iterations. It is worth noting that the objective function's value increases rapidly initially, and satisfactory results can be achieved after approximately 3000 iterations. The DC operator can judiciously determine the number of iterations based on specific requirements.

To illustrate the impact of dynamic meta-model updating, we compare the differential evolution-based static KPLS meta-model with HA-DEDKM in Table 4.4. The static KPLS meta-model algorithm requires solving only the lower-level problem during meta-model construction, resulting in fewer calls to the lower-level model. However, the meta-model's low prediction accuracy leads to weaker final results. On the other hand, the HA-DEDKM algorithm dynamically updates the meta-model

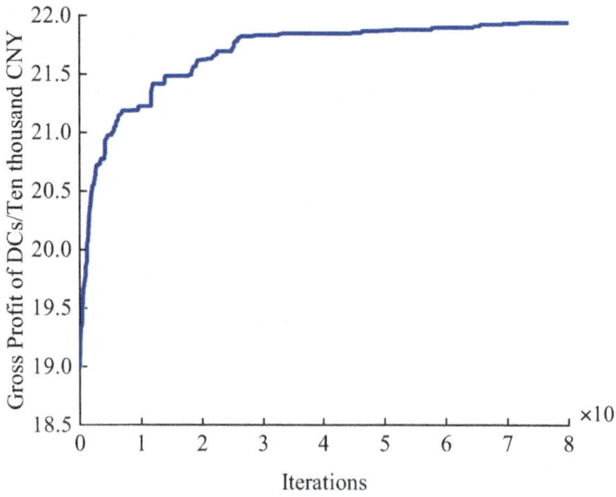

Fig. 4.10 Electricity trading results of networked DCs

Table 4.4 Performance comparison between the two algorithms

	The number of iterations	The number of lower-layer model calls	Total profit of DCs (CNY)
Static KPLS meta-model	8000	500	199,572.68
HA-DEDKM	8000	8500	219,491.60

based on prediction errors, increasing the number of lower-level model calls and leading to better solutions.

4.7 Conclusions

In this chapter, an optimal trading strategy of networked DCs in multiple LEMs is proposed to allocate workloads among multiple DCs optimally. The simulation results with three geographically networked DCs show:

(1) LEMs enable prosumers to trade freely, enhancing their profits. Compared to trading with ToU prices, total market welfare increases by 10.48%. The ADMM algorithm clears the LEM while ensuring prosumers' privacy. It takes about 507 iterations to achieve maximum social welfare in the LEMs, which is acceptable.
(2) The proposed model shifts workload from DCs with higher ToU electricity prices to those with lower prices in LEMs, increasing networked DCs' profits by 23.54%. This shift also improves the utilization of distributed resources, especially renewable energies, and raises total social welfare in LEMs.
(3) The proposed trading strategy, based on HA-DEDKM, does not access the operating information of LEMs and prosumers, preserving their privacy. Optimal workload allocation among DCs can be achieved with fewer than 8000 interactions between the DC operator and LEMs.

Appendix 4.1

First, the LEM clearing problem is converted into the standard form of the ADMM algorithm and then the iterative format of the optimization problem is obtained. Finally, the termination conditions of the iteration process are given.

1. Transformation of the optimization problem

Let $\boldsymbol{x}_{\xi}^{\mathrm{T}} = [\boldsymbol{x}_{\xi,1}^{\mathrm{T}}, \ldots, \boldsymbol{x}_{\xi,k}^{\mathrm{T}}, \ldots, \boldsymbol{x}_{\xi,K_{\xi}}^{\mathrm{T}}]$ represent the decision variables of all prosumers, where the decision variables of the k-th prosumer are described as follows:

$$x_{\xi,k}^{T}=\begin{cases}[m_{\xi,1,t}^{\text{IW}},p_{\xi,1,t}^{\text{DC}},p_{\xi,1,t}^{\text{grid},s},p_{\xi,1,t}^{\text{grid},b},p_{\xi,1m,t}^{\text{LEM}}]_{m\in K_{\xi},t\in T} & k=1\\ [p_{\xi,k,t}^{\text{DG}},p_{\xi,k,t}^{\text{ESS},ch},p_{\xi,k,t}^{\text{ESS},dis},p_{\xi,k,t}^{\text{FL}},p_{\xi,k,t}^{\text{grid},s},p_{\xi,k,t}^{\text{grid},b},p_{\xi,km,t}^{\text{LEM}}]_{m\in K_{\xi},t\in T} & k=2,3,\ldots,K_{\xi}\end{cases} \tag{4.40}$$

(a) Convert the maximization problem to the minimization problem.

Define the function $f_{\xi,k}$ as follows:

$$f_{\xi,k}\left(x_{\xi,k}\right)=\begin{cases}\sum_{t=1}^{T}(-B_{\xi,1,t}^{\text{DC}}-B_{\xi,1,t}^{\text{grid}}+C_{\xi,1,t}^{\text{net}}), & k=1\\ \sum_{t=1}^{T}(C_{\xi,k,t}^{\text{DG}}+C_{\xi,k,t}^{\text{ESS}}-U_{\xi,k,t}^{\text{FL}}-B_{\xi,k,t}^{\text{grid}}+C_{\xi,k,t}^{\text{net}}), & k=2,\ldots,K_{\xi}\end{cases} \tag{4.41}$$

The optimization problem of LEM clearing is equivalent to

$$\begin{aligned}&\min\sum_{k=1}^{K_{\xi}}f_{\xi,k}(x_{\xi,k})\\ &\text{s.t.}\ \ (4.15)–(4.31)\end{aligned} \tag{4.42}$$

(b) Remove inequality constraints.

Define the extended real-valued function of $f_{\xi,k}$

$$\tilde{f}_{\xi,k}\left(x_{\xi,k}\right)=\begin{cases}f_{\xi,k}(x_{\xi,k}),\ x_{\xi,k} & \text{constraints satisfied}\\ +\infty,\ x_{\xi,k} & \text{constraints not satisfied}\end{cases} \tag{4.43}$$

Putting the prosumers' constraints implicitly in the objective function, the market optimization problem is equivalent to

$$\tilde{f}_{\xi,k}\left(x_{\xi,k}\right)=\begin{cases}f_{\xi,k}(x_{\xi,k}),\ x_{\xi,k} & \text{constraints satisfied}\\ +\infty,\ x_{\xi,k} & \text{constraints not satisfied}\end{cases} \tag{4.44}$$

(c) Split local variables and global variables.

The $p_{\xi,km,t}^{\text{LEM}}$ in the above model is the local variable of the k-th prosumer. In the constraints, it is coupled with the local variable $p_{\xi,mk,t}^{\text{LEM}}$ of the m-th prosumer. In order to decompose the problem, the global variable $h_{\xi,km,t}$ is introduced, and the market power balance constraint are rewritten as:

$$\frac{h_{\xi,km,t}-h_{\xi,mk,t}}{2}-p_{\xi,km,t}^{\text{LEM}}=0 \tag{4.45}$$

By multiplying Δt, the market power balance constraint is changed into the electric energy balance constraint. Then the market optimization problem is finally expressed as follows:

$$\begin{aligned} &\min \sum_{k=1}^{K_\xi} \tilde{f}_{\xi,k}(x_{\xi,k}) \\ &\text{s.t.} \left(\frac{h_{\xi,km,t} - h_{\xi,mk,t}}{2} - p_{\xi,km,t}^{\text{LEM}} \right) \Delta t = 0 \end{aligned} \tag{4.46}$$

2. Iterative format for optimization problems.

Define $\boldsymbol{H}_\xi^{\text{T}} = [h_{\xi,km,t}](m \in K_\xi, t \in T, k \in K_\xi)$.

Let $\boldsymbol{\Pi}_\xi^{\text{T}} = [\boldsymbol{\Pi}_{\xi,1}^{\text{T}}, \ldots, \boldsymbol{\Pi}_{\xi,k}^{\text{T}}, \ldots, \boldsymbol{\Pi}_{\xi,K_\xi}^{\text{T}}]$ denote the dual variable of all constraints, where $\boldsymbol{\Pi}_{\xi,k}^{\text{T}} = [\pi_{\xi,km,t}](m \in K_\xi, t \in T, k \in K_\xi)$.

The augmented Lagrange function of the optimization problem is

$$L_{\rho,\xi}(\boldsymbol{x}_\xi, \boldsymbol{H}_\xi, \boldsymbol{\Pi}_\xi) = \sum_{k=1}^{K_\xi} L_{\rho,\xi,k}(\boldsymbol{x}_{\xi,k}, \boldsymbol{H}_\xi, \boldsymbol{\Pi}_{\xi,k}) \tag{4.47}$$

where

$$\begin{aligned} L_{\rho,\xi,k}(\boldsymbol{x}_{\xi,k}, \boldsymbol{H}_\xi, \boldsymbol{\Pi}_{\xi,k}) = \tilde{f}_{\xi,k}(\boldsymbol{x}_{\xi,k}) + \sum_{t=1}^{T} \sum_{m=1}^{K_\xi} \Bigg(& \pi_{\xi,km,t} \left(\frac{h_{\xi,km,t} - h_{\xi,mk,t}}{2} - p_{\xi,km,t}^{\text{LEM}} \right) \Delta t \\ & + \frac{\rho_\xi}{2} \left(\frac{h_{\xi,km,t} - h_{\xi,mk,t}}{2} - p_{\xi,km,t}^{\text{LEM}} \right)^2 \Delta t^2 \Bigg) \end{aligned} \tag{4.48}$$

The iterative process of the ADMM algorithm is as follows

$$\begin{aligned} \boldsymbol{x}_{\xi,k}^{l+1} &= \arg\min_{x_{\xi,k}} L_{\rho,\xi,k}(\boldsymbol{x}_{\xi,k}, \boldsymbol{H}_\xi^l, \boldsymbol{\Pi}_{\xi,k}^l) h_{\xi,km,t}^{l+1} \\ &= \frac{p_{\xi,km,t}^{\text{LEM},l+1} - p_{\xi,mk,t}^{\text{LEM},l+1}}{2} - \frac{\pi_{\xi,km,t}^l - \pi_{\xi,mk,t}^l}{2\rho_\xi \Delta t} \pi_{\xi,km,t}^{l+1} \\ &= \pi_{\xi,km,t}^l + \rho_\xi \left(\frac{h_{\xi,km,t}^{l+1} - h_{\xi,mk,t}^{l+1}}{2} - p_{\xi,km,t}^{\text{LEM},l+1} \right) \Delta t \end{aligned} \tag{4.49}$$

Combining the 2nd and 3rd of the above formulas gives

$$\pi_{\xi,km,t}^{l+1} = -\rho_\xi \frac{p_{\xi,km,t}^{\text{LEM},l+1} + p_{\xi,mk,t}^{\text{LEM},l+1}}{2} \Delta t + \frac{\pi_{\xi,km,t}^l + \pi_{\xi,mk,t}^l}{2} \tag{4.50}$$

Thus

$$\pi_{\xi,km,t}^{l+1} - \pi_{\xi,mk,t}^{l+1} = 0 \tag{4.51}$$

Choose an appropriate initial value so that $\pi_{\xi,km,t}^{0} - \pi_{\xi,km,t}^{0} = 0$, then

$$\pi_{\xi,km,t}^{l} - \pi_{\xi,mk,t}^{l} = 0 \tag{4.52}$$

$$h_{\xi,km,t}^{l+1} = \frac{p_{\xi,km,t}^{\mathrm{LEM},l+1} - p_{\xi,mk,t}^{\mathrm{LEM},l+1}}{2} \tag{4.53}$$

Therefore, the iterative process of ADMM algorithm can be expressed as

$$\begin{aligned} \boldsymbol{x}_{\xi,k}^{l+1} &= \arg\min_{\boldsymbol{x}_{\xi,k}} \Big(\tilde{f}_{\xi,k}\left(\boldsymbol{x}_{\xi,k}\right) \\ &\quad + \sum_{t=1}^{T}\sum_{m=1}^{K_\xi} (\pi_{\xi,km,t}^{l}(\frac{p_{\xi,km,t}^{\mathrm{LEM},l} - p_{\xi,mk,t}^{\mathrm{LEM},l}}{2} - p_{\xi,km,t}^{\mathrm{LEM}})\Delta t \\ &\quad + \frac{\rho_\xi}{2}\left(\frac{p_{\xi,km,t}^{\mathrm{LEM},l} - p_{\xi,mk,t}^{\mathrm{LEM},l}}{2} - p_{\xi,km,t}^{\mathrm{LEM}}\right)^2 \Delta t^2\Big)\pi_{\xi,km,t}^{l+1} \\ &= \pi_{\xi,km,t}^{l} - \rho_\xi \frac{p_{\xi,km,t}^{\mathrm{LEM},l+1} + p_{\xi,mk,t}^{\mathrm{LEM},l+1}}{2}\Delta t \end{aligned} \tag{4.54}$$

3. The iteration termination condition of ADMM algorithm.

$$\begin{cases} \left\|\boldsymbol{r}^{l+1}\right\|_2 \le \varepsilon^{\mathrm{pri}} \\ \left\|\boldsymbol{s}^{l+1}\right\|_2 \le \varepsilon^{\mathrm{dual}} \end{cases} \tag{4.55}$$

Appendix 4.2

See Fig. 4.11 and Table 4.5.

Appendix 4.3

See Figs. 4.12, 4.13, 4.14 and 4.15.

Table 4.5 Prosumer's electricity purchase/selling price from/to distribution system

	Period	Time	Buying price (CNY/kWh)	Selling price (CNY/kWh)
Market1	Peak	9:00–12:00 16:00–21:00	0.9306	0.1
	Normal	7:00–9:00 12:00–16:00 21:00–23:00	0.6204	
	Valley	23:00–7:00	0.2854	
Market2	Peak	7:00–9:00 17:00–23:00	0.6798	0.1
	Normal	23:00–7:00	0.4532	
	Valley	9:00–17:00	0.2266	
Market3	Peak	8:00–11:00 17:00–22:00	1.0263	0.1
	Normal	11:00–17:00 22:00–24:00	0.5968	
	Valley	0:00–8:00	0.2498	

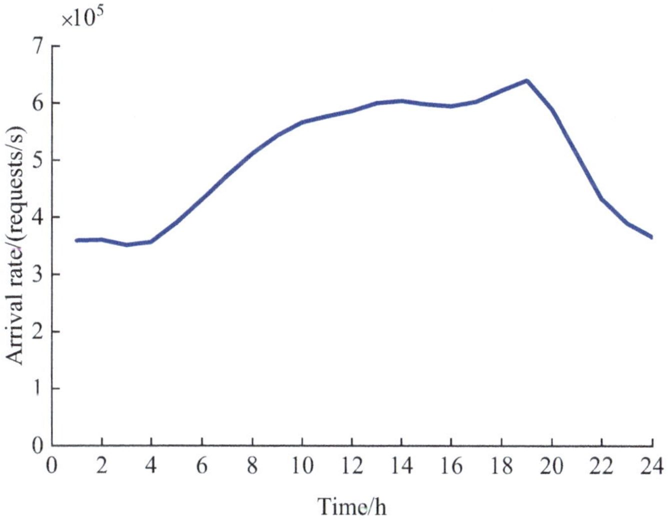

Fig. 4.11 Mean arrival rate of workloads

Appendix 4.4

See Fig. 4.16.

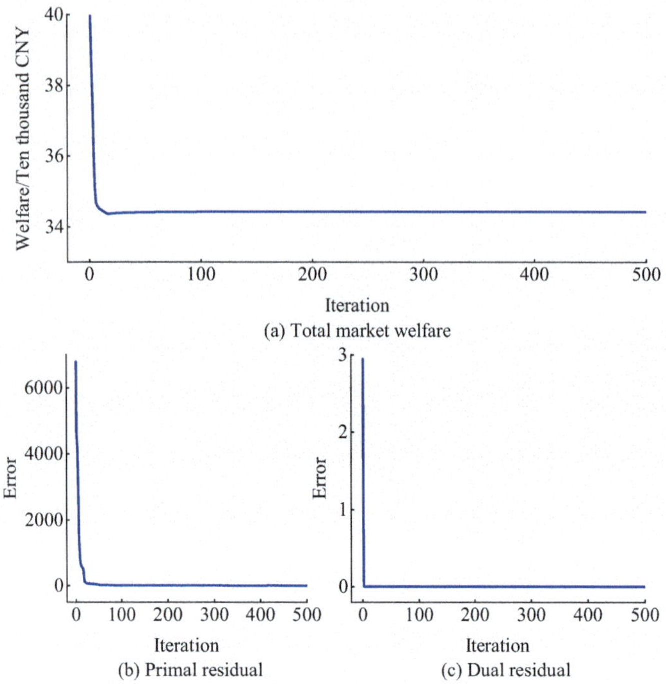

Fig. 4.12 Convergence curve of ADMM algorithm

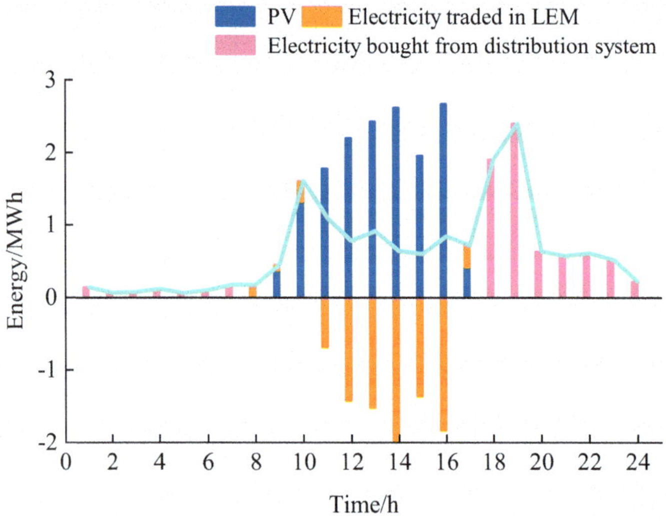

Fig. 4.13 Operation optimization scheme of PV and FL at node 6

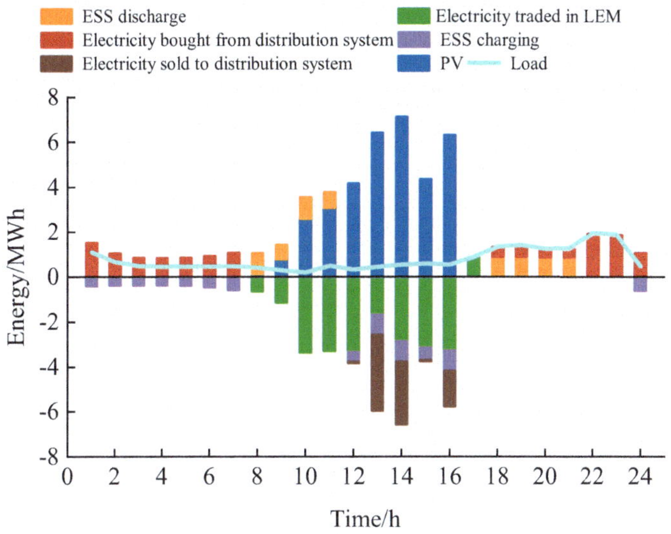

Fig. 4.14 Operation optimization scheme of PV, ESS, and FL at node 8

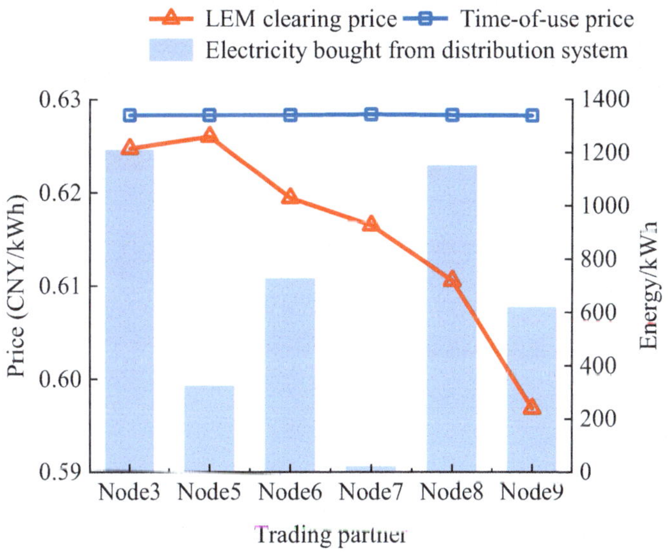

Fig. 4.15 LEM trading power and price of DC during 10:00–11:00

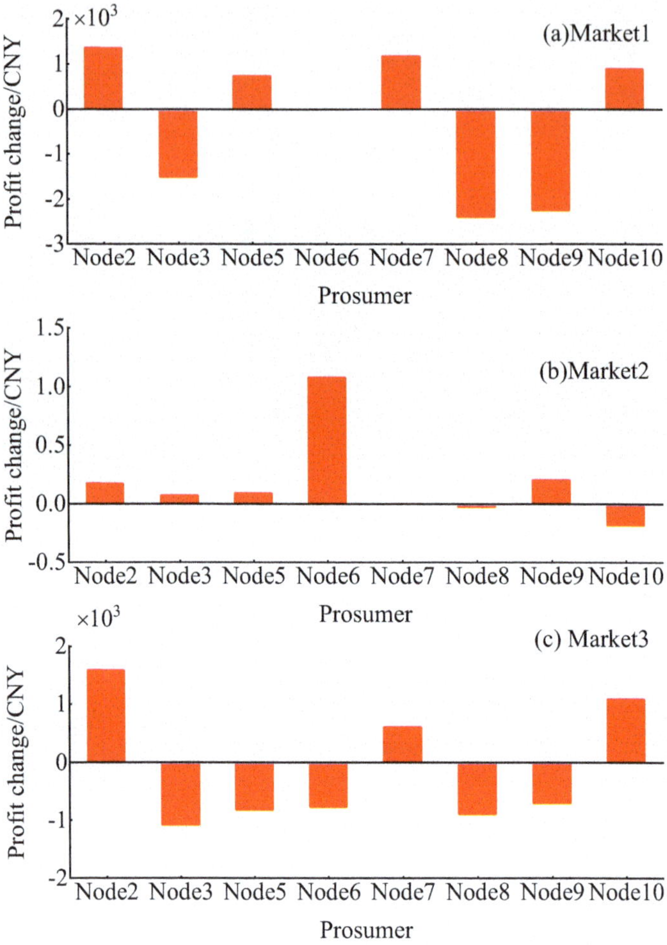

Fig. 4.16 Profit changes of conventional prosumers with the proposed bi-level optimization

References

1. Cedillo, M. H., Sun, H., Jiang, J., & Cao, Y. (2022). Dynamic pricing and control for EV charging stations with solar generation. *Applied Energy, 326*, 119920.
2. Li, Z., & Ma, C. (2022). A temporal–spatial charging coordination scheme incorporating probability of EV charging availability. *Applied Energy, 325*, 119838.
3. Song, M., & Gao, C. (2022). *Integration of distributed resources in smart grids for demand response and transactive energy: A case study of TCLs*. Springer.
4. Chen, M., Gao, C., Shahidehpour, M., Li, Z., Chen, S., & Li, D. (2020). Internet data center load modeling for demand response considering the coupling of multiple regulation methods. *IEEE Transactions on Smart Grid, 12*(3), 2060–2076.

5. Melton, R. B. (2013). Gridwise transactive energy framework (draft version) (No. PNNL-SA-22946). Pacific Northwest National Lab (PNNL), Richland, WA (United States).
6. Song, M., Cai, Y., Gao, C., Chen, T., Yao, Y., & Ming, H. (2022). Transactive energy in power distribution systems: Paving the path towards cyber-physical-social system. *International Journal of Electrical Power & Energy Systems, 142*, 108289.
7. Baroche, T., Pinson, P., Latimier, R. L. G., & Ahmed, H. B. (2019). Exogenous cost allocation in peer-to-peer electricity markets. *IEEE Transactions on Power Systems, 34*(4), 2553–2564.
8. Khorasany, M., Mishra, Y., & Ledwich, G. (2019). A decentralized bilateral energy trading system for peer-to-peer electricity markets. *IEEE Transactions on industrial Electronics, 67*(6), 4646–4657.
9. Song, M., Sun, W., Wang, Y., Shahidehpour, M., Li, Z., & Gao, C. (2019). Hierarchical scheduling of aggregated TCL flexibility for transactive energy in power systems. *IEEE Transactions on Smart Grid, 11*(3), 2452–2463.
10. Yang, Q., & Wang, H. (2021). Privacy-preserving transactive energy management for IoT-aided smart homes via blockchain. *IEEE Internet of Things Journal, 8*(14), 11463–11475.
11. Johanning, S., & Bruckner, T. (2019, September). Blockchain-based peer-to-peer energy trade: A critical review of disruptive potential. In *2019 16th International Conference on the European Energy Market (EEM)* (pp. 1–8), Ljubljana, Slovenia.
12. S&P Global Platts. (2020). *World's first high-frequency decentralized energy market helps drive Port of Rotterdam's energy transition*. https://www.spglobal.com/platts/en/about-platts/media-center/press-releases/2020/051020-world-s-first-high-frequency-decentralized-energy-market-driveport-of-rotterdam-energy-transition. Accessed May 10, 2020.
13. Power Technology. (2017). *The Brooklyn microgrid: Blockchain-enabled community power*. https://www.power-technology.com/features/featurethe-brooklyn-microgrid-blockchain-enabled-community-power-5783564/. Accessed January 12, 2017.
14. Share&Charge. (2020). *Open charging network: The next level of OCPI-Based ERoaming*. https://shareandcharge.com/. Accessed July 10, 2020.
15. News Desk. (2018). *Energo boost clean energy production through Qtum Blockchain implementation in Philippines*. https://www.geospatialworld.net/news/energo-clean-energy-production-qtum-blockchain/. Accessed February 7, 2018.
16. Zhu, S., Song, M., Lim, M. K., Wang, J., & Zhao, J. (2020). The development of energy blockchain and its implications for China's energy sector. *Resources Policy, 66*, 101595.
17. News Desk. (2024). Industry Leaders Adopt NVIDIA Robotics for Development of AI-Powered Autonomous Machines. https://www.geospatialworld.net/news/industry-leaders-nvidia-robotics-ai-powered-autonomous-machines/, Accessed June 10, 2024.
18. The Government of the People's Republic of China. (2021). *Outline of the fourteenth five-year plan for national economic and social development of the People's Republic of China and Vision 2035*. https://www.ndrc.gov.cn/xxgk/zcfb/ghwb/202103/P020210323538797779059.pdf. Accessed April 21, 2021.
19. IESPLZA. (2018). Supporting data centres to control carbon without controlling electricity. https://www.iesplaza.com/article-6006-1.html. Accessed November 25, 2018.
20. Mahbod, M. H. B., Chng, C. B., Lee, P. S., & Chui, C. K. (2022). Energy saving evaluation of an energy efficient data center using a model-free reinforcement learning approach. *Applied Energy, 322*, 119392.
21. Zhang, Q., Zeng, W., Lin, Q., Chng, C. B., Chui, C. K., & Lee, P. S. (2023). Deep reinforcement learning towards real-world dynamic thermal management of data centers. *Applied Energy, 333*, 120561.
22. Iqbal, W., Berral, J. L., Erradi, A., & Carrera, D. (2019). Real-time data center's telemetry reduction and reconstruction using Markov chain models. *IEEE Systems Journal, 13*(4), 4039–4050.

23. Lu, X., Kong, F., Liu, X., Yin, J., Xiang, Q., & Yu, H. (2017). Bulk savings for bulk transfers: Minimizing the energy-cost for geo-distributed data centers. *IEEE Transactions on Cloud Computing, 8*(1), 73–85.
24. Yu, L., Jiang, T., & Zou, Y. (2016). Distributed real-time energy management in data center microgrids. *IEEE Transactions on Smart Grid, 9*(4), 3748–3762.
25. Zhang, Y., Wang, Y., & Wang, X. (2012). Electricity bill capping for cloud-scale data centers that impact the power markets. In *2012 41st International Conference on Parallel Processing (ICPP)* (pp. 440–449), Pittsburgh, PA, USA.
26. Li, S., Brocanelli, M., Zhang, W., & Wang, X. (2014). Integrated power management of data centers and electric vehicles for energy and regulation market participation. *IEEE Transactions on Smart Grid, 5*(5), 2283–2294.
27. Yu, L., Jiang, T., & Zou, Y. (2016). Distributed online energy management for data centers and electric vehicles in smart grid. *IEEE Internet of Things Journal, 3*(6), 1373–1384.
28. Cuffe, P., & Keane, A. (2015). Visualizing the electrical structure of power systems. *IEEE Systems Journal, 11*(3), 1810–1821.
29. Fan, X., Weber, W. D., & Barroso, L. A. (2007). Power provisioning for a warehouse-sized computer. *ACM SIGARCH Computer Architecture news, 35*(2), 13–23.
30. Chen, M., Gao, C., Chen, S., Li, D., & Liu, Q. (2019). Bi-level economic dispatch modeling considering the load regulation potential of internet data centers. *Proceedings of the CSEE, 39*(5), 1301–1313.
31. Ye, Y., Qiu, D., Sun, M., Papadaskalopoulos, D., & Strbac, G. (2019). Deep reinforcement learning for strategic bidding in electricity markets. *IEEE Transactions on Smart Grid, 11*(2), 1343–1355.
32. Sinha, A., & Shaikh, V. (2021). Solving bilevel optimization problems using kriging approximations. *IEEE Transactions on Cybernetics, 52*(10), 10639–10654.
33. Storn, R., & Price, K. (1997). Differential evolution—a simple and efficient heuristic for global optimization over continuous spaces. *Journal of Global Optimization, 11*, 341–359.
34. Opara, K. R., & Arabas, J. (2019). Differential evolution: A survey of theoretical analyses. *Swarm and Evolutionary Computation, 44*, 546–558.
35. Han, Z. H. (2016). Kriging surrogate model and its application to design optimization: A review of recent progress. *Acta Aeronautica et Astronautica Sinica, 37*(11), 3197–3225.
36. Xiao, H., Pei, W., Dong, Z., Pu, T., Chen, N., & Li, K. (2018). Reactive power optimization of distribution system with distributed generation using metamodel-based global optimization method. *Proceedings of the Chinese Society of Electrical Engineering, 38*(3), 5751–5762.
37. Bouhlel, M. A., Hwang, J. T., Bartoli, N., Lafage, R., Morlier, J., & Martins, J. R. (2019). A Python surrogate modeling framework with derivatives. *Advances in Engineering Software, 135*, 102662.
38. Santner, T. J., Williams, B. J., Notz, W. I., & Williams, B. J. (2003). *The design and analysis of computer experiments* (Vol. 1). Springer.
39. Bouhlel, M. A., Bartoli, N., Otsmane, A., & Morlier, J. (2016). Improving kriging surrogates of high-dimensional design models by Partial Least Squares dimension reduction. *Structural and Multidisciplinary Optimization, 53*, 935–952.
40. Tang, J., Wang, D., Jia, H., Zhang, Y., & Xiong, J. (2018). Optimal economic operation of active distribution networks based on hybrid algorithm of surrogate model and particle swarm optimization. *Automation of Electric Power Systems, 42*(4), 95–103.
41. Thereska, E., Donnelly, A., & Narayanan, D. (2011, April). Sierra: practical power-proportionality for data center storage. In *Proceedings of the Sixth Conference on Computer Systems* (pp. 169–182).
42. Baghzouz, Y., & Ertem, S. (1990). Shunt capacitor sizing for radial distribution feeders with distorted substation voltages. *IEEE Transactions on Power Delivery, 5*(2), 650–657.
43. Ratnam, E. L., Weller, S. R., Kellett, C. M., & Murray, A. T. (2017). Residential load and rooftop PV generation: An Australian distribution network dataset. *International Journal of Sustainable Energy, 36*(8), 787–806.

44. IESO. (2022). *Generator output and capability.* https://www.ieso.ca/en/Power-Data/Data-Directory. Accessed July 17, 2022.
45. Jazzbin. (2018). *The genetic and evolutionary algorithm toolbox with high performance in python.* http://geatpy.com. Accessed February 12, 2022.

Part II
Local Energy Market Design Considering the Impacts of Uncertainties

Chapter 5
Prospect Incentive-Compatible Local Energy Market Design with Adaptive Robust Approaches

Abstract The local energy market (LEM) has recently been introduced to facilitate peer-to-peer (P2P) energy trading within an integrated energy market. Nevertheless, sharing energy can result in profit loss for participants, reducing their incentive to join the coalition. This chapter introduces a trading mechanism for the market organizer to allocate cooperative surplus based on the market contribution of each participant. This model ensures key market mechanism attributes: social welfare maximization, individual rationality, cost recovery, and incentive compatibility, allowing participants to receive their fair share and avoid profit reductions. Additionally, the unpredictable and intermittent nature of power outputs from distributed renewable energy sources presents a significant challenge for organizing a LEM. Thus, an adaptive robust optimization (ARO) algorithm is used to identify scenarios that enhance conservativeness and economy, improving the efficiency of the market-clearing process. We transform the original problem into a tri-level model and solve it through iterative internal and external formulations. Numerical results demonstrate a social welfare increase of $2751 compared to the independent mode, with a 71.55% improvement in calculation efficiency using the ARO algorithm.

5.1 Introduction

The integrated energy system has recently gained significant traction for various applications such as energy coupling, cost reduction, and enhancing energy efficiency [1]. In an integrated energy system, the Combined Cooling, Heating, and Power (CCHP) can be utilized as central hubs to integrate electrical power systems, natural gas systems, and heating/cooling demands. In addition, another facet of integrated energy applications has been explored through the local energy market (LEM) structure [2]. O. Abrishambaf has presented an extensive analysis of the latest developments in LEM [3]. This chapter provides an overview of methods to integrate small-scale LEM into networks [4]. We establish a LEM structure to enable distributed energy trading among multiple CCHPs. Traditionally, in such a market, participants could only trade energy with an external power system [5]. This trading

M. Song et al., *Local Energy Markets*, https://doi.org/10.1007/978-981-97-9750-9_5

mode often results in some participants' underutilization of surplus renewable energy due to net metering caps restricting the excess energy that can be sent back to the grids [6]. Consequently, this can lead to renewable energy loss and reduced profits. Furthermore, the variability in renewable energy sources (RES) introduces a significant level of uncertainty, affecting the operation of Distributed Integrated Energy Management (DIEM) by compromising the reliability of energy supply for CCHPs and increasing operational risks [7]. To address these challenges, we propose developing an algorithm to manage stochastic RES, promote resource conservation, and enhance social welfare by efficiently handling large-scale stochastic generation.

In this chapter, we devise an incentive-compatible trading mechanism for integrated energy hubs, like CCHPs, known as the DIEM. Recent literature has proposed various trading models for LEM. For instance, a regional energy transaction market incorporating a ladder carbon trading mechanism to minimize economic costs is discussed in [8]. A non-cooperative game model is presented in [9], ensuring fair and stable peer-to-peer (P2P) trading among LEM owners. Furthermore, a contribution-based energy trading mechanism that allocates surplus energy to market participants is introduced in [10].

Most existing studies do not effectively model incentive compatibility, a crucial aspect of trading mechanisms, because the objectives of maximizing individual market member benefits and overall social welfare often conflict, and the incentive standards are not reliably designed [11]. The study in [8] examines minimizing global economic costs without considering individuals with multi-energy equipment. Participants in [9] optimize their own trading strategies without aiming to maximize social welfare. In [10], energy users are incentivized by a distribution factor, but this factor depends on the distributor's preferences rather than an objective standard, reducing its reliability. Therefore, this work implements an efficient mechanism with an incentive-compatible attribute, promoting rational and fair distribution of cooperative surplus among participants and encouraging their participation in market transactions.

Recent research frequently utilizes the Vickrey-Clark-Groves (VCG) theory to achieve market incentive compatibility [12]. Nevertheless, VCG's requirement of excluding each participant from the market in every computation step to determine their market contribution leads to significant computational complexity, limiting its scalability [13]. For instance, VCG is applied to optimize users' energy consumption one renewable source at a time [12]. A scene-reduction technique limits the calculation scope when applying VCG to clear the market [13]. Moreover, the VCG algorithm struggles to address fairness and incentive challenges in a DIEM context. Therefore, this chapter introduces an incentive compatibility mechanism based on market contribution identification. The integrated energy service provider (IESP) encourages active participation in DIEM by redistributing the cooperative surplus without diminishing member profits. The IESP acts as an independent participant in the distributed energy market, offering various energy forms to consumers and engaging in transactions with the external energy grid. The operational planning of a regional IESP is analyzed by creating a risk-averse day-ahead model to handle price uncertainty [14]. A day-ahead economic dispatch strategy for IESP is proposed in the electricity-gas market [15]. A reliability-constrained decision model incorporating demand response programs is also developed to demonstrate how to allocate demand in the smart distribution network [16].

With the widespread deployment of renewable energy resources owned by various market participants across multiple geographic locations, uncertainties and volatility increasingly challenge the system's operation. This raises concerns about the reliability of DIEM. Several stochastic approaches have been utilized to optimize operation strategies, including chance-constrained optimization [17], stochastic optimization [18], and robust optimization [19]. Chance-constrained optimization involves deriving solutions based on specified probability ranges, which may occasionally lead to constraint violations [20]. Stochastic optimization decomposes the complex problem into simpler subproblems, yet it encounters computational challenges when handling numerous sampling scenarios [21]. Robust optimization, on the other hand, seeks a conservative solution to mitigate uncertainties, but it risks producing infeasible solutions [22]. Adaptive robust optimization (ARO) has gained traction in addressing uncertainties by offering conservative solutions that minimize potential performance degradation in worst-case scenarios. This approach typically divides the optimization problem into day-ahead [13] and real-time [23] stages, focusing on identifying the worst-case scenario regarding distributed renewable energy outputs during the real-time stage. By identifying these scenarios, ARO guides for system operations to mitigate potential profit reductions under adverse conditions. ARO formulations have been extensively applied in various optimization contexts within the power sector, including tie-line dispatch [24], unit commitment [25], and collaborative optimization for multi-energy systems [26], among others.

Currently, the primary algorithms for ARO include Benders decomposition [27], column constraint generation [28], random scene-based methods [29], and linear decision rules [30]. Benders decomposition and column constraint generation algorithms use the Karush–Kuhn–Tucker (KKT) conditions to transform the original problem into a single-level model, necessitating the introduction of numerous intermediate variables. The scene-based approach in [29] introduces an auxiliary variable for the objective function of the inner max–min problem. While these methods can address ARO models, they all encounter low-efficiency issues when handling large-scale random scenarios. Bilinear terms arise from the duality or the transformation of the inner max–min problem into a single-level equivalent in both Benders decomposition [27] and column constraint generation [28]. Although big-M methods can linearize these terms, the resulting bounding parameters and binary variables are computationally demanding. The scene-based algorithm often employs scene-reduction techniques to mitigate computational burdens [29], compromising sample uniqueness and simulation accuracy. Linear decision rules reformulate the robust problem into a computationally tractable model [30], yet the solutions tend to be somewhat conservative. We introduce an enhanced ARO solution algorithm based on selected scenario identification to address these challenges. This iterative solution involves a set of selected scenarios encompassing all stochastic information from distributed renewable energy sources, simplifying the initial stochastic scenarios and thereby improving ARO solution efficiency.

5.3 Design of Incentive-Compatible DIEM Mechanism

5.3.1 Operation Modes of DIEM

Based on the transaction status between market participants, the DIEM operation mode is classified into Disagreement Point (DP) mode and Agreement Point (AP) mode, which disable and enable P2P trading among CCHPs. In DP mode, the IESP needs to optimize resource management for each CCHP individually, without interactions between CCHPs. This mode reflects the maximized benefit of each CCHP. On the other side, AP mode allows P2P trading among CCHPs, with the IESP optimizing the entire system's performance. Generally, AP mode is expected to outperform DP mode due to mutual support among CCHPs, leading to more efficient use of internal energy resources and reduced operational costs. However, implementing AP mode presents a significant challenge: creating an effective and fair mechanism for benefit allocation among CCHPs. Then, two-stage ARO-based optimization models are proposed to represent the AP and DP mechanisms.

5.3.2 DIEM Trading Model in DP Mode

In DP mode, IESP achieves market clearing to minimize the cost of all participants. Under the ARO structure, the proposed model can be demonstrated as a tri-level min–max–min optimization problem.

(1) Objective function

The objective function of DP is formulated in (5.1), representing the negative expected social welfare to be minimized. The first term corresponds to the net cost in the day-ahead stage, which equals the operational cost minus the energy demand utility. The second term corresponds to the net cost in the real-time stage. Note that the superscript DP is omitted for the variables here, and the superscript index d indicates the variables in the day-ahead stage. Additionally, it is important to note that stochastic scenarios, denoted by s, refer to cases with different wind power outputs.

$$\min c_i^{DP} \Rightarrow \min \sum_{t\in\Omega_T} \left(F_{i,t}^{D} - F_{i,t}^{U}\right) + \max_{s\in\Omega_O} \min \sum_{t\in\Omega_T} F_{i,t,s}^{R} \tag{5.1}$$

where

$$\begin{aligned} F_{t,i}^{\mathrm{D}} = {} & \left(S_{\mathrm{PE},i,t}^{d} + S_{\mathrm{PG},i,t}^{d}\right) + \left(c_{\mathrm{G},i}^{\mathrm{U}} R_{\mathrm{G},i,t}^{\mathrm{U},d} + c_{\mathrm{G},i}^{\mathrm{D}} R_{\mathrm{G},i,t}^{\mathrm{D},d}\right) \\ & + \left(c_{\mathrm{E},i}^{\mathrm{U}} R_{\mathrm{E},i,t}^{\mathrm{U},d} + c_{\mathrm{E},i}^{D} R_{\mathrm{E},i,t}^{D,d}\right), \quad \forall i, t \end{aligned} \tag{5.2}$$

$$S_{\mathrm{PE},i,t}^{d} = \lambda_{\mathrm{E},t}^{\mathrm{R}} P_{\mathrm{NEL},i,t}^{d} - \lambda_{\mathrm{E},t}^{\mathrm{S}} P_{\mathrm{NES},i,t}^{d}, \quad \forall i, t \tag{5.3}$$

$$S_{\mathrm{PG},i,t}^{d} = \lambda_{\mathrm{G},t}^{\mathrm{R}} Q_{\mathrm{NGL},i,t}^{d}, \quad \forall i, t \tag{5.4}$$

$$F_{t,i}^{\mathrm{U}} = u_{\mathrm{E},i} P_{\mathrm{Le},i,t}^{d} + u_{\mathrm{H},i} Q_{\mathrm{Lh},i,t}^{d} + u_{\mathrm{C},i} Q_{\mathrm{Lc},i,t}^{d}, \quad \forall i, t \tag{5.5}$$

$$\begin{aligned} F_{t,i,s}^{\mathrm{R}} = {} & \lambda_{\mathrm{E},t}^{\mathrm{R}} \left(r_{i,t,s}^{\mathrm{ER,U}} - r_{i,t,s}^{\mathrm{ER,D}} \right) - \lambda_{\mathrm{E},t}^{\mathrm{S}} \left(r_{i,t,s}^{\mathrm{ES,U}} - r_{i,t,s}^{\mathrm{ES,D}} \right) \\ & + \lambda_{\mathrm{G},t}^{\mathrm{R}} \left(r_{i,t,s}^{\mathrm{GR,U}} - r_{i,t,s}^{\mathrm{GR,D}} \right) + \delta_{\mathrm{E},i} \left(P_{\mathrm{Le},i,t}^{d} - P_{\mathrm{Le},i,t,s}^{r} \right) \\ & + \delta_{\mathrm{H},i} \left(Q_{\mathrm{Lh},i,t}^{d} - Q_{\mathrm{Lh},i,t,s}^{r} \right) + \delta_{\mathrm{C},i} \left(Q_{\mathrm{Lc},i,t}^{d} - Q_{\mathrm{Lc},i,t,s}^{r} \right), \quad \forall i, t, s \end{aligned} \tag{5.6}$$

The day-ahead stage costs are detailed in Eqs. (5.2)–(5.4), which encompasses the net expenses for purchasing electricity/gas and maintaining reserve capacities. Specifically, the formulations within the brackets denote the net costs associated with trading electricity and gas with the external energy network and the reserve costs for gas and electricity. The total utility demands of participants are formulated in Eq. (5.5), which include utilities for electricity, heat, and cold demands. In addition, the real-time stage costs are given by Eq. (5.6), incorporating energy balancing costs and demand shedding costs. Specifically, the first term pertains to the balancing regulation cost of electricity purchased by CCHP i, while the second represents the revenue from sales. The third term denotes the gas regulation cost, and the last three refer to the costs associated with shedding electricity, heat, and cold demands. It is noteworthy that CCHPs are assumed incapable of producing natural gas, resulting in a one-way gas transaction between IESP and CCHPs, where IESP does not repurchase natural gas from CCHPs.

(2) Constraints of day-ahead stage

The operation devices in CCHPs are restricted by (5.7)–(5.16).

$$P_{\mathrm{MT},i,t}^{d} = W_{\mathrm{MT},i,t}^{d} \eta_{\mathrm{E,MT}}, \quad \forall i, t \tag{5.7}$$

$$Q_{\mathrm{re},i,t}^{d} = P_{\mathrm{MT},i,t}^{d} \eta_{\mathrm{HE,MT},i}, \quad \forall i, t \tag{5.8}$$

$$-R_{\mathrm{r},i} P_{\mathrm{MT},i}^{\max} \le P_{\mathrm{MT},i,t}^{d} - P_{\mathrm{MT},i,t-1}^{d} \le R_{\mathrm{r},i} P_{\mathrm{MT},i}^{\max}, \quad \forall i, t \tag{5.9}$$

$$Q_{\mathrm{EHB},i,t}^{d} = Q_{\mathrm{re},i,t}^{d} \eta_{\mathrm{EHB},i}, \quad \forall i, t \tag{5.10}$$

$$Q_{\mathrm{EHB},i,t}^{d} = Q_{\mathrm{EHB\_HE},i,t}^{d} + Q_{\mathrm{EHB\_AR},i,t}^{d}, \quad \forall i, t \tag{5.11}$$

$$Q_{\mathrm{HE},i,t}^{d} = Q_{\mathrm{EHB\_HE},i,t}^{d} \eta_{\mathrm{HE},i}, \quad \forall i, t \tag{5.12}$$

$$Q_{\mathrm{GB},i,t}^{d} = W_{\mathrm{GB},i,t}^{d} \eta_{\mathrm{GB},i}, \quad \forall i, t \tag{5.13}$$

$$Q_{\mathrm{AR},i,t}^{d} = Q_{\mathrm{EHB\_AR},i,t}^{d} \eta_{\mathrm{AR},i}, \quad \forall i, t \tag{5.14}$$

$$Q_{\mathrm{ER},i,t}^{d} = P_{\mathrm{ER},i,t}^{d} \eta_{\mathrm{ER},i}, \quad \forall i, t \tag{5.15}$$

$$\begin{gathered} X_{\mathrm{dev}}^{\min} \le X_{\mathrm{dev},i,t}^{d} \le X_{\mathrm{dev}}^{\max}, \quad \forall i, t \\ X_{\mathrm{dev}}^{d} \in \left[P_{\mathrm{MT},t}^{d}, W_{\mathrm{GB},t}^{d}, Q_{\mathrm{EHB},t}^{d}, Q_{\mathrm{AR},t}^{d}, Q_{\mathrm{ER},t}^{d}\right] \end{gathered} \tag{5.16}$$

where Eqs. (5.7)–(5.8) describe the operation of micro-turbines (MTs), detailing their electricity and exhaust heat generation, respectively. The heat generation and distribution of Exhaust-heat Boilers (EHBs) are addressed in Eqs. (5.9)–(5.10), Specifically, (5.9) specifies the exhaust heat conversion capability, ensuring energy conservation following the conversion process described in (5.10). Equations (5.11)–(5.15) denote the operation specifics for heat exchangers (HEs), gas boilers (GBs), absorption refrigerators (ARs), and electric refrigerators (ERs). Additionally, inequality constraints governing the operational behavior of these devices are presented in (5.16), which imposes restrictions on the ramp rates of MTs to maintain operational stability across consecutive periods.

$$R_{\mathrm{G},i,t}^{\mathrm{U},\min} \le R_{\mathrm{G},i,t}^{\mathrm{U},d}, R_{\mathrm{G},i,t}^{\mathrm{D},\min} \le R_{\mathrm{G},i,t}^{\mathrm{D},d}, \quad \forall i, t \tag{5.17}$$

$$Q_{\mathrm{NGL},i,t}^{d} + R_{\mathrm{G},i,t}^{\mathrm{U},d} \le Q_{\mathrm{G},i}^{\max}, \quad \forall i, t \tag{5.18}$$

$$Q_{\mathrm{NGL},i,t}^{d} - R_{\mathrm{G},i,t}^{\mathrm{D},d} \ge 0, \quad \forall i, t \tag{5.19}$$

$$R_{\mathrm{E},i,t}^{\mathrm{U},\min} \le R_{\mathrm{E},i,t}^{\mathrm{U},d}, R_{\mathrm{E},i,t}^{\mathrm{D},\min} \le R_{\mathrm{E},i,t}^{\mathrm{D},d}, \quad \forall i, t \tag{5.20}$$

$$P_{\mathrm{NEL},i,t}^{d} + R_{\mathrm{E},i,t}^{\mathrm{U},d} \le P_{\mathrm{E},i}^{\max}, \quad \forall i, t \tag{5.21}$$

$$P_{\mathrm{NEL},i,t}^{d} - R_{\mathrm{E},i,t}^{\mathrm{D},d} \ge 0, \quad \forall i, t \tag{5.22}$$

$$P_{\mathrm{NEL},i,t}^{d} - P_{\mathrm{NES},i,t}^{d} = P_{\mathrm{Le},i,t}^{d} - P_{\mathrm{MT},i,t}^{d} - P_{\mathrm{DRES},i,t}^{d} + P_{\mathrm{ER},i,t}^{d}, \quad \forall i, t \tag{5.23}$$

$$Q_{\mathrm{HE},i,t}^{d} + Q_{\mathrm{GB},i,t}^{d} = Q_{\mathrm{Lh},i,t}^{d}, \quad \forall i, t \tag{5.24}$$

$$Q_{\mathrm{NGL},i,t}^{d} = W_{\mathrm{MT},i,t}^{d} + W_{\mathrm{GB},i,t}^{d}, \quad \forall i, t \tag{5.25}$$

$$Q_{\mathrm{Lc},i,t}^{d} = Q_{\mathrm{AR},i,t}^{d} + Q_{\mathrm{ER},i,t}^{d}, \quad \forall i, t \tag{5.26}$$

$$0 \leq P_{\mathrm{NEL},i,t}^{d} \leq P_{\mathrm{NEL},i}^{\max}, 0 \leq P_{\mathrm{NES},i,t}^{d} \leq P_{\mathrm{NES},i}^{\max}, \quad \forall i, t \tag{5.27}$$

$$0 \leq Q_{\mathrm{NGL},i,t}^{d} \leq Q_{\mathrm{NGL},i}^{\max}, \quad \forall i, t \tag{5.28}$$

$$\begin{aligned} & X_{\mathrm{L}}^{\min} \leq X_{\mathrm{L},i,t}^{d} \leq X_{\mathrm{L}}^{\max}, \\ & X_{\mathrm{L},i,t}^{d} \in \left[P_{\mathrm{Le},i,t}^{d}, Q_{\mathrm{Lh},i,t}^{d}, Q_{\mathrm{Lc},i,t}^{d}\right], \quad \forall i, t \end{aligned} \tag{5.29}$$

$$0 \leq P_{\mathrm{DRES},i,t}^{d} \leq P_{\mathrm{FRES},i,t}^{d}, \quad \forall i, t \tag{5.30}$$

The reserve capacity constraints for electricity and natural gas purchases are formulated in (5.17)–(5.22). Specifically, (5.17) and (5.20) define the upper and lower limits for gas and electricity reserves, respectively. The relationships between net gas demand and day-ahead reserve capacity are described in (5.18)–(5.19), mirroring the constraints for electricity demand detailed in (5.21)–(5.22). Equations (5.23)–(5.26) establish the balancing constraints for electricity, heat, gas, and cold. The net energy demand and surplus energy boundaries are outlined in (5.27)–(5.28). Energy demands encompassing electricity, heat, and cold are specified in (5.29). Equation (5.30) characterizes the contribution of renewable energy sources in the day-ahead stage.

(3) Constraints of real-time stage

The real-time stage is a pivotal mechanism for balancing energy production and consumption, as illustrated below. Specifically, (5.31)–(5.33) establish the relationships between net electricity demand, surplus, and net gas demand during the real-time stage and their counterparts from the day-ahead stage. (5.34)–(5.35) ensure that dispatched renewable energy in real-time remains within a reasonable range, not exceeding forecasted generation values from both the day-ahead and real-time stages. It's important to note that previously outlined constraints, such as device operations (5.7)–(5.16) and balancing constraints (5.22)–(5.29) and (5.30), are not reiterated here, with corresponding variable superscripts now denoted as r.

$$P_{\mathrm{NEL},i,t,s}^{r} = P_{\mathrm{NEL},i,t}^{d} + r_{i,t,s}^{\mathrm{ER},\mathrm{U},r} - r_{i,t,s}^{\mathrm{ER},\mathrm{D},r}, \quad \forall i, s, t \tag{5.31}$$

$$P_{\mathrm{NES},i,t,s}^{r} = P_{\mathrm{NES},i,t}^{d} + r_{i,t,s}^{\mathrm{ES},\mathrm{U},r} - r_{i,t,s}^{\mathrm{ES},\mathrm{D},r}, \quad \forall i, s, t \tag{5.32}$$

$$Q_{\mathrm{NGL},i,t,s}^{r} = Q_{\mathrm{NGL},i,t}^{d} + r_{i,t,s}^{\mathrm{GR},\mathrm{U},r} - r_{i,t,s}^{\mathrm{GR},\mathrm{D},r}, \quad \forall i, s, t \tag{5.33}$$

$$0 \leq P_{\mathrm{DRES},i,t,s}^{r} \leq P_{\mathrm{FRES},i,t}^{d}, \quad \forall i, s, t \tag{5.34}$$

$$0 \leq P^{r}_{\mathrm{DRES},i,t,s} \leq P^{r}_{\mathrm{FRES},i,t}, s \in \Omega_{\mathrm{O}}, \quad \forall i, s, t \tag{5.35}$$

5.3.3 *Incentive-Compatible DIEM Trading Model in AP*

In this section, the trading model concerning AP mode is presented.

(1) DIEM trading model in AP

Here we define social welfare as the difference between the total revenue of IESP and the total cost.

$$r_{\mathrm{SW}} = \left(r_{\mathrm{IESP}} - \sum_{i \in \Omega_{\mathrm{I}}} R_{\mathrm{ES},i} \right) - \sum_{i \in \Omega_{\mathrm{I}}} \left(c_i - R_{\mathrm{ES},i} \right) = r_{\mathrm{IESP}} - \sum_{i \in \Omega_{\mathrm{I}}} c_i \tag{5.36}$$

where $R_{\mathrm{ES},i}$ is the incentive fee paid by IESP to each CCHP, which is offset in (5.36).

The objective function described in (5.37) aims to maximize the social welfare of AP within the ARO framework. The first term minimizes negative social welfare by incorporating all CCHPs and IESP in the day-ahead stage. The second term maximizes the minimized global cost across all market participants under various scenarios during the real-time stage. For simplicity, the superscript AP is omitted for variables in this context.

$$\begin{aligned} \max r_{\mathrm{SW}} \Rightarrow \min(-r_{\mathrm{SW}}) \Rightarrow \min\left(\sum_{t \in \Omega_{\mathrm{T}}} \sum_{i \in \Omega_{\mathrm{I}}} \left(F^{\mathrm{D}}_{t,i} - F^{\mathrm{U}}_{t,i} \right) - r^{d}_{\mathrm{IESP}} \right) \\ + \max_{s \in \Omega_{\mathrm{O}}} \min \left(\sum_{t \in \Omega_{\mathrm{T}}} \sum_{i \in \Omega_{\mathrm{I}}} F^{\mathrm{R}}_{t,i,s} - r^{r}_{\mathrm{IESP},s} \right) \end{aligned} \tag{5.37}$$

where the net revenues of IESP in the day-ahead stage as

$$r^{d}_{\mathrm{IESP}} = \sum_{t \in \Omega_{T}} \left(B^{d}_{\mathrm{E},t} + B^{d}_{\mathrm{G,t}} + S^{d}_{\mathrm{E},t} + S^{d}_{\mathrm{G},t} \right). \tag{5.38}$$

The net benefits of IESP originate from trading energy with external energy grids and CCHPs. The first two terms in (5.39)–(5.40) quantify the net benefits of trading electricity and gas with external energy grids. The last two terms, as described in (5.41)–(5.42), represent the net benefits of interactions with CCHPs.

$$B^{d}_{\mathrm{E},t} = \lambda^{\mathrm{out}}_{\mathrm{E},t} P^{d}_{\mathrm{TNES},t} - \lambda^{\mathrm{in}}_{\mathrm{E},t} P^{d}_{\mathrm{TNEL},t}, \quad \forall t \tag{5.39}$$

$$B^{d}_{\mathrm{G},t} = -\lambda^{\mathrm{in}}_{\mathrm{G},t} Q^{d}_{\mathrm{TNGL},t}, \quad \forall t \tag{5.40}$$

$$S_{E,t}^{d} = \lambda_{E,t}^{R} \sum_{i \in \Omega_I} P_{NEL,i,t}^{d} - \lambda_{E,t}^{S} \sum_{i \in \Omega_I} P_{NES,i,t}^{d}, \quad \forall t \tag{5.41}$$

$$S_{G,t}^{d} = \lambda_{G,t}^{S} \sum_{i \in \Omega_I} Q_{NGL,i,t}^{d}, \quad \forall t \tag{5.42}$$

where

$$P_{TNEL,t}^{d} - P_{TNES,t}^{d} = \sum_{i \in \Omega_I} P_{NEL,i,t}^{d} - \sum_{i \in \Omega_I} P_{NES,i,t}^{d}, \quad \forall t \tag{5.43}$$

$$P_{TNEL,t}^{d} \geq 0, P_{TNES,t}^{d} \geq 0, \quad \forall t \tag{5.44}$$

$$Q_{TNGL,t}^{d} = \sum_{i \in \Omega_I} Q_{NGL,i,t}^{d}, \quad \forall t. \tag{5.45}$$

In (5.43), the energy balance encompasses the total calculated value and the energy aggregate of all CCHPs. (5.45) ensures non-negativity in the total net surplus and demand. Although (5.45) resembles (5.43), it introduces a constraint on the gas energy balance.

The real-time stage net benefits of IESP are detailed as follows, omitting the reiterated constraints from (5.43)–(5.45). Specifically, (5.46) defines these net benefits akin to (5.38), encompassing gains from energy trading with external grids and CCHPs. Notably, calculations in (5.47)–(5.50) exclude day-ahead transactions to prevent the double counting.

$$r_{IESP,s}^{r} = \sum_{t \in \Omega_T} \left(B_{E,t,s}^{r} + B_{G,t,s}^{t} + S_{E,t,s}^{r} + S_{G,t,s}^{r} \right) \tag{5.46}$$

$$B_{E,t,s}^{r} = \lambda_{E,t}^{out} \left(P_{TNES,t,s}^{r} - P_{TNES,t}^{d} \right) - \lambda_{E,t}^{in} \left(P_{TNEL,t,s}^{r} - P_{TNEL,t}^{d} \right), \quad \forall t, s \tag{5.47}$$

$$B_{G,t,s}^{r} = -\lambda_{G,t}^{in} \left(Q_{TNGL,t,s}^{r} - Q_{TNGL,t}^{d} \right), \quad \forall t, s \tag{5.48}$$

$$S_{E,t,s}^{r} = \lambda_{E,t}^{R} \sum_{i \in \Omega_I} \left(P_{NEL,i,t,s}^{r} - P_{NEL,i,t}^{d} \right) - \lambda_{E,t}^{S} \sum_{i \in \Omega_I} \left(P_{NES,i,t,s}^{r} - P_{NES,i,t}^{d} \right), \quad \forall t, s \tag{5.49}$$

$$S_{G,t,s}^{r} = \lambda_{G,t}^{S} \sum_{i \in \Omega_I} \left(Q_{NGL,i,t,s}^{r} - Q_{NGL,i,t}^{d} \right), \quad \forall t, s \tag{5.50}$$

For the balance constraints of AP, we need to take into consideration energy trading variables, i.e. P_{EES}, P_{EHS}, and P_{ECS}.

$$P_{NEL,i,t} - P_{NES,i,t} - P_{EES,i,t} = P_{Le,i,t} - P_{MT,i,t} - P_{DRES,i,t} + P_{ER,i,t}, \quad \forall i, t \tag{5.51}$$

$$Q_{\mathrm{HE},i,t} + Q_{\mathrm{GB},i,t} - Q_{\mathrm{EHS},i,t} = Q_{\mathrm{Lh},i,t}, \quad \forall i, t \tag{5.52}$$

$$Q_{\mathrm{Lc},i,t} - Q_{\mathrm{ECS},i,t} = Q_{\mathrm{AR},i,t} + Q_{\mathrm{ER},i,t}, \quad \forall i, t \tag{5.53}$$

$$\sum_{i \in \Omega_{\mathrm{I}}} P_{\mathrm{EES},i,t} = \sum_{i \in \Omega_{\mathrm{I}}} Q_{\mathrm{EHS},i,t} = \sum_{i \in \Omega_{\mathrm{I}}} Q_{\mathrm{ECS},i,t} = 0, \quad \forall i, t \tag{5.54}$$

$$\begin{gathered} -X_{\mathrm{Share}}^{\max} \le X_{\mathrm{Share},i,t} \le X_{\mathrm{Share}}^{\max}, \quad \forall i, t, \\ X_{\mathrm{Share},i,t} \in \left[Q_{\mathrm{ECS},i,t}, P_{\mathrm{EES},i,t}, Q_{\mathrm{EHS},i,t}\right] \end{gathered} \tag{5.55}$$

where Eqs. (5.51)–(5.53) establish the balancing constraints for integrated energy, encompassing P2P trading among participants. Equation (5.54) ensures that the total P2P trading across all participants must sum to zero.

(2) Market contribution of CCHPs in AP

As previously discussed, a crucial aspect of ensuring the participation of CCHPs in DIEM is a fair and effective benefit allocation mechanism. Here, the proposed allocation is based on the market contribution of each CCHP, defined as the economic value derived from participating in P2P trading. This value is calculated as the product of the shared energy and its unit value. In DIEM, the market contribution of a CCHP can be expressed by Eq. (5.56), which incorporates the contributions from shared electricity, heat, and cold in both day-ahead and real-time stages.

$$\begin{aligned} C_{\mathrm{EV},i,t} = & \left|\lambda_{\mathrm{EES},t}^{d} P_{\mathrm{EES},i,t}^{d} + \lambda_{\mathrm{EES},t}^{\mathrm{ws}}\left(P_{\mathrm{EES},i,t}^{\mathrm{ws}} - P_{\mathrm{EES},i,t}^{d}\right)\right| \\ & + \left|\lambda_{\mathrm{EHS}}^{d} Q_{\mathrm{NHS},i,t}^{d} + \lambda_{\mathrm{EHS},t}^{\mathrm{ws}}\left(Q_{\mathrm{EHS},i,t}^{\mathrm{ws}} - Q_{\mathrm{EHS},i,t}^{d}\right)\right| \\ & + \left|\lambda_{\mathrm{ECS}}^{d} Q_{\mathrm{NCS},i,t}^{d} + \lambda_{\mathrm{ECS},t}^{\mathrm{ws}}\left(Q_{\mathrm{ECS},i,t}^{\mathrm{ws}} - Q_{\mathrm{ECS},i,t}^{d}\right)\right|, \quad \forall i, t \end{aligned} \tag{5.56}$$

where variables with the superscript ws represent those in the worst-case scenario of the real-time stage. The locational marginal prices for electricity, heat, and cold (denoted as λ_{EES}, λ_{EHS}, and λ_{ECS}) can be derived from the dual variables of the energy trading balance (5.54) for both stages. It is important to note that both supply and demand are considered contributions to DIEM. The market contribution of each participant is calculated as

$$\omega_i = \frac{(1-\sigma)\sum_{t \in \Omega_{\mathrm{T}}} C_{\mathrm{EV},i,t}}{\sum_{t \in \Omega_{\mathrm{T}}} \sum_{i \in \Omega_{\mathrm{I}}} C_{\mathrm{EV},i,t}}, \quad \forall i \tag{5.57}$$

where the constant σ, defined within the range [0, 1], represents the ratio of management fees charged by IESP under the AP mode to the total market contribution.

(3) Cooperative surplus allocation based on incentive compatibility

According to economic principles [32], the cooperative surplus is the total market cost reduction achieved when comparing the two operating modes, DP and AP.

$$\Delta = \left(-r_{\mathrm{IESP}}^{\mathrm{DP}} + \sum_{i\in\Omega_{\mathrm{I}}} c_i^{\mathrm{DP}}\right) - \left(-r_{\mathrm{IESP}}^{\mathrm{AP}} + \sum_{i\in\Omega_{\mathrm{I}}} c_i^{\mathrm{AP}}\right) \tag{5.58}$$

Nevertheless, energy trading will cause CCHPs to deviate from their optimal performance, increasing costs for some CCHPs and violating their rational requirements. In market mechanism design theory, individual rationality is crucial for a mechanism to be accepted by all participants, meaning that the original benefits of the participants should not be compromised [13]. To address this, the cooperative surplus is calculated and reallocated by the IESP based on the judgment and evaluation of each CCHP in the market, as specified by formula (5.56). This process ensures that a certain incentive fee is paid to each CCHP by the IESP, thereby maintaining the profits and individual rational attributes of all participants.

In DIEM, we need to set the following constraints to ensure individual rationality.

$$\kappa_{\mathrm{IESP}} = \left(r_{\mathrm{IESP}}^{\mathrm{AP}} - R_{\mathrm{IESP}}\right) - r_{\mathrm{IESP}}^{\mathrm{DP}} \geq 0 \tag{5.59}$$

$$\kappa_{\mathrm{ES},i} = c_i^{\mathrm{DP}} - \left(c_i^{\mathrm{AP}} - R_{\mathrm{ES},i}\right) \geq 0, \quad \forall i \tag{5.60}$$

where R_{IESP} and $R_{\mathrm{ES},i}$ satisfy the following constraint

$$\sum_{i\in\Omega_{\mathrm{I}}} R_{\mathrm{ES},i} - R_{\mathrm{IESP}} = 0 \tag{5.61}$$

When Eq. (5.59) through (5.61) are satisfied, $R_{\mathrm{ES},i}$, defined in Eq. (5.36), and R_{IESP} can be determined using the Nash bargaining model [32]. The objective function maximizes the Cobb–Douglas utility, which is the product of the profit increase of IESP and the cost reduction of all CCHPs.

$$\max f^{\mathrm{CD}} = \kappa_{\mathrm{IESP}}^{\sigma} \prod_{i\in\Omega_{\mathrm{I}}} \left(\kappa_{\mathrm{ES},i}^{\omega_i}\right). \tag{5.62}$$

Then, we can take the log of both sides to obtain

$$\max \ln f^{\mathrm{CD}} = \sigma \ln \kappa_{\mathrm{IESP}} + \sum_{i\in\Omega_{\mathrm{I}}} \omega_i \ln \kappa_{\mathrm{ES},i} \tag{5.63}$$

It is proved that the Eq. (5.62) can be a concave function. Thus, we can leverage the first-order KKT condition [19] to obtain its analytical solution.

$$\partial \ln f^{\mathrm{CD}} \partial^{-1} R_{\mathrm{ES},i} = \omega_j \kappa_{\mathrm{ES},i}^{-1} - \sigma \kappa_{\mathrm{IESP}}^{-1} = 0, \quad \forall i \tag{5.64}$$

Solving the N equations in (5.64)

$$\sum_{i \in \Omega_{\mathrm{I}}} R_{\mathrm{ES},i} = (1-\sigma)\left[-r_{\mathrm{IESP}}^{\mathrm{DP}} - r_{\mathrm{IESP}}^{\mathrm{AP}} + \sum_{i \in \Omega_{\mathrm{I}}} \left(c_i^{\mathrm{DP}} - c_i^{\mathrm{AP}} \right) \right] - \sum_{i \in \Omega_{\mathrm{I}}} \left(c_i^{\mathrm{DP}} - c_i^{\mathrm{AP}} \right), \quad \forall i. \tag{5.65}$$

Then, $R_{\mathrm{ES},i}$ can be described as

$$\begin{aligned} R_{\mathrm{ES},i} &= \omega_i \left[r_{\mathrm{IESP}}^{\mathrm{AP}} - r_{\mathrm{IESP}}^{\mathrm{DP}} + \sum_{i \in \Omega_{\mathrm{I}}} \left(c_i^{\mathrm{DP}} - c_i^{\mathrm{AP}} \right) \right] - c_i^{\mathrm{DP}} + c_i^{\mathrm{AP}} \\ &= \omega_i \Delta + c_i^{\mathrm{AP}} - c_i^{\mathrm{DP}}, \quad \forall i. \end{aligned} \tag{5.66}$$

where $R_{\mathrm{ES},i}$ consists of two components. The first component $\omega_i \Delta$ represents the profit of each CCHP derived from the cooperative surplus based on market contributions. The remaining terms signify the cost increase incurred by CCHPs due to energy trading. Economically, $R_{\mathrm{ES},i}$ reflects that IESP allocates Δ to CCHPs based on their market contributions and also covers the cost adjustments arising from energy trading.

Afterward, substitute $R_{\mathrm{ES},}$ into (5.63)

$$\max \ln f^{CD} = \sigma^2 \ln \Delta + \sum_{i \in \Omega_I} \omega_i^2 \ln \Delta \propto \ln \Delta \tag{5.67}$$

Because the natural logarithmic function is monotonically increasing within its domain, maximizing the Nash bargaining target is equivalent to maximizing the cooperative surplus.

Once $r_{\mathrm{SW}}^{\mathrm{AP}}$ it is given, we can get

$$\arg\max f^{\mathrm{CD}} \Leftrightarrow \arg\max \Delta \Leftrightarrow \arg\max r_{\mathrm{SW}}^{\mathrm{AP}} \tag{5.68}$$

which proves that the proposed model can maximize the whole social welfare.

Total cost reduction can be calculated as

$$\kappa_{\mathrm{IESP}} + \sum_{i \in \Omega_I} \kappa_{\mathrm{ES},i} = \left(\sigma + \sum_{i \in \Omega_I} \omega_i \right) \Delta = \Delta \tag{5.69}$$

We can summarize that the proposed incentive-compatible cooperative surplus allocation mechanism meets the individual rationality criterion. Additionally, it satisfies all four fundamental attributes of mechanism design theory [13], demonstrating that the mechanism is both reasonable and effective.

5.3.4 DIEM Trading and Settlement Process

IESP begins by formulating the DIEM settlement rules and signing energy hosting and trading agreements. As presented in Algorithm 1, we list the detailed transaction process of DIEM.

Algorithm 1: The trading process of DIEM

1:Each CCHP reports the schedule for the next day's integrated energy demands.
2: $c_i^{\mathrm{DP}}, r_{\mathrm{IESP}}^{\mathrm{DP}}, X_i^{\mathrm{DP}}$ ←Independent ARO decisions are made by IESP for each CCHP in DP.
3: $c_i^{\mathrm{AP}}, r_{\mathrm{IESP}}^{\mathrm{AP}}, X_i^{\mathrm{AP}}$ ←ARO decisions are made by IESP for DIEM including all CCHPs in AP.
4:According to the decision results, energy transactions are conducted by IESP with the external energy grid to ensure energy supply and demand balance.
5:IESP calculates ω_i and Δ.
6:$R_{\mathrm{ES},i}$←Cobb-Douglas utility model is solved by IESP.
7:Settlements for DIEM are made by IESP.

5.4 Solutions

To solve the tri-level optimization model presented in Sect. 5.3, the ARO solution algorithm is employed, transforming the original problem into two single-level problems: the external problem (EP) and the internal problem (IP). Using AP as an example, EP first addresses the outer min–max problem with an initially selected subset, obtaining decision variables for the day-ahead stage. Then, IP iterates to find the worst-case scenarios in the real-time stage based on EP results, adding them to the selected subset, encompassing all uncertain information from the original sets.

(a) The external problem

We can convert Eq. (5.37) into two single-level problems. Specifically, the external problem can be expressed as

$$\min\left(\sum_{t\in\Omega_T}\sum_{i\in\Omega I}\left(F_{t,i}^{\mathrm{D}}-F_{t,i}^{\mathrm{U}}\right)-r_{\mathrm{IESP}}^{d}\right)+\varphi_{\mathrm{EP}}, \tag{5.70}$$

$$\varphi_{\mathrm{EP}}\geq\sum_{t\in\Omega_T}\sum_{i\in\Omega I}F_{t,i,s}^{\mathrm{R}}-r_{\mathrm{IESP},s}^{r},\quad s\in\Omega_S. \tag{5.71}$$

The variable φ_{EP} denotes the real-time balancing cost in the worst-case scenario, representing the optimal solution of the internal problem. Both day-ahead and real-time constraints constrain the objective function of EP. Consequently, the decision

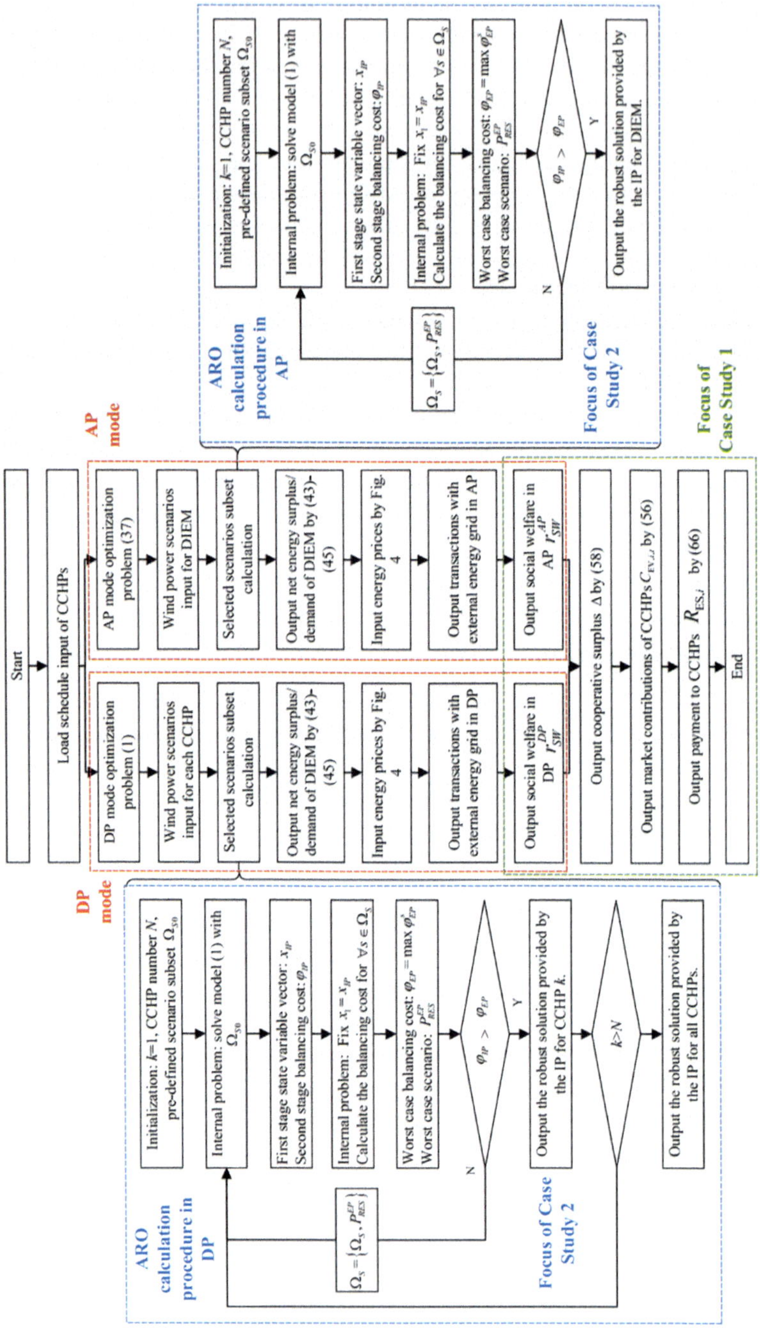

Fig. 5.3 Simulation framework

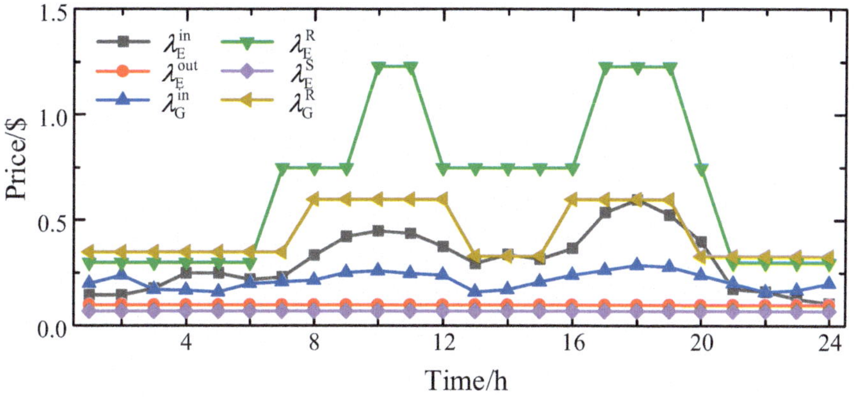

Fig. 5.4 Energy price

DIEM, which include one IESP and five CCHPs, are compared. Each mode is tested across 30 scenarios characterized by a 24-h wind power output profile. The specific operational procedures for each mode are outlined as follows:

Mode 1: The market operates under the DP mode.
Mode 2: The market operates under the AP mode without reallocation of the cooperative surplus.
Mode 3: The market operates under the AP mode with the cooperative surplus equally distributed among all CCHPs.
Mode 4: The market operates under the AP mode with incentive compatibility used for reallocating the cooperative surplus.

Table 5.1 details the net benefit of the IESP under the four modes previously described. In Mode 2, the IESP's net benefit is lower than in Mode 1 because the shared energies are utilized by CCHPs to meet their energy demands, reducing the amount of energy the IESP sells to them. This underscores the importance of developing a fair and effective surplus allocation mechanism. When a reallocation mechanism is implemented, the IESP's profit increases as its individual rationality is ensured within the constraints of the reallocation model. It is important to note that different surplus allocation mechanisms only affect the distribution of total market profit among the CCHPs, not the overall market net benefit.

Table 5.1 Net benefit of IESP under different modes

	Mode 1	Mode 2	Mode 3	Mode 4
Net benefit (k$)	22.70	22.36	22.97	22.97
Profit growth compared to Mode 1 (k$)	N/A	– 0.34	0.27	0.27

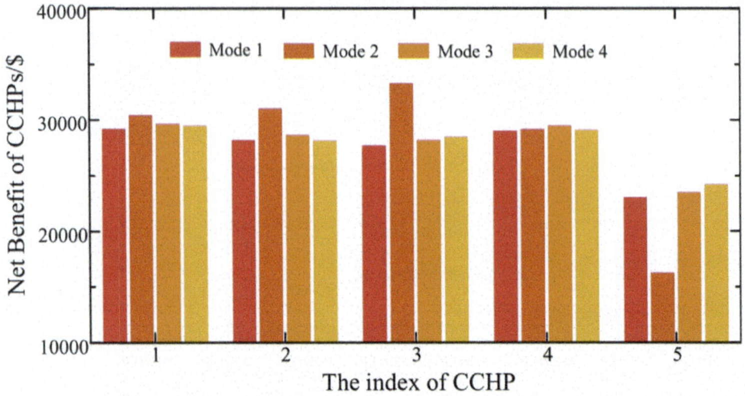

Fig. 5.5 The net benefit of each CCHP under different modes

To further explore the impact of different modes on individual CCHPs, Fig. 5.5 shows the net benefits of CCHPs under different modes. It is evident that Modes 3 and 4 result in improved net benefits for all CCHPs, satisfying the individual rationality attribute. However, in Mode 2, the net benefit of CCHP5 decreases compared to Mode 1, which could reduce its incentive to participate in the market coalition.

The following subsection will analyze the comparison between Modes 3 and 4, as well as market contributions and cost savings.

Figure 5.6 illustrates the market contributions and cost savings of CCHPs under Modes 3 and 4, examining the fairness of their respective cooperative surplus allocation mechanisms. The results reveal inconsistencies between market contributions and allocated cooperative surplus for CCHPs under Mode 3. Notably, despite CCHP5 making the highest contribution, the surplus allocated to CCHP5 is the same as for other participants. Such an unfair allocation mechanism could discourage further participation in the coalition. In contrast, Mode 4's surplus allocation strategy achieves optimal consistency between market contribution and benefit allocation, as depicted in Fig. 5.6. Implementing this method would enhance market fairness, particularly incentivizing CCHPs with lower energy production costs to participate in the energy-sharing market. Table 5.2 summarizes detailed transactions between CCHPs and the IESP organizer for cooperative surplus reallocation under Modes 3 and 4. A positive value indicates that the IESP pays a CCHP to recognize higher market contributions, while a negative value indicates that a CCHP reimburses part of the profits to the IESP to compensate participants with higher contributions. In Table 5.2, the change ratio B; is defined as

$$\beta_i = \frac{R_{\mathrm{ES},i}^{\mathrm{mode4}} - R_{\mathrm{ES},i}^{\mathrm{mode3}}}{R_{\mathrm{ES},i}^{\mathrm{mode3}}} \tag{5.73}$$

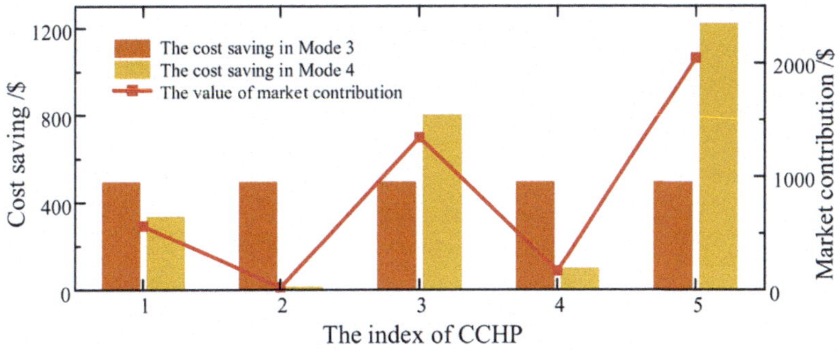

Fig. 5.6 CCHPs' market contribution and cost saving under different modes

Table 5.2 Allocations paid by IESP to CCHP under different modes

CCHP index	Mode 3 ($)	Mode 4 ($)	β_i (%)
1	− 744.40	− 901.82	− 21.15
2	− 2349.34	− 2829.84	− 20.45
3	− 5119.17	− 4812.07	5.99
4	329.55	− 66.46	− 120.16
5	7267.95	7994.77	10.00

Table 5.3 The social welfare in different modes

Mode index	1	2	3	4
Social welfare ($)	159,739	162,490	162,490	162,490

From Table 5.3, it is evident that the allocation mechanism designed in Mode 4 does not impact the overall social welfare of the market, which remains at 162,490, matching the maximum social welfare derived in (5.37).

In summary, the DIEM market mechanism designed in this chapter aligns with the principles of mechanism design theory. Additionally, in this study, seven scenarios are iterated as the selected subset of AP in the ARO formulation, as depicted in Fig. 5.7. A detailed discussion of the ARO formulation's capability to manage uncertainties will be provided in the following subsection.

5.5.3 *Case Study 2*

In Case study 2, we concentrate on a large-scale DIEM with the initial random scenarios increasing to 50, as shown in Fig. 5.8. The primary objective of the subsequent tests is to evaluate the performance of the ARO formulation in terms

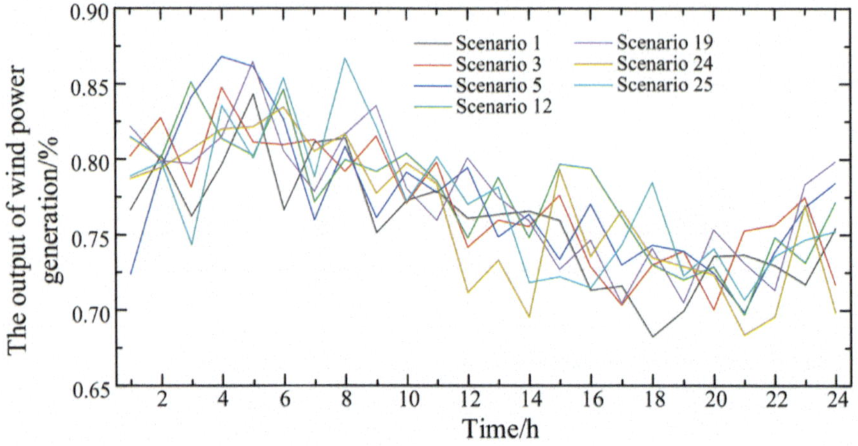

Fig. 5.7 The iteration results in the selected subset in the ARO formulation

of scenario selection generality, computational efficiency, and the robustness and conservativeness of the results, in comparison with the conventional TSO method.

To verify that the chosen scenarios accurately represent the entire set of random information, it is important to demonstrate that the scenarios not selected do not influence the results of the optimization problem. Thus, using the outcomes derived from the ARO formulation, the RO formulation developed in [29] is utilized to confirm correctness through a comparison of the following four approaches.

Approach 2A: All scenarios are included in the ARO model.
Approach 2B: Only the selected scenarios calculated by the ARO are incorporated into the RO (Robust Optimization) model.

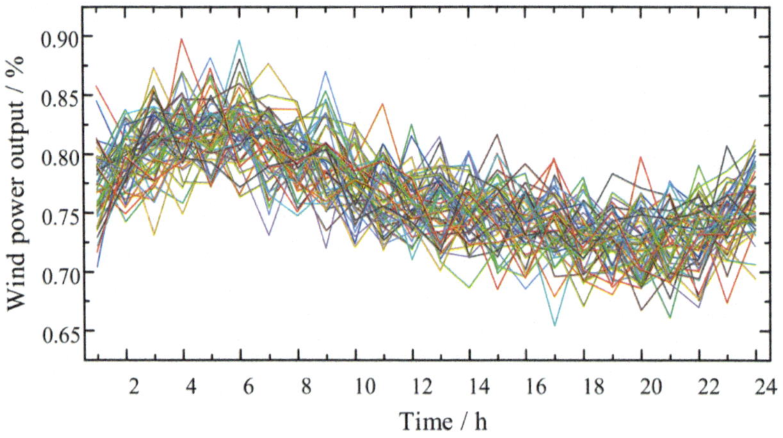

Fig. 5.8 50 random scenarios in Case Study 2

Approach 2C: The selected scenarios calculated by the ARO + half of the unselected scenarios are included in the RO model.
Approach 2D: The selected scenarios calculated by the ARO + all unselected scenarios, are included in the RO model.

Table 5.4 presents a summary of the results and time required to solve the model using the four approaches previously mentioned. The consistency in model outcomes confirms that the scenarios which cannot affect the results, thereby validating the proposed scenario selection method. Additionally, omitting unnecessary scenarios greatly enhances computational efficiency.

Additionally, hybrid, state-of-the-art, and PSO-oriented algorithms have proven to be effective tools for solving complex optimization problems, including FBICLPSO [33], CHPSA [34], SPSO [35], and HGWO-PSO [36], among others. The computational efficiency of the proposed method is compared with these PSO-oriented algorithms, as shown in Table 5.5. Due to the simulation scale, the calculation times vary. The proposed ARO method, applied to a more complex model, requires more computation time compared to others.

In addition, to assess the robustness and conservativeness of the ARO algorithm, four cases are designed to compare ARO with the conventional TSO formulation developed in [29]. The standard deviation of the normal distribution of 50 random scenarios represents the randomness of the four cases, set at 5%, 10%, 15%, and 20%

Table 5.4 Results comparison in different approaches

	DP (k$)	AP (k$)	Auxiliary (k$)	Time (s)
Approach 2A	313.44	320.11	156.35	598.9
Approach 2B	313.44	320.11	156.35	248.1
Approach 2C	313.44	320.11	156.35	492.8
Approach 2D	313.44	320.11	156.35	872.2

Table 5.5 Performance of the ARO method compared with PSO-oriented algorithms

	Simulation scale	Proposed algorithm (s)	Traditional method (s)	Time difference (s)	Efficiency improvement (%)
ARO in this chapter	1-IESP, 10-CCHPs, 30-scenarios	248.1	872.2	624.1	71.6
FBICLPSO in [33]	30-bus, with FACTS	75.41	400	324.59	81.1
CHPSA in [34]	24-bus	19.65	86.18	66.53	77.2
SPSO in [35]	30-bus	21.124	152.32	131.2	86.1
HGWO-PSO in [36]	40-unit, 4-area	12.1	24.0	11.9	49.6

Table 5.6 Comparison of TSO and ARO results

		Case2D	Case2E	Case2F	Case2G
r_{SW}^{AP} (k$)	TSO	320.00	319.90	319.87	\
	ARO	320.11	319.96	319.91	319.88
φ_{IP} ($)	TSO	161.12	182.33	206.62	\
	ARO	156.35	169.35	187.87	232.66

for cases 2D-2G. Their performance in terms of robustness and conservativeness is quantified by social welfare in energy-sharing mode and the balancing cost of the internal problem.

Table 5.6 presents the comparison results of TSO and ARO. In high randomness conditions (Approach 2G), a feasible solution cannot be obtained using the TSO model, while the optimal solution is achievable with the ARO model. This indicates that TSO has limitations in handling large-scale random problems, whereas the ARO formulation improves robustness by explicitly considering extreme scenarios.

For the remaining approaches, the social welfare obtained from the ARO model is higher than that from the TSO model. Additionally, the ARO model demonstrates better performance in terms of balancing costs in worst-case scenarios. For example, in Approaches 2D-2F, the φ_{IP} values for ARO are consistently lower than those for TSO. This improvement is due to ARO's optimization of the worst-case internal problem, which adaptively adjusts for uncertain parameters. In contrast, TSO weights each scenario without independently optimizing the worst case. These results further confirm that compared to TSO, the ARO formulation enhances both conservativeness and economic efficiency when determining energy consumption and market clearing decisions.

5.6 Conclusions

To resolve the issues concerning the lack of incentives for CCHPs to participate in distributed energy markets and enable energy trading, this chapter presents a comprehensive DIEM trading mechanism to maximize social welfare while maintaining individual rationality, incentive compatibility, and cost recovery characteristics. Generally, we develop an optimization problem to maximize social welfare by scheduling the energy interaction between CCHPs. A cooperative surplus allocation strategy based on market contributions is proposed to enhance LEM fairness and attract more participants. The proposed mechanism is demonstrated to satisfy individual rationality, incentive compatibility, and cost recovery. Furthermore, an ARO formulation is adopted to determine energy consumption and market-clearing decisions for all participants. This is solved using a selected-scenario-based algorithm. Numerical results show that the proposed model improves conservatism compared to conventional robust optimization formulations.

Given the premise that the LEM sets prices received from the external power system, input electricity prices are unchangeable. Adjustments to the market clearing strategy are necessary for the integrated energy service provider when external network prices change. Future work could consider incorporating randomness from multiple energy sources, including uncertainties in prices and power output, into the model. Additionally, integrating network architecture and physical transmission constraints could improve the model's accuracy. Addressing profit distribution in cases of transmission congestion would further refine market operations. To enhance computational efficiency for handling complex models involving transmission and energy flows, algorithms oriented toward particle swarm optimization could be explored.

Appendix

1. Proof of social welfare maximization

To demonstrate that the proposed incentive-compatible market mechanism maximizes social welfare, we first establish that the Nash bargaining model in Sect. 5.3.3 is a convex problem. This entails deriving the analytical optimal solution of the settlement mechanism. To facilitate this calculation, we begin by computing the analytical optimal solution of the logarithmic Nash bargaining model. Central to this is proving that the logarithmic Nash bargaining model is convex, which hinges on demonstrating that its objective function is concave. Given that the natural logarithm function monotonically increases over the domain $\{X > 0\}$, we apply it to the objective function (5.62). This transforms the Nash bargaining model into

$$\max \ln f^{\mathrm{CD}} = \sigma \ln \kappa_{\mathrm{IESP}} + \sum_{i \subset \Omega_{\mathrm{I}}} \omega_i \ln \kappa_{\mathrm{ES},i} \tag{5.74}$$

s.t. (5.55), (5.61).

Notably, the individual rational constraints of IESP and CCHP are inherently included within the domain. To proceed with calculating the Hessian matrix of the logarithmic objective function, we begin by computing the second-order partial derivatives with respect to the decision variables:

$$\frac{\partial^2 \ln f^{\mathrm{CD}}}{\partial X_i^2} = -\frac{\omega_i}{\left(c_i^{\mathrm{DP}} - c_i^{\mathrm{AP}} + R_{\mathrm{ES},i}\right)^2}\left(\frac{\partial c_i^{\mathrm{AP}}}{\partial X_i}\right)^2 - \frac{\omega_i}{c_i^{\mathrm{DP}} - c_i^{AP} + R_{\mathrm{ES},i}}\frac{\partial^2 c_i^{\mathrm{AP}}}{\partial X_i^2} \tag{5.75}$$

$$\frac{\partial^2 \ln f^{\mathrm{CD}}}{\partial X_i X_j} = 0 \tag{5.76}$$

$$\frac{\partial^2 \ln f^{\mathrm{CD}}}{\partial R_{\mathrm{ES},i}^2} = -\frac{\omega_i}{\left(c_i^{\mathrm{DP}} - c_i^{\mathrm{AP}} + R_{\mathrm{ES},i}\right)^2} \tag{5.77}$$

$$\frac{\partial^2 \ln f^{\mathrm{CD}}}{\partial R_{\mathrm{ES},i} R_{\mathrm{ES},j}} = 0 \tag{5.78}$$

Due to the convexity of the CCHPs' objective function, we can obtain

$$\frac{\partial^2 c_i^{\mathrm{AP}}}{\partial X_i^2} \geq 0 \tag{5.79}$$

Therefore, the value of (5.75) cannot be greater than zero, which corresponds to Eq. (5.77). Actually, the Hessian matrix belongs to a diagonal matrix, and all elements in it cannot be greater than 0. These above features enable the Hessian matrix as non-positive definite and log objective function is convex. Therefore, we can calculate the Karush–Kuhn–Tucker Condition as:

$$\partial \ln f^{\mathrm{CD}} \partial^{-1} R_{\mathrm{ES},i} = \omega_j \kappa_{\mathrm{ES},i}^{-1} - \sigma \kappa_{\mathrm{IESP}}^{-1} = 0, \quad \forall i. \tag{5.80}$$

Although Eq. (5.80) has N equations of CCHPs, we can solve these parallelly as follows.

$$\sum_{i \in \Omega_1} R_{\mathrm{ES},i} = (1-\sigma)\left[-r_{\mathrm{IESP}}^{\mathrm{DP}} - r_{\mathrm{IESP}}^{\mathrm{AP}} + \sum_{i \in \Omega_1} \left(c_i^{\mathrm{DP}} - c_i^{\mathrm{AP}} \right) \right] - \sum_{i \in \Omega_1} \left(c_i^{\mathrm{DP}} - c_i^{\mathrm{AP}} \right), \quad \forall i \tag{5.81}$$

Further, (5.82) can be derived

$$R_{\mathrm{ES},i} = \omega_i \left[r_{\mathrm{IESP}}^{\mathrm{AP}} - r_{\mathrm{IESP}}^{\mathrm{DP}} + \sum_{i \in \Omega_1} \left(c_i^{\mathrm{DP}} - c_i^{\mathrm{AP}} \right) \right] - c_i^{\mathrm{DP}} + c_i^{\mathrm{AP}} = \omega_i \Delta + c_i^{\mathrm{AP}} - c_i^{\mathrm{DP}}, \quad \forall i \tag{5.82}$$

From Eq. (5.82), we can observe that the payment received by CCHP largely depends on the cost reduction of DIEM and the change of cost after participating in the market. Thereafter, we can substitute (5.82) into (5.74) to gain the following optimization function.

$$\max \ln f^{\mathrm{CD}} = \sigma^2 \ln \Delta + \sum_{i \in \Omega_I} \omega_i^2 \ln \Delta \propto \ln \Delta \tag{5.83}$$

Therefore, the Nash bargaining model is akin to the DIEM model in that they both aim to minimize the total operational cost of the system, thus achieving the goal of maximizing social welfare, namely

$$X^* \in \arg\max f^{\mathrm{CD}} \Leftrightarrow \arg\max \Delta \Leftrightarrow \arg\max r_{SW}^{\mathrm{AP}} \tag{5.84}$$

Therefore, the attribute of maximizing social welfare is proved.

2. Proof of individual rationality

If the Nash bargaining model has a feasible solution, the constraints (5.59)–(5.60) in the model naturally ensure that the proposed trading mechanism satisfies individual rationality. Therefore, the feasibility of the solution is demonstrated.

From constraints (5.59)–(5.60) in the model, it is evident that the existence of a feasible solution implies the satisfaction of individual rationality constraints for IESP and CCHP. Therefore, the necessary and sufficient condition for satisfying these individual rationality constraints is that the cooperative surplus is non-negative. In other words, the condition that there exists $\{R_{\mathrm{ES},i} \geq 0, \forall i\}$ which ensures both CCHP and IESP satisfy individual rationality is equivalent to having a non-negative cooperative surplus.

$$\begin{cases} (r_{\mathrm{IESP}}^{\mathrm{AP}} - R_{\mathrm{IESP}}) - r_{\mathrm{IESP}}^{\mathrm{DP}} \geq 0 \\ c_i^{\mathrm{DP}} - (c_i^{\mathrm{AP}} - R_{\mathrm{ES},i}) \geq 0 \end{cases} \Leftrightarrow \Delta c \geq 0 \tag{5.85}$$

Therefore, it suffices to demonstrate that the cooperative surplus generated by DIEM is non-negative.

Given that the optimization variables for CCHPs are independent of each other, the model where each CCHP operates independently before engaging in the market is equivalent to the following model:

$$\begin{cases} \min c_1^{\mathrm{AP}} \text{s.t. } \mathrm{X}_1 \in \chi_1 \\ \min c_2^{\mathrm{AP}} \text{s.t. } \mathrm{X}_2 \in \chi_2 \\ \cdots \\ \min c_N^{AP} \text{s.t. } \mathrm{X}_N \in \chi_N \end{cases} \Leftrightarrow \min \sum_{i=1}^{N} c_i^{\mathrm{AP}} \text{s.t. } \mathrm{X} \in \chi \tag{5.86}$$

The total cost incurred by IESP and CCHP in this independent operation model must be higher than the total cost under DIEM, as DIEM is designed to minimize the total cost for IESP and CCHP. Therefore, $\Delta c \geq 0$.

To sum up, there exists a feasible solution to the Nash bargaining model, and the incentive-compatible mechanism satisfies the individual rationality attribute.

3. Proof of cost recovery

In terms of cost recovery, we can distribute the cooperative to market participators, like IESP and CCHPs. Based on Eqs. (5.59)–(5.60), we can calculate the sum of expenditure reduction as

$$\kappa_{\mathrm{IESP}} + \sum_{i \in \Omega_I} \kappa_{\mathrm{ES},i} \left(\sigma + \sum_{i \in \Omega_I} \omega_i \right) \Delta = \Delta \tag{5.87}$$

It is noted that Eq. (5.87) represents the cooperation surplus generated by the DIEM transaction, which proves that the incentive-compatible mechanism also has the cost recovery attribute.

4. Proof of incentive compatibility

To demonstrate that the proposed trading mechanism satisfies incentive compatibility, it is crucial to show that individual objectives align with the overall system goal. For instance, consider CCHP J. The income for CCHP J within the trading mechanism is calculated as Eq. (5.88):

$$\prod_{J} = -c_J^{\mathrm{AP}} + R_{\mathrm{ES},J} \tag{5.88}$$

Equation (5.88), $-c_J^{\mathrm{AP}}$ represents the net income under the AP model, and the other term $R_{\mathrm{ES},J}$ denotes the fee paid by the IESP for participating in energy trading. It is assumed that CCHP J would take strategic actions while optimizing its energy trading. Therefore, we can combine Eq. (5.82) to get the optimal shared energy quantity of CCHP J. Specifically, it can be derived from the following optimization problem.

$$\max\{-c_J^{\mathrm{AP}} + R_{\mathrm{ES},J}\} = \max\{\omega_J \Delta - c_J^{\mathrm{DP}}\} \tag{5.89}$$

On the right side of Eq. (5.89), the second term is determined by the DP mode and is independent of the energy trading, which means this term can be ignored when determining the energy trading amount. Consequently, the objective function is simplified to optimize the first term. Combined with Eq. (5.84), it is clear that the first term is directly proportional to the total social welfare. Therefore, the goal of maximizing social welfare has been internalized in the optimal goal of CCHP J, encouraging it to adopt a strategy of sharing energy, which ultimately leads to maximizing social welfare.

In summary, the objective of individual benefit aligns with the global objective, thus proving the character of incentive compatibility.

References

1. Zhang, Z., Jing, R., Lin, J., Wang, X., Dam, K. H., Wang, M., et al. (2020). Combining agent-based residential demand modeling with design optimization for integrated energy systems planning and operation. *Applied Energy, 263*, 114623.
2. Gridwise Architecture Council. (2020). *Transactive energy*. https://www.gridwiseac.org/about/transactive_energy.aspx
3. Abrishambaf, O., Lezama, F., Faria, P., & Vale, Z. (2019). Towards transactive energy systems: An analysis on current trends. *Energy Strategy Reviews, 26*, 100418.
4. Guerrero, J., Gebbran, D., Mhanna, S., Chapman, A. C., & Verbič, G. (2020). Towards a transactive energy system for integration of distributed energy resources: Home energy management,

distributed optimal power flow, and peer-to-peer energy trading. *Renewable and Sustainable Energy Reviews, 132*, 110000.
5. Wang, Y., Wang, Y., Huang, Y., Yu, H., Du, R., Zhang, F., et al. (2018). Optimal scheduling of the regional integrated energy system considering economy and environment. *IEEE Transactions on Sustainable Energy, 10*(4), 1939–1949.
6. Wang, J., Wei, Z., Yang, B., Yong, Y., Xue, M., Sun, G., et al. (2019). Two-stage integrated electricity and heat market clearing with energy stations. *IEEE Access, 7*, 44928–44938.
7. Chen, S., Wei, Z., Sun, G., Cheung, K. W., & Sun, Y. (2016). Multi-linear probabilistic energy flow analysis of integrated electrical and natural-gas systems. *IEEE Transactions on Power Systems, 32*(3), 1970–1979.
8. Wang, Y. L., Wang, Y., Huang, Y., Yand, J., Ma, Y., Yu, H., et al. (2019). Operation optimization of regional integrated energy system based on the modeling of electricity-thermal-natural gas network. *Applied Energy, 251*, 113410.
9. Jadhav, A. M., Patne, N. R., & Guerrero, J. M. (2018). A novel approach to neighborhood fair energy trading in a distribution network of multiple microgrid clusters. *IEEE Transactions on Industrial Electronics, 66*(2), 1520–1531.
10. Park, S., Lee, J., Bae, S., Hwang, G., & Choi, J. K. (2016). Contribution-based energy-trading mechanism in microgrids for future smart grid: A game theoretic approach. *IEEE Transactions on Industrial Electronics, 63*(7), 4255–4265.
11. Myerson, R. B., Eatwell, J., Milgate, M., & Peter, N. (1989). *Allocation, information and markets*.
12. Mediwaththe, C. P., Shaw, M., Halgamuge, S., Smith, D. B., & Scott, P. (2019). An incentive-compatible energy trading framework for neighborhood area networks with shared energy storage. *IEEE Transactions on Sustainable Energy, 11*(1), 467–476.
13. Exizidis, L., Kazempour, J., Papakonstantinou, A., Pinson, P., De Greve, Z., & Vallée, F. (2019). Incentive-compatibility in a two-stage stochastic electricity market with high wind power penetration. *IEEE Transactions on Power Systems, 34*(4), 2846–2858.
14. Yang, L., Jian, J., Xu, Y., Dong, Z., & Ma, G. (2017). Multiple perspective-cuts outer approximation method for risk-averse operational planning of regional energy service providers. *IEEE Transactions on Industrial Informatics, 13*(5), 2606–2619.
15. Guo, L., Wang, L., Wei, W., & Wu, J. (2019). Day ahead economic dispatch of district integrated energy service providers under energy market. In *2019 IEEE 3rd International Electrical and Energy Conference* (pp. 1878–1883).
16. Mahboubi-Moghaddam, E., Nayeripour, M., & Aghaei, J. (2016). Reliability constrained decision model for energy service provider incorporating demand response programs. *Applied Energy, 183*, 552–565.
17. Wang, Q., Guan, Y., & Wang, J. (2011). A chance-constrained two-stage stochastic program for unit commitment with uncertain wind power output. *IEEE Transactions on Power Systems, 27*(1), 206–215.
18. Lu, Z., Liu, M., Lu, W., & Deng, Z. (2019). Stochastic optimization of economic dispatch with wind and photovoltaic energy using the nested sparse grid-based stochastic collocation method. *IEEE Access, 7*, 91827–91837.
19. Qiu, H., Zhao, B., Gu, W., & Bo, R. (2018). Bi-level two-stage robust optimal scheduling for AC/DC hybrid multi-microgrids. *IEEE Transactions on Smart Grid, 9*(5), 5455–5466.
20. Cao, Y., Tan, Y., Li, C., & Rehtanz, C. (2013). Chance-constrained optimization-based unbalanced optimal power flow for radial distribution networks. *IEEE Transactions on Power Delivery, 28*(3), 1855–1864.
21. Seddig, K., Jochem, P., & Fichtner, W. (2019). Two-stage stochastic optimization for cost-minimal charging of electric vehicles at public charging stations with photovoltaics. *Applied energy, 242*, 769–781.
22. Xie, S., Hu, Z., & Wang, J. (2020). Two-stage robust optimization for expansion planning of active distribution systems coupled with urban transportation networks. *Applied Energy, 261*, 114412.

23. Li, P., Wu, Z., Wang, Y., Dou, X., Hu, M., & Hu, J. (2018). Adaptive robust optimal reactive power dispatch in unbalanced distribution networks with high penetration of distributed generation. *IET Generation, Transmission & Distribution, 12*(6), 1382–1389.
24. Li, Z., Wu, W., Shahidehpour, M., & Zhang, B. (2015). Adaptive robust tie-line scheduling considering wind power uncertainty for interconnected power systems. *IEEE Transactions on Power Systems, 31*(4), 2701–2713.
25. Bertsimas, D., Litvinov, E., Sun, X. A., Zhao, J., & Zheng, T. (2012). Adaptive robust optimization for the security constrained unit commitment problem. *IEEE Transactions on Power Systems, 28*(1), 52–63.
26. Chen, S., Wei, Z., Sun, G., Cheung, K. W., Wang, D., & Zang, H. (2017). Adaptive robust day-ahead dispatch for urban energy systems. *IEEE Transactions on Industrial Electronics, 66*(2), 1379–1390.
27. Jamalzadeh, R., & Hong, M. (2018). Microgrid optimal power flow using the generalized benders decomposition approach. *IEEE Transactions on Sustainable Energy, 10*(4), 2050–2064.
28. Soares, J., Canizes, B., Ghazvini, M. A. F., Vale, Z., & Venayagamoorthy, G. K. (2017). Two-stage stochastic model using benders' decomposition for large-scale energy resource management in smart grids. *IEEE Transactions on Industry Applications, 53*(6), 5905–5914.
29. Morales, J. M., Conejo, A. J., Madsen, H., Pinson, P., & Zugno, M. (2013). *Integrating renewables in electricity markets: Operational problems*. Springer Science & Business Media.
30. Bertsimas, D., & de Ruiter, F. J. (2016). Duality in two-stage adaptive linear optimization: Faster computation and stronger bounds. *INFORMS Journal on Computing, 28*(3), 500–511.
31. Gu, W., Wang, Z., Wu, Z., Luo, Z., Tang, Y., & Wang, J. (2016). An online optimal dispatch schedule for CCHP microgrids based on model predictive control. *IEEE Transactions on Smart Grid, 8*(5), 2332–2342.
32. Gregory, M. N. (2008). *The principles of microeconomics*. South-Western.
33. Naderi, E., Pourakbari-Kasmaei, M., & Abdi, H. (2019). An efficient particle swarm optimization algorithm to solve optimal power flow problem integrated with FACTS devices. *Applied Soft Computing, 80*, 243–262.
34. Naderi, E., Pourakbari-Kasmaei, M., & Lehtonen, M. (2020). Transmission expansion planning integrated with wind farms: A review, comparative study, and a novel profound search approach. *International Journal of Electrical Power & Energy Systems, 115*, 105460.
35. Naderi, E., Pourakbari-Kasmaei, M., Cerna, F. V., & Lehtonen, M. (2021). A novel hybrid self-adaptive heuristic algorithm to handle single-and multi-objective optimal power flow problems. *International Journal of Electrical Power & Energy Systems, 125*, 106492.
36. Azizivahed, A., Arefi, A., Naderi, E., Narimani, H., Fathi, M., & Narimani, M. R. (2020). An efficient hybrid approach to solve bi-objective multi-area dynamic economic emission dispatch problem. *Electric Power Components and Systems, 48*(4–5), 485–500.

Chapter 6
Peer-to-Peer Energy Trading of Microgrids Considering Renewable Energy Uncertainty

Abstract As subsidies for distributed renewable energy are reduced, market-based measures are essential to promote the integration of renewable energy into the grid. However, renewable energy generation's intermittent and volatile nature creates significant decision-making challenges. To mitigate uncertainty from inaccurate renewable energy forecasts, this chapter presents a peer-to-peer (P2P) energy trading strategy for multiple microgrids (MGs) using distributionally robust optimization (DRO). Firstly, a fuzzy set based on Wasserstein distance is developed to model prediction errors for each MG's renewable energy. Secondly, a day-ahead P2P energy trading model utilizing DRO is proposed to handle power fluctuations. Thirdly, dual theory converts the nonlinear model into a linear and convex programming problem. A distributed strategy based on the alternating direction method of multipliers (ADMM) is applied to maintain MG independence and privacy. Finally, the case study demonstrates that the proposed trading strategy can enhance MG revenue from P2P transactions, protect MG privacy, and support renewable energy development. The DRO approach also ensures the economic and reliable execution of transactions for real-time implementation.

6.1 Introduction

Global green and low-carbon development adoption is accelerating due to energy crises and climate change [1]. This shift necessitates a transition in the power system to renewable energy for decarbonization. As noted in [2], distributed energy resources (DERs) like distributed power generation, energy storage (ES), and flexible loads are rapidly expanding in the distribution grid, a trend expected to persist with the growing economic viability of small-scale distributed RE [3]. However, integrating DERs on a large scale decentralizes the distribution network and complicates decision-making due to RE generation uncertainties, potentially undermining power system reliability if not managed properly [4, 5]. Multiple microgrids (MGs) effectively manage DERs by reducing dependence on the distribution network through local power complementation.

M. Song et al., *Local Energy Markets*, https://doi.org/10.1007/978-981-97-9750-9_6

Additionally, the local energy market (LEM) at the distribution grid level supports DER market participation, facilitating price-driven regulation, especially with peer-to-peer (P2P) energy [6, 7]. The National Energy Administration of China has highlighted P2P trading at the distribution network level as a focus of future market reform. However, the uncertainty of distributed RE generation poses a significant challenge for MG operators, who must balance economic considerations with intraday implementation reliability based on day-ahead RE forecasts. To date, few studies have addressed P2P energy while accounting for this uncertainty, this chapter aims to fill a gap.

In LEM, DERs can share surplus energy with various entities like MGs, virtual power plants, and load aggregators. The MG-based energy-sharing mechanism has been a major research area recently. For example, a model for MGs with price-based demand response is proposed in [8], utilizing a dynamic pricing model based on supply and demand to enhance local PV consumption. In [9], an energy-sharing cloud mechanism for MGs with renewable energy sources and storage is introduced. A Stackelberg game approach for MG energy sharing management is discussed in [10]. Authors in [11] developed a co-optimization model for multiple virtual power plants using demand response and game theory to manage DERs. In [12], a P2P energy trading platform was created, using game theory to simulate P2P trading and presenting a framework for identifying essential components and technological aspects. In [13], a DC MG system with P2P energy sharing is proposed using nonlinear programming. These studies, however, rely on a central agent for global optimization, requiring data from all participants, which raises privacy concerns and communication pressure.

To address these issues, distributed algorithm-based P2P energy trading has become a research focus. Distributed algorithms decompose the global optimization problem, allowing participants to independently determine their strategies, thereby protecting privacy. Common algorithms include the primal–dual gradient descent method, subgradient method, distributed consensus algorithm, and alternating direction method of multipliers (ADMM). In [14], a P2P trading model based on consumer preferences is developed and solved using an improved relaxation consensus method. To prevent line congestion, authors in [15] use the power transmission distribution factor in the P2P trading model, solved with the primal–dual gradient descent method. In [16], a consensus ADMM algorithm addresses the P2P energy trading model, incorporating network usage fees to calculate power transmission costs, helping to maintain and update infrastructure to avoid congestion. Simulation results show good convergence and solution speed for the ADMM method. Although distributed P2P energy trading protects aggregator privacy, it rarely addresses uncertainty. Typically, uncertain parameters are pre-studied while the model can be converted to a deterministic problem, such as dynamic power reserve [17]. Considering uncertainty's significant impact on decision-making, exploring P2P energy trading accounting for uncertainty is crucial.

When scheduling decisions are made, DER aggregators face uncertainties such as RE generation, load demand, and electricity prices [18]. Research over the past decade has explored various strategies for managing these uncertainties, including stochastic programming (SP) and robust optimization (RO). For example, SP simulates uncertainties in RE output and load, developing a demand response model to mitigate fluctuations [19], and a multi-stage SP approach optimizes aggregators' intraday market participation [20]. RO models, meanwhile, are used to develop optimal scheduling strategies for short- and medium-term markets by addressing uncertainties in electricity load, heat load, and RE generation [21, 22]. Information gap decision theory (IGDT) is another method for handling uncertainty, although it is considered deterministic [17, 23]. SP requires accurate probability distributions of uncertain parameters, whereas RO is often seen as too conservative, focusing only on worst-case scenarios.

Distributionally robust optimization (DRO) is gaining popularity in power systems optimization for handling uncertainty [24, 25]. It balances conservatism and feasibility by combining SP and RO benefits and requires only historical data. DRO has been used to address wind power uncertainty in unit commitment problems [25], optimal power flow issues [26, 27], and demand response uncertainty in electricity-gas coupled systems [28]. However, uncertainty in P2P energy trading remains underexplored. Applying the DRO method to manage RE prediction uncertainty in P2P energy trading for MGs is a promising research area.

Building on the previous analysis, this chapter explores the energy-sharing mechanism among multiple MGs while accounting for RE uncertainty. Figure 6.1 depicts the multi-MG framework in the distribution network, where each MG includes various DERs. MGs can buy and sell power from retailers and participate in P2P trading with other MGs. This chapter focuses on the P2P energy trading of multiple MGs, addressing RE uncertainty. Unlike previous studies that do not consider uncertainty, this chapter employs the Wasserstein distance-based DRO method to manage RE uncertainty in multi-MG decision-making. Additionally, an ADMM-based algorithm is proposed for P2P energy trading to protect MGs' privacy. Case studies confirm the reliability and economic efficiency of the proposed method.

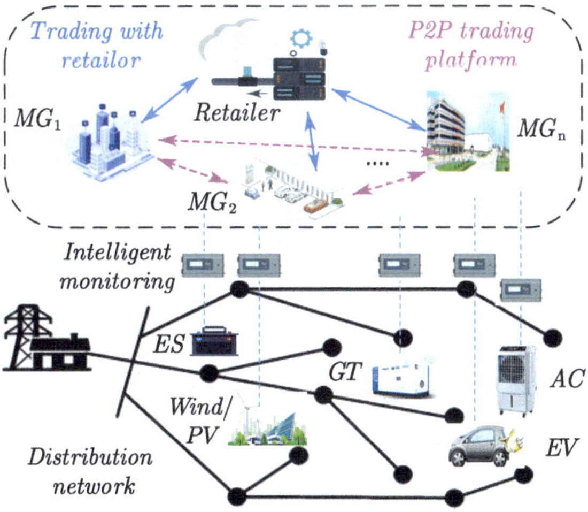

Fig. 6.1 Schematic diagram of the energy-sharing framework of multi-MGs

6.2 Basic Models of Multi-MGs in Different Time Scales

Each MG is assumed to contain gas turbine (GT) units, ES, demand response (DR) loads, and distributed RE units.

6.2.1 Day-Ahead Optimization Model

1. *Operating Cost*

In each MG i, its operating cost includes GT unit cost, DR cost, electricity purchase and selling cost with the retailer, P2P trading cost, and power transmission cost.

(a) *GT unit operating cost.*

The capacity of GT units in the distribution network is typically small, and it is assumed that these units are always online, without considering start-up and shutdown costs. Therefore, a linear function is used to calculate their costs as in Eq. (6.1)

$$C_{i,t}^{\mathrm{gt}} = a_i p_{i,t}^{\mathrm{gt}} + b_i \tag{6.1}$$

where a_i and b_i are the GT cost coefficients, and $p_{i,t}^{gt}$ is the power output of the GT unit in period t.

(b) *ES operating cost.*

$$C_{i,t}^{\mathrm{es}} = c_i^{\mathrm{es}} (p_{i,t}^{\mathrm{cha}} + p_{i,t}^{\mathrm{dis}}) \tag{6.2}$$

where c_i^{es} is the depreciation cost factor of ES. $p_{i,t}^{\mathrm{cha}}$ and $p_{i,t}^{\mathrm{dis}}$ are the charging power and discharging power of ES in the period t, respectively. Both of them are positive. Only one of them can be positive at time t to prevent arbitrage.

(c) *DR cost.*

$$C_{i,t}^{\mathrm{dr}} = c_{i,t}^{\mathrm{dr}} p_{i,t}^{\mathrm{dr}} \tag{6.3}$$

where $c_{i,t}^{\mathrm{dr}}$ is the subsidy price for reducing load demand during period t. $p_{i,t}^{\mathrm{dr}}$ is the capacity of DR power.

(d) *Revenue from internal electricity sales.*

$$C_{i,t}^{\mathrm{sell}} = \lambda_{i,t}^{\mathrm{ins}} p_{i,t}^{\mathrm{load}} \tag{6.4}$$

where $\lambda_{i,t}^{\mathrm{ins}}$ is the price of MG i selling electricity to internal load users. $p_{i,t}^{\mathrm{load}}$ is the total internal load demand at time step t.

(e) *Cost of trading with distribution grid.*

$$C_{i,t}^{\text{grid}} = \lambda_{i,t}^{b} p_{i,t}^{b} - \lambda_{i,t}^{s} p_{i,t}^{s} \tag{6.5}$$

where $\lambda_{i,t}^{b}$ and $p_{i,t}^{b}$ are the price and electricity purchased by MG i from the retailer in the period t, respectively. $\lambda_{i,t}^{s}$ and $p_{i,t}^{s}$ are the price and power selling to the grid, respectively. Usually, $\lambda_{i,t}^{s}$ is much lower than $\lambda_{i,t}^{b}$ which is the basis for MGs to increase profits or reduce costs through P2P trading.

(f) *P2P energy trading cost.*

MG i can trade with any MG j participating in the energy-sharing transaction. Its cost is as follows

$$C_{i,t}^{p2p} = \sum_{j \in N \backslash i} \lambda_{i,t}^{j} p_{i,t}^{j} \tag{6.6}$$

where N is the set of MGs participating in the P2P trading. $\lambda_{i,t}^{j}$ is the transaction price between MG i and MG j in period t, which should be between $\lambda_{i,t}^{b}$ and $\lambda_{i,t}^{s}$. $p_{i,t}^{j}$ is the transaction power between them in period t. It indicates that MG i purchases power from MG j when it is positive, and vice versa.

(g) *Power transmission cost.*

MGs must pay a network usage fee to the distribution network operator based on their P2P trading power. This fee, approved by the operator, is then allocated to each MG for maintaining and upgrading the distribution grid. Additionally, this cost encourages P2P trading between nearby MGs, helping to minimize the risk of network congestion. The cost of P2P trading is shared between both parties and is calculated as follows:

$$\phi_{ij} = c^{\text{dist}} d_{ij} / 2 \tag{6.7}$$

where ϕ_{ij} is the transmission cost per unit of energy transactions between MG i and MG j. This cost is proportional to the electrical distance d_{ij}, which is determined by the distribution grid structure. It can be obtained by calculating the Thevenin impedance distance based on the shortest path algorithm. c^{dist} is the transmission cost per unit distance set by the distribution network operator. Thus, the power transmission cost of P2P trading for MG i is as follows.

$$C_{i,t}^{tr} = \sum_{j \in N \backslash i} \phi_{ij} |p_{i,t}^{j}| \tag{6.8}$$

2. *Constraints*

The following constraints must be satisfied by MG *i* for the day-ahead trading model.

(a) Generation constraints of distributed resources.

$$0 \le p_{i,t}^{\mathrm{re}} \le p_i^{\mathrm{re\_max}} \tag{6.9}$$

$$p_i^{\mathrm{gt\_min}} \le p_{i,t}^{gt} \le p_i^{\mathrm{gt\_max}} \tag{6.10}$$

$$0 \le p_{i,t}^{\mathrm{cha}} \le p_i^{\mathrm{cha\_max}} \tag{6.11}$$

$$0 \le p_{i,t}^{\mathrm{dis}} \le p_i^{\mathrm{dis\_max}} \tag{6.12}$$

$$0 \le p_{i,t}^{\mathrm{dr}} \le p_i^{\mathrm{dr\_max}} \tag{6.13}$$

$$\mathrm{SoC}_{i,t} = \mathrm{SoC}_{i,t-1} + (\eta_i^c \cdot p_{i,t}^{\mathrm{cha}} - p_{i,t}^{\mathrm{dis}}/\eta_i^d)/E_i^{\max} \tag{6.14}$$

$$\mathrm{SoC}_{i,0} = \mathrm{SoC}_{i,T} \tag{6.15}$$

$$\mathrm{SoC}_i^{\min} \le \mathrm{SoC}_{i,t} \le \mathrm{SoC}_i^{\max} \tag{6.16}$$

where $p_{\mathrm{i}}^{\mathrm{re\_max}}$, $p_i^{\mathrm{gt\_max}}$, $p_i^{\mathrm{cha\_max}}$, $p_i^{\mathrm{dis\_max}}$, and $p_i^{\mathrm{dr\_max}}$ are the upper limit of RE generation, GT power, ES charging power, ES discharging power, and DR power, respectively. $p_i^{\mathrm{gt\_min}}$ is the lower limit of GT power. $E_i^{\max}$ is the maximum capacity of the ES. $\mathrm{SoC}_{i,t}$ is the state of charge (SoC) of the ES. $\mathrm{SoC}_{i,0}$ and $\mathrm{SoC}_{i,T}$ are respectively the SoC at the beginning and end of the dispatch cycle. $\mathrm{SoC}_{\mathrm{i}}^{\max}$ and $\mathrm{SoC}_{\mathrm{i}}^{\min}$ are the upper and lower limits of the ES SoC, respectively.

(b) P2P energy trading constraints.

For any period *t*, the electricity sold by MG *i* to MG *j* should be equal to the electricity purchased by MG *j* from MG *i*, as Eq. (6.17).

$$p_{i,t}^{j} = -p_{j,t}^{i}, \quad \forall t \in T, \forall i \in N, j \in N \backslash i \tag{6.17}$$

(c) Power balance.

For any period *t*, the sum of electricity production and purchase by MG *i* is equal to the sum of load demand and electricity sales, as shown in Eq. (6.18). $p_{i,t}^{\mathrm{load}}$ is the base load in MG *i* at period *t*.

$$p_{i,t}^{\mathrm{re}} + p_{i,t}^{\mathrm{gt}} + p_{i,t}^{\mathrm{dr}} + p_{i,t}^{\mathrm{dis}} + p_{i,t}^{b} - p_{i,t}^{\mathrm{cha}} - p_{i,t}^{s} + \sum_{i \in N, j \in N \setminus i} p_{i,t}^{j} = p_{i,t}^{\mathrm{load}} \tag{6.18}$$

3. *Optimization Objective Function*

MG i participates in market transactions to make decisions that minimize its operating cost or maximize total profit. The objective function is depicted in Eq. (6.19).

$$F_i = \sum_{t=1}^{T} \left(C_{i,t}^{\mathrm{gt}} + C_{i,t}^{\mathrm{es}} + C_{i,t}^{\mathrm{dr}} + C_{i,t}^{\mathrm{grid}} + C_{i,t}^{\mathrm{p2p}} + C_{i,t}^{\mathrm{tr}} - C_{i,t}^{\mathrm{sell}} \right) \tag{6.19}$$

The multi-MGs energy sharing alliance's optimization goal is to reduce total operational costs. It should be noticed that the signs of the transaction power between two MGs are opposite (see Eq. (6.17), with the purchase being positive and the sale being negative), even though the transaction prices are equal. Thus, the P2P trading costs (see Eq. (6.6)) between MGs cancel out ($\sum_{i \in N} \sum_{j \in N \setminus i} \lambda_{i,t}^{j} p_{i,t}^{j} = 0$). The objective function of market clearing is shown as Eq. (6.1), with the constraints set $\Phi = \{(6.9)\text{–}(6.18)\}$.

$$F = \min \sum_{i \in N} F_i = \min \sum_{i}^{N} \sum_{t}^{T} \left(C_{i,t}^{\mathrm{gt}} + C_{i,t}^{\mathrm{es}} + C_{i,t}^{\mathrm{dr}} + C_{i,t}^{\mathrm{grid}} + C_{i,t}^{\mathrm{tr}} - C_{i,t}^{\mathrm{sell}} \right) \tag{6.20}$$

6.2.2 Real-Time Scheduling Model

1. *Re-scheduling cost*

Day-ahead scheduling might fail during the intraday stage due to variations in RE output. MGs must reschedule controllable resources like GT and ES based on day-ahead plans to avoid high real-time market balancing costs. Therefore, MGs need to reserve space for intraday rescheduling when making day-ahead decisions. An affine strategy is used by the MG operator to adjust GT and ES power to handle RE prediction errors. By assigning participation factors, each resource takes on a portion of the rescheduling power, as detailed in Eqs. (6.21)–(6.26).

$$\Delta p_{i,t}^{\mathrm{gt}} = -\alpha_{i,t}^{\mathrm{gt}} \cdot \omega_{i,t}^{\mathrm{re}} \tag{6.21}$$

$$\Delta p_{i,t}^{\mathrm{cha}} = \alpha_{i,t}^{\mathrm{cha}} \cdot \omega_{i,t}^{\mathrm{re}} \tag{6.22}$$

$$\Delta p_{i,t}^{\mathrm{dis}} = -\alpha_{i,t}^{\mathrm{dis}} \cdot \omega_{i,t}^{\mathrm{re}} \tag{6.23}$$

$$p_{i,t}^{\text{gt\_}r} = p_{i,t}^{\text{gt}} + \Delta p_{i,t}^{\text{gt}} \tag{6.24}$$

$$p_{i,t}^{\text{cha\_}r} = p_{i,t}^{\text{cha}} + \Delta p_{i,t}^{\text{cha}} \tag{6.25}$$

$$p_{i,t}^{\text{dis\_}r} = p_{i,t}^{\text{dis}} + \Delta p_{i,t}^{\text{dis}} \tag{6.26}$$

where $\omega_{i,t}^{\text{re}}$ is the RE output deviation in MG i in period t. $\Delta p_{i,t}^{\text{gt}}$, $\Delta p_{i,t}^{\text{cha}}$, and $\Delta p_{i,t}^{\text{dis}}$ are the rescheduling power of GT, ES charging, and ES discharging in real-time stages, respectively. $\alpha_{i,t}^{\text{gt}}$, $\alpha_{i,t}^{\text{gt}}$, and $\alpha_{i,t}^{\text{dis}}$ are, respectively, their participation factors. They are also decision variables with constraints Eq. (6.27). $p_{i,t}^{\text{gt\_}r}$, $p_{i,t}^{\text{cha\_}r}$, and $p_{i,t}^{\text{dis\_}r}$ are, respectively, their rescheduled powers.

$$\begin{cases} 0 \le \alpha_{i,t}^{\text{gt}} \le 1 \\ 0 \le \alpha_{i,t}^{\text{cha}} \le 1 \\ 0 \le \alpha_{i,t}^{\text{dis}} \le 1 \\ \alpha_{i,t}^{\text{gt}} + \alpha_{i,t}^{\text{dis}} - \alpha_{i,t}^{\text{cha}} = 1 \end{cases} \tag{6.27}$$

The rescheduling cost C_i^{real} of MG i can be expressed as Eq. (6.28).

$$C_i^{\text{real}} = \sum_{t=1}^{T} [a_i \Delta p_{i,t}^{\text{gt}} + b_i + c_i^{\text{es}} (\Delta p_{i,t}^{\text{cha}} + \Delta p_{i,t}^{\text{dis}})] \tag{6.28}$$

2. *Constraints*

In addition to Eq. (6.27), the rescheduling power of MG i in the real-time stage must meet the operating constraints $\Theta = \{(6.29)\text{–}(6.35)\}$.

$$p_{i,t}^{\text{gt\_ min}} \le p_{i,t}^{\text{gt\_}r} \le p_{i,t}^{\text{gt\_ max}} \tag{6.29}$$

$$0 \le p_{i,t}^{\text{cha\_}r} \le p_i^{\text{cha\_ max}} \tag{6.30}$$

$$0 \le p_{i,t}^{\text{dis\_}r} \le p_i^{\text{dis\_ max}} \tag{6.31}$$

$$\text{SoC}_{i,t}^r = \text{SoC}_{i,t-1}^r + (\eta_i^c \cdot p_{i,t}^{\text{cha\_}r} - p_{i,t}^{\text{dis\_}r} / \eta_i^d) / E_i^{\text{max}} \tag{6.32}$$

$$\text{SoC}_{i,0}^r = \text{SoC}_{i,T}^r \tag{6.33}$$

$$\text{SoC}_i^{\text{min}} \le \text{SoC}_{i,t}^r \le \text{SoC}_i^{\text{max}} \tag{6.34}$$

$$p_{i,t}^{\text{re}} + p_{i,t}^{\text{gt}\_r} + p_{i,t}^{\text{dr}} + p_{i,t}^{\text{dis}\_r} + p_{i,t}^{b} - p_{i,t}^{\text{cha}\_r} - p_{i,t}^{s} + \sum_{i \in N, j \in N \backslash i} p_{i,t}^{j} = p_{i,t}^{\text{load}} \tag{6.35}$$

where the superscript r indicates the variables in the real-time stage. Equations (6.29)–(6.31) represent the rescheduled power constraints of GT, ES charging, and ES discharging power respectively. Equations (6.32)–(6.34)are the SoC constraints of ES in the real-time stage. Equation (6.35) is the power balance constraint considering the real-time rescheduling.

6.3 DRO-Based Trading Models of Multiple MGs Considering RE Uncertainty

6.3.1 Principle of DRO

The objective function of DRO is generally the min–max problem shown in Eq. (6.36).

$$\min_{x \in X} \sup_{P \in \Gamma} \{E_P[f(\boldsymbol{x}, \omega)]\} \tag{6.36}$$

where $\boldsymbol{x}$ is the decision variable vector and X is the decision variable set. ω is the uncertain parameter, and P is its probability distribution. Γ is a fuzzy dataset constructed from the historical data of ω. First, according to the historical data of ω, construct a fuzzy set including its true distribution. Then, under the worst distribution scenario of ω in Γ, find the solution of $\boldsymbol{x}$ that optimizes the expected value of the objective function $f(\boldsymbol{x}, \omega)$.

Two primary methods for constructing fuzzy sets are based on moment information and probability distance, with their pros and cons discussed in [29]. This chapter uses the Wasserstein probability distance to create the RE uncertainty set, offering several advantages: it doesn't need prior knowledge of parameter distribution, avoiding uncertainties in statistical inference, and the DRO model using Wasserstein distance can be converted into a solvable convex optimization problem through dual theory.

To create a fuzzy set using probability distance, the process involves estimating the empirical distribution of uncertain parameters from historical data. Then, assuming the real distribution lies within a certain range, a probability sphere is formed with the empirical distribution at its center. Mathematically, fuzzy sets can be described as shown in Eq. (6.37).

$$\Gamma := \{P \in \Psi | d(P_K, P) \le \varepsilon\} \tag{6.37}$$

$$\begin{cases} \sup\limits_{P\in\Gamma} E_P[h(\boldsymbol{x},\omega)] = \sup\limits_{P_k\in\Gamma} \frac{1}{K}\sum\limits_{k=1}^{K}\int\limits_{\omega\in\Xi} h(\boldsymbol{x},\omega)P_k(\mathrm{d}\omega) \\ \text{s.t. } \frac{1}{K}\sum\limits_{k=1}^{K}\int\limits_{\omega\in\Xi} \|\omega_k-\omega\|P_k(\mathrm{d}\omega) \le \varepsilon \end{cases} \tag{6.45}$$

The decision variable in the aforementioned model is the probability distribution, which is challenging to solve. The model is, therefore, further converted into Eq. (6.46), in accordance with the strong duality theory. γ is the dual variable. The pre-and post-transformation models are convex optimization problems. However, the latter is easier to resolve.

$$\sup_{P\in\Gamma} E_P[h(\boldsymbol{x},\omega)] = \inf_{\gamma\ge 0}\left\{\gamma\varepsilon + \frac{1}{K}\sum_{k=1}^{K}\sup_{\omega\in\Xi}(h(\boldsymbol{x},\omega)-\gamma\|\omega_k-\omega\|)\right\} \tag{6.46}$$

Further introducing auxiliary variables τ_k, the above formula can be reshaped as:

$$\begin{aligned} &\inf_{\gamma\ge 0}\left\{\gamma\varepsilon + \frac{1}{K}\sum_{k=1}^{K}\tau_k\right\} \\ &\text{s.t. } \sup_{\omega\in\Xi}(h(\boldsymbol{x},\omega)-\gamma\|\omega_k-\omega\|) \le \tau_k,\quad \forall k\in K \end{aligned} \tag{6.47}$$

The supremum of the constraints Eq. (6.47) can only be obtained at the boundary of the uncertain variable or ω_k. Accordingly, the above model is transformed into a linear programming problem Eq. (6.48).$\overline{\omega}$ and $\underline{\omega}$ are the upper and lower bounds of RE forecast uncertainty ω, respectively.

$$\inf_{\gamma\ge 0}\left\{\gamma\varepsilon + \frac{1}{K}\sum_{k=1}^{K}\tau_k\right\}\text{s.t.}\begin{cases} h(\boldsymbol{x},\overline{\omega}) - \gamma||\overline{\omega}-\omega_k|| \le \tau_k, & \forall k\in K \\ h(\boldsymbol{x},\underline{\omega}) - \gamma||\underline{\omega}-\omega_k|| \le \tau_k, & \forall k\in K \\ h(\boldsymbol{x},\omega_k) \le \tau_k, \quad \forall k =\in K \end{cases} \tag{6.48}$$

From the model in Sect. 6.3, it is clear that $l(\boldsymbol{x},\omega)$ is also a linear polynomial whose optimal solution must be found at the boundary of ω. Accordingly, the MG optimal scheduling model is finally transformed into the following form:

$$\min \gamma\varepsilon + \frac{1}{K}\sum_{k=1}^{K}\tau_k \tag{6.49}$$

$$\text{s.t.}\ \begin{cases} Ax = \boldsymbol{b} \\ h(\boldsymbol{x}, \overline{\omega}) - \gamma \|\overline{\omega} - \omega_k\| \le \tau_k, & \forall k \in K \\ h(\boldsymbol{x}, \underline{\omega}) - \gamma \|\omega - \omega_k\| \le \tau_k, & \forall k \in K \\ h(\boldsymbol{x}, \omega_k) \le \tau_k, \quad \forall k \in K \\ \gamma \ge 0 \\ El(\boldsymbol{x}, \overline{\omega}) - \boldsymbol{d} \le 0 \\ El(\boldsymbol{x}, \underline{\omega}) - \boldsymbol{d} \le 0 \end{cases} \tag{6.50}$$

So far, the DRO model has been transformed into an easy-to-solve linear convex optimization problem that can be solved with commercial software solvers.

6.4 Distributed Algorithms

The optimization model for one MG is extended to multi-MGs, and the overall optimization model is as (6.51).

$$\min_{x} \sum_{i \in N} \sup_{P_i \in \Gamma_i} E_{P_i}[h_i(\boldsymbol{x}_i, \omega_i)] = \min F + \sum_{i \in N} \left(\boldsymbol{\gamma}_i \boldsymbol{\varepsilon}_i + 1/K \sum_{k \in K} \tau_{i,k} \right) \tag{6.51}$$

The model constraints include Eq. (6.17). and Eq. (6.50). for P2P trading between multiple MGs. While Eq. (6.51). can be solved centrally, it requires extensive information exchange. Moreover, the model ignores internal P2P trading costs (Sect. 6.5.2), challenging cost allocation. Additionally, since each MG operates independently, ensuring the privacy and security of MGs is difficult.

An ADMM-based distributed solution is proposed to address these issues, allowing each MG to make independent decisions. In each iteration, only the P2P trading price and power are exchanged between MGs, preserving privacy and reducing communication load.

The ADMM algorithm's principle and convergence proof are detailed in [33]. It is used to solve the multi-MG trade problem presented in this chapter. Except for the global coupling constraints in Eq. (6.17). Related to P2P energy trading, all other constraints pertain only to each MG's decision variables. Thus, auxiliary variables are introduced to decouple the constraints, as shown in Eqs. (6.52)–(6.53), aligning with the ADMM algorithm format.

$$p_{i,t}^{j,\text{aux}} + p_{j,t}^{i,\text{aux}} = 0, \quad \forall t \in T, i \in N, j \in N \backslash i \tag{6.52}$$

$$p_{i,t}^{j} = p_{i,t}^{j,\text{aux}}, \quad \forall t \in T, i \in N, j \in N \backslash i \tag{6.53}$$

where $p_{i,t}^{j,\text{aux}}$ and $p_{j,t}^{i,\text{aux}}$ are the introduced auxiliary variables. From an economic viewpoint, the Lagrange multipliers related to the P2P trading constraints represent the shadow price of the trading power. Let $\boldsymbol{y}_{i,t} := \{p_{i,t}^{j,\text{aux}}\}_{j\in N\backslash i}$ the Lagrange multiplier corresponding to the constraint Eq. (6.53).is $\boldsymbol{\lambda}_{i,t} := \{\lambda_{i,t}^{j}\}_{j\in N\backslash i}$. Then, the augmented Lagrangian function of the optimization problem is shown in Eq. (6.54).

$$L(\boldsymbol{x}, \boldsymbol{y}, \boldsymbol{\lambda}) = \sum_{i\in N}\sum_{t\in T} L_{i,t}(\boldsymbol{x}_{i,t}, \boldsymbol{y}_{i,t}, \boldsymbol{\lambda}_{i,t}) \tag{6.54}$$

$$\begin{aligned} L_{i,t}(\boldsymbol{x}_{i,t}, \boldsymbol{y}_{i,t}, \boldsymbol{\lambda}_{i,t}) &= F_i + \gamma_i \varepsilon_i + 1/K \sum_{k\in K} \tau_{i,k} \\ &+ \sum_{i\in N\backslash j} \left[(\lambda_{i,t}^{j})^T (p_{i,t}^{j} - p_{i,t}^{j,\text{aux}}) + \frac{\rho}{2} \left\| p_{i,t}^{j} - p_{i,t}^{j,\text{aux}} \right\|_2^2 \right] \end{aligned} \tag{6.55}$$

where ρ is the penalty factor. Let v be the iteration times. Then the iterative process of the ADMM algorithm can be decomposed into the following three steps, as presented in Algorithm 1.

Algorithm 1

Inputs: RE and load forecasts $(\boldsymbol{p}_i^{re}, \boldsymbol{p}_i^{load})$, electricity price of retailer $(\boldsymbol{\lambda}^b, \boldsymbol{\lambda}^s)$, and other preset parameters for MGs.
Initialize $\boldsymbol{y}_i^{(1)} = 0, \boldsymbol{\lambda}_i^{(1)} = 0, \rho = 1.3, \xi_1 = \xi_2 = 1e^{-3}, \mathrm{V} = 100$.
for $v = 1$ *to* V **do**
  (1) **Step 1**: Each MG independently optimizes locally in parallel for $\boldsymbol{x}_i^{(v+1)}$ with $\boldsymbol{y}_i^{(v)}$ and $\boldsymbol{\lambda}_i^{(v)}$.
  (2) **Step 2**: MGs update auxiliary variable $\boldsymbol{y}_i^{(v+1)}$ locally.
  (3) If the convergence conditions are met?
    if $\|r^{(v)}\|_2^2 \leq \xi_1$, and $\|s^{(v)}\|_2^2 \leq \xi_2$ **then**
      Problem is solved, and stop iteration.
    else
      if $v > \mathrm{V}$ **then**
        The model is unsolvable. The MGs will have to trade with the retailors with $(\boldsymbol{\lambda}^s, \boldsymbol{\lambda}^b)$.
      end if
    end if
  (4) **Step 3**: Update $\boldsymbol{\lambda}_i^{(v+1)}$ with $\boldsymbol{y}_i^{(v+1)}$.
  (5) $v \leftarrow v+1$, and return to (1).
end for

Step 1: MG i updates local decision variables.

Each MG i independently updates the local decision variable $\boldsymbol{x}$ by solving Eq. (6.56). according to the auxiliary variable $\boldsymbol{y}_{i,t}^{(v)}$ and the Lagrange multiplier $\boldsymbol{\lambda}_{i,t}^{(v)}$ obtained from the previous iteration v.

$$\boldsymbol{x}_{i,t}^{(v+1)} := \arg\min_{x_{i,t}} \sum_{t\in T} L_{i,t}(\boldsymbol{x}_{i,t}, \boldsymbol{y}_{i,t}^{(v)}, \lambda_{i,t}^{(v)}) p_{i,t}^{j(v)} \tag{6.56}$$

The constraints are shown in Eq. (6.50). After updating the local variables $\boldsymbol{x}$, MG i sends the new P2P trading volume $p_{i,t}^{j}(v)$ to MG j via the energy-sharing platform.

Step 2: Update auxiliary variables.

The update of the auxiliary variable $\boldsymbol{y}_{i,t}$ needs to solve the global optimization problem Eq. (6.57). with constraints Eq. (6.52).

$$\min_{\boldsymbol{y}} L(\boldsymbol{x}^{(v+1)}, \boldsymbol{y}, \boldsymbol{\lambda}^{(v)}) \tag{6.57}$$

The above optimization problem should be shifted to further achieve decentralized transactions between MGs and secure their privacy. Since $p_{i,t}^{j,\mathrm{aux}}$ is only coupled to $p_{j,t}^{i,\mathrm{aux}}$, the objective function is decomposable. Therefore, they can be obtained by solving Eq. (6.58). with constraints Eq. (6.52).

$$\min_{p_{i,t}^{j,\mathrm{aux}}} -(\lambda_{i,t}^{j})^{T} p_{j,t}^{i,\mathrm{aux}} - (\lambda_{j,t}^{i})^{T} p_{i,t}^{j,\mathrm{aux}} + \frac{\rho}{2}\left\| p_{i,t}^{j,\mathrm{aux}} - p_{i,t}^{j} \right\|_2^2 + \frac{\rho}{2}\left\| p_{j,t}^{i,\mathrm{aux}} - p_{j,t}^{i} \right\|_2^2 \tag{6.58}$$

The analytical solution of $p_{i,t}^{j,\mathrm{aux}}$ is achieved with the Karush–Kuhn–Tucker (KKT) condition to improve the solving speed of the ADMM algorithm. During the *(v + 1)th* iteration, it can be calculated with Eq. (6.59). With this transformation, after receiving $p_{i,t}^{j(v+1)}$ and $\lambda_{i,t}^{(v)}$, MG i can update $p_{i,t}^{j,\mathrm{aux}}$ independently without relying on the platform for global optimization, realizing a fully distributed solution.

$$p_{i,t}^{j,\mathrm{aux},(v+1)} = -p_{j,t}^{i,\mathrm{aux},(v+1)} = \frac{p_{i,t}^{j,(v+1)} - p_{j,t}^{i,(v+1)}}{2} + \frac{\lambda_{i,t}^{j,(v)} - \lambda_{j,t}^{i,(v)}}{2\rho} \tag{6.59}$$

Step 3: Update Lagrange Multiplier $\lambda_{i,t}$.

MG i updates $\lambda_{i,t}^{j}$ independently and sends the price signal to the corresponding MG j via the platform.

$$\lambda_{i,t}^{j,(v+1)} = \lambda_{i,t}^{j,(v)} + \rho\left(p_{i,t}^{j,(v+1)} - p_{i,t}^{j,\mathrm{aux},(v+1)}\right) \tag{6.60}$$

Equations (6.61) and (6.62) show the convergence criteria of the ADMM algorithm: primal residual and dual residual, respectively. ξ_1 and ξ_2 are the set error accuracy.

$$r^{(v)} = \sqrt{\sum_{i\in N}\sum_{j\in N\backslash i}\sum_{t\in T}(p_{i,t}^{j,(v)} - p_{i,t}^{j,\mathrm{aux},(v)})^2} \le \xi_1 \tag{6.61}$$

$$s^{(v)} = \sqrt{\sum_{i\in N}\sum_{j\in N\backslash i}\sum_{t\in T}(p_{i,t}^{j,\mathrm{aux},(v)} - p_{i,t}^{j,\mathrm{aux},(v-1)})^2} \le \xi_2 \tag{6.62}$$

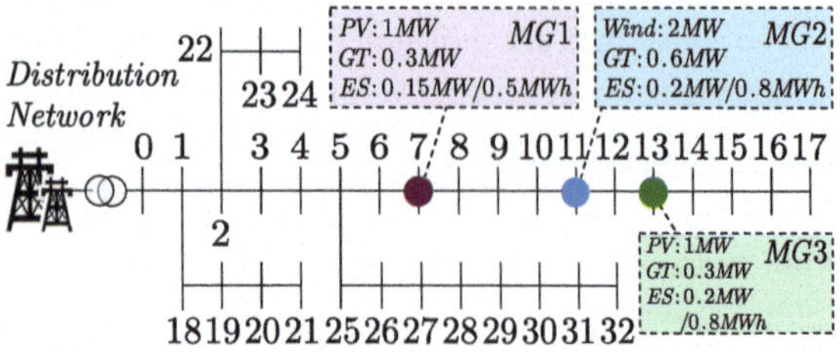

Fig. 6.2 Resource configuration of three MGs

The $p_{i,t}^{j}$ and $\lambda_{i,t}^{j}$ obtained after satisfying the convergence criterion are the P2P trading power and price between MG i and MG j in the period t.

6.5 Case Studies

6.5.1 *System Modeling and Parameter Setting*

Using the IEEE-33 node distribution network as an example, three MGs are positioned at nodes 7, 11, and 13. Figure 6.2 shows their unit capacity configurations, and Fig. 6.3 depicts their RE output and load prediction curves. The GT unit has a minimum power output of 30 kW, and the maximum load reduction is 10% of the predicted load per period. ES units have charge and discharge efficiencies of 0.95. The initial ES SoC is 60%, ranging from 10 to 90% of the rated capacity. The adjustment costs for GT and ES are $90/MWh and $10/MWh, respectively. Internal power sale prices for MG1, MG2, and MG3 are $82/MWh, $90/MWh, and $80/MWh, respectively. The demand response subsidy is 20% of the electricity sales price. Time-of-use retail prices for the distribution network are $80/MWh during off-peak hours (23:00–6:00) and $100/MWh during peak hours. The feed-in tariff for MG power surplus is $40/MWh. Electrical distances between the three nodes are listed in Table 6.1, with a unit distance transmission fee of $10/MWh.

Each MG is assumed to have historical data on RE output, load demand, electricity pricing, and various meteorological data stored in a SCADA system. The SCADA system described in [34] is suitable for predicting and analyzing DERs in MGs.

6.5.2 *P2P Energy Trading Results*

Figure 6.4 illustrates the day-ahead trading results for MG1, MG2, and MG3. The complementary outputs of wind turbines and PV units enable the three MGs to benefit from P2P exchanges. At night, when wind turbine output is high, MG2 sells excess power to MG1 and MG3 while meeting its own load. During the day, when PV output is high, MG1 and MG3 sell excess electricity to MG2. MGs trade with the

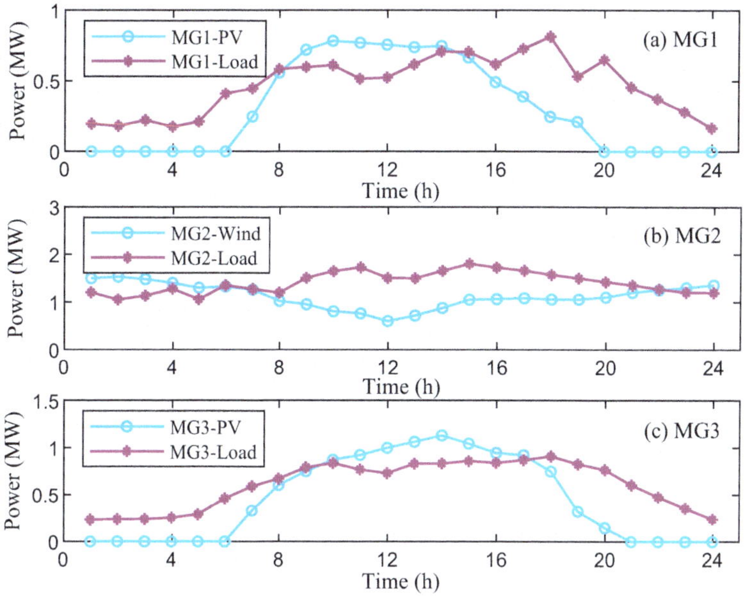

Fig. 6.3 Day-ahead predicted RE output and load demand in three MGs

Table 6.1 Electrical distance between three MGs

	MG1-MG2	MG2-MG3	MG1-MG3
Distance (km)	1.11	0.79	1.9

distribution network only when there is a surplus or shortfall in electricity after P2P trading to balance supply and demand. For instance, in periods 1–3 h, MG2 has extra power to sell to retailers after fulfilling MG1 and MG3's load demands through P2P trading. In period 6 h, P2P trading alone is insufficient to meet MG1 and MG3's load demands, so they purchase electricity from retailers. This indicates a preference for P2P trading over trading with retailers.

Figure 6.5 shows the P2P energy trading price, the prices of electricity bought, and sold from the retailer using the ADMM algorithm proposed in Sect. 6.5.4. The distribution network's Retail electricity price and feed-in tariff are labeled A and B, respectively, and the P2P price between MG i and MG j is labeled C. As observed, there is no P2P trading between MG1 and MG3 because their supply and demand roles are similar, resulting in no P2P price.

Figure 6.5 also illustrates the relationship between P2P trading price trends and the quantity of supply and demand within the energy-sharing alliance. When internal energy is ample, the P2P price is lower, and vice versa, with the price ranging between D and E. Thus, participants in P2P trading can achieve higher returns than direct retailer trading. This supports the effectiveness of the proposed pricing strategy in facilitating energy trading within the distribution grid.

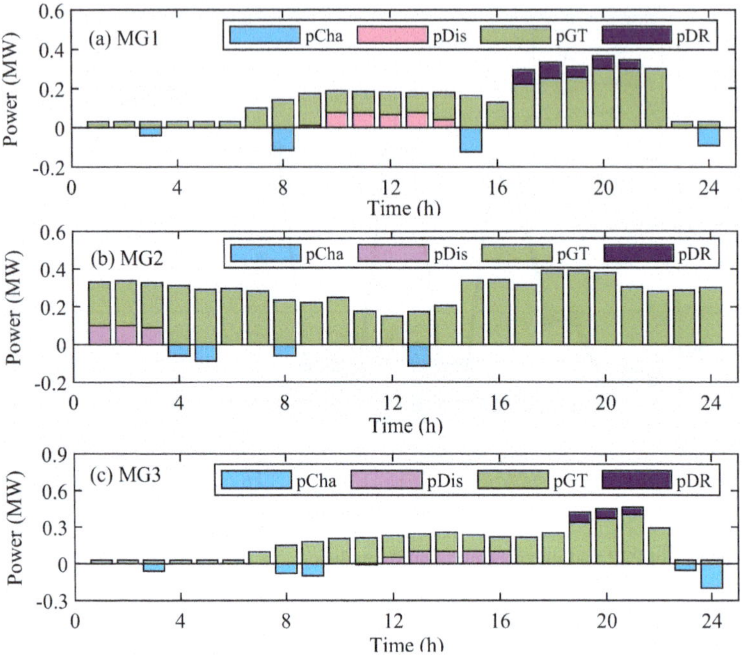

Fig. 6.6 Day-ahead optimal scheduling results of DERs in MGs

6.5.4 Performance Analysis of the ADMM Algorithm

Figure 6.7 displays the ADMM algorithm results. The iterative process for P2P trading power and price between MG2 and MG3 at 1 h is shown in Fig. 6.7a and b, respectively, converging after 40 iterations. The convergence of primal and dual residuals is shown in Fig. 6.7c and d, respectively, achieving high accuracy at 90 iterations and meeting the error requirements. Optimizing one MG in a loop takes about 2.8 s, resulting in a solution time of 15 min in a serial computing environment and less than 5 min in a parallel environment. Thus, the proposed ADMM algorithm demonstrates good performance and convergence speed, suitable for practical applications.

6.5.5 Effectiveness of the DRO Model

To verify the DRO model's effectiveness, 5000 additional sample data sets were used as a test set to check the model's solution adaptability. Figure 6.8 shows the total social welfare for in-sample and out-of-sample performance of the DRO model under different sample sizes.

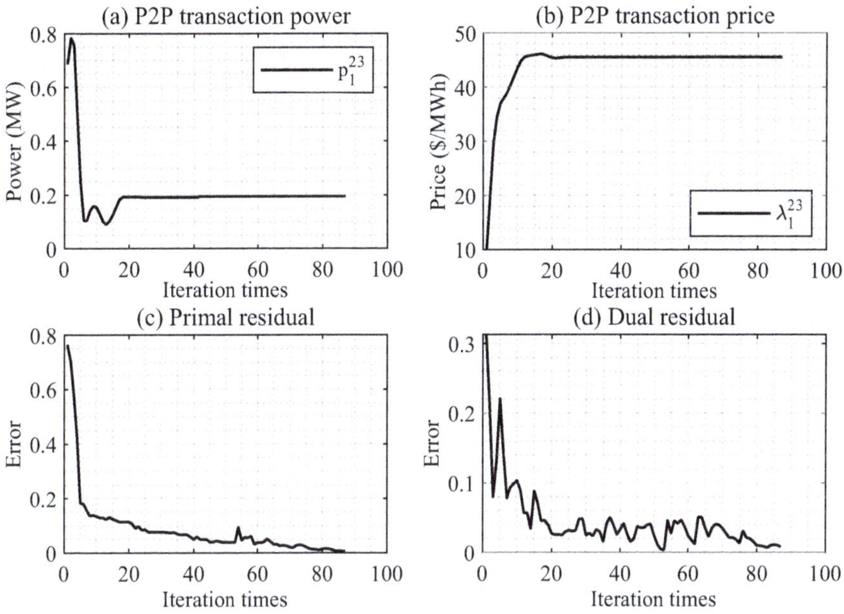

Fig. 6.7 ADMM algorithm results: **a** P2P trading power; **b** P2P trading price; **c** convergence of primal residual; **d** convergence of dual residual

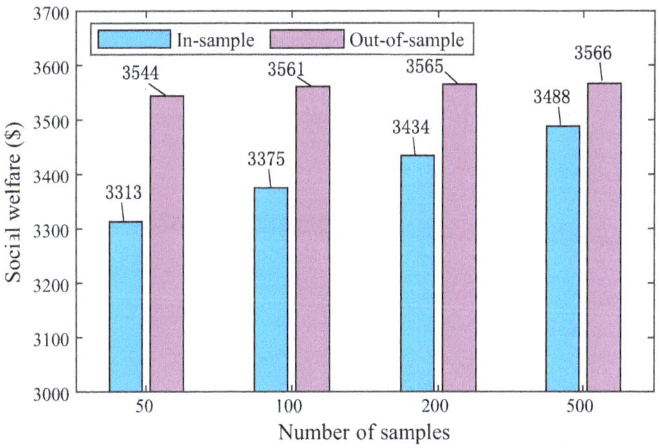

Fig. 6.8 Effectiveness of the DRO Model for various sample sizes

As the number of samples increases, the model's conservatism decreases, overall social welfare improves, and in-sample welfare approaches out-of-sample results. Optimization results based on in-sample data are always lower than out-of-sample performance because the DRO model considers the worst-case distribution within the fuzzy set. This proves the DRO model's robustness. As the sample size grows, the DRO results converge more smoothly, stabilizing with over 100 samples. This highlights the advantage of the Wasserstein distance-based DRO method, which can make good decisions even with limited historical data by constructing fuzzy sets.

6.5.6 Comparison of DRO Models with SP and RO

Figure 6.9 compares total social welfare calculated using RO, DRO, and SP methods for different sample sizes. DRO results consistently fall between RO and SP, aligning with the discussed theory. RO's conclusions are overly conservative, focusing on the worst-case scenario. SP aims to maximize expected profit, often producing overly optimistic estimates due to its disregard for extreme events. DRO balances probability distribution and worst-case scenarios, yielding conservative yet cost-effective results.

Table 6.3 shows total social welfare for various sample sizes using different methods. While SP's in-sample result (e.g., \$3499 with 500 samples) is better than DRO (\$3488), its out-of-sample performance is weaker (\$3482 compared to \$3567 for DRO). This discrepancy arises because SP optimizes based solely on empirical data, leading to poor outcomes with smaller samples. As sample sizes increase, in-sample and out-of-sample results converge. Thus, without knowledge of uncertain parameters' distribution, DRO performs better.

Table 6.4 compares the solution times for a single ADMM algorithm iteration. RO has the shortest iteration time, unaffected by sample size increase, as it optimizes the worst-case scenario with limited variables and constraints. However, Figure 6.9 shows that RO's overly conservative decisions significantly reduce revenue in practice. SP requires the longest time to solve, with efficiency declining as sample sizes

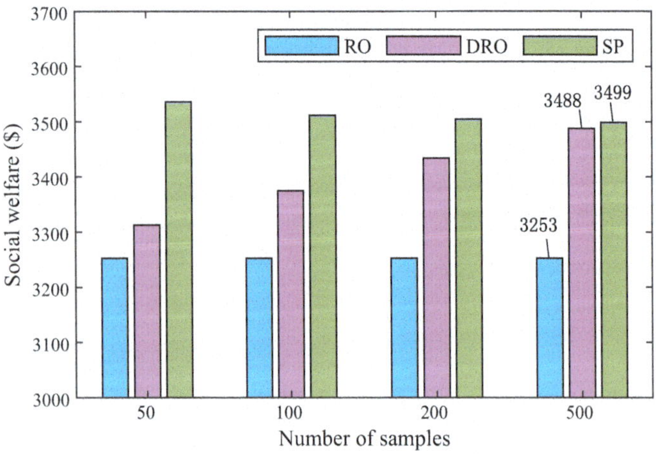

Fig. 6.9 Comparison of DRO Models with SP and RO for various sample sizes

Table 6.3 Optimized profit of DRO and SP with different sample sizes

Sample size	Methods			
	DRO ($)		SP ($)	
	In-sample	Out-of-sample	In-sample	Out-of-sample
50	3313	3544	3536	3421
100	3375	3561	3512	3458
200	3434	3565	3505	3472
500	3488	3567	3499	3482

Table 6.4 Solution time for a single iteration with different methods

Methods	Samples			
	Solution time for a single iteration (s)			
	50	100	200	500
RO	1.12	1.51	1.81	2.32
DRO	2.12	2.83	4.71	6.35
SP	2.15	2.95	6.89	12.38

grow, due to the necessity to evaluate all scenarios and proportional increase in variables and constraints. DRO's solution time falls between RO and SP. According to the transformation of the DRO model in Sect. 6.5.5, only some variables and constraints increase with more samples, ensuring high solution efficiency.

In conclusion, the proposed DRO model is shown to be conservative and economical through comparative analysis in terms of conservatism, economy, and solution efficiency. It maintains good solution efficiency even with a large number of samples.

6.6 Conclusions

This chapter explored P2P energy trading among multiple MGs, factoring in renewable energy uncertainty. Initially, a day-ahead P2P trading model for MGs was developed. A Wasserstein distance-based DRO method for optimal scheduling was proposed to handle RE prediction errors. The ADMM algorithm was used to solve the energy-sharing model while preserving MG privacy. Numerical case studies revealed that: (1) P2P energy trading enhances MG profits, with trading prices influenced by internal supply and demand; lower prices occur when internal energy is abundant and higher when scarce. (2) Network usage costs promote trading RE with nearby consumers. (3) The ADMM algorithm achieves results comparable to a centralized solution after a few iterations. (4) The DRO method provides transaction results that balance economy and reliability. Finally, as the uncertainty set of parameters significantly impacts the model, future research will focus on improving the accuracy of the RE uncertainty set and the economic efficiency of the scheduling.

References

1. Schaeffer, R., Szklo, A. S., de Lucena, A. F. P., Borba, B. S. M. C., Nogueira, L. P. P., Fleming, F. P., & Boulahya, M. S. (2012). Energy sector vulnerability to climate change: A review. *Energy, 38*(1), 1–12.
2. Mielczarski, W. (2020, September). Is the climate neutrality an illusion? In *2020 17th International Conference on the European Energy Market* (EEM) (pp. 1–4). IEEE.
3. Gielen, D., Gorini, R., Wagner, N., Leme, R., Gutierrez, L., Prakash, G., et al. (2019). Global energy transformation: a roadmap to 2050.
4. Xu, T., Gao, W., Qian, F., & Li, Y. (2022). The implementation limitation of variable renewable energies and its impacts on the public power grid. *Energy, 239*, 121992.
5. Quint, R., Dangelmaier, L., Green, I., Edelson, D., Ganugula, V., Kaneshiro, R., Pigeon, J., Quaintance, B., Riesz, J., & Stringer, N. (2019). Transformation of the grid: The impact of distributed energy resources on bulk power systems. *IEEE Power and Energy Magazine, 17*(6), 35–45.
6. Adeyemi, A., Yan, M., Shahidehpour, M., Bahramirad, S., & Paaso, A. (2020). Energy markets for managing energy exchanges in power distribution systems. *The Electricity Journal, 33*(9), 106868.
7. Song, M., Cai, Y., Gao, C., Chen, T., Yao, Y., & Ming, H. (2022). Energy in power distribution systems: Paving the path towards cyber-physical-social system. *International Journal of Electrical Power & Energy Systems, 142*, 108289.
8. Liu, N., Yu, X., Wang, C., Li, C., Ma, L., & Lei, J. (2017). Energy-sharing model with price-based demand response for microgrids of peer-to-peer prosumers. *IEEE Transactions on Power Systems, 32*(5), 3569–3583.
9. Li, S., Zhu, J., & Dong, H. (2021). A novel energy sharing mechanism for smart microgrid. *IEEE Transactions on Smart Grid, 12*(6), 5475–5478.
10. Erol, Ö., & Filik, Ü. B. (2022). A Stackelberg game approach for energy sharing management of a microgrid providing flexibility to entities. *Applied Energy, 316*, 118944.
11. Wang, Y., Ai, X., Tan, Z., Yan, L., & Liu, S. (2015). Interactive dispatch modes and bidding strategy of multiple virtual power plants based on demand response and game theory. *IEEE Transactions on Smart Grid, 7*(1), 510–519.
12. Zhang, C., Wu, J., Zhou, Y., Cheng, M., & Long, C. (2018). Peer-to-peer energy trading in a microgrid. *Applied Energy, 220*, 1–12.
13. Iqbal, S., Nasir, M., Zia, M. F., Riaz, K., Sajjad, H., Khan, H. A., & Member, I. E. E. E. (2021). A novel approach for system loss minimization in a peer-to-peer energy sharing community DC microgrid. *International Journal of Electrical Power & Energy Systems, 129*, 106775.
14. Sorin, E., Bobo, L., & Pinson, P. (2018). Consensus-based approach to peer-to-peer electricity markets with product differentiation. *IEEE Transactions on Power Systems, 34*(2), 994–1004.
15. Khorasany, M., Mishra, Y., & Ledwich, G. (2019). A decentralized bilateral energy trading system for peer-to-peer electricity markets. *IEEE Transactions on Industrial Electronics, 67*(6), 4646–4657.
16. Baroche, T., Pinson, P., Latimier, R. L. G., & Ahmed, H. B. (2019). Exogenous cost allocation in peer-to-peer electricity markets. *IEEE Transactions on Power Systems, 34*(4), 2553–2564.
17. Yan, X., Gao, C., Song, M., Chen, T., Ding, J., Guo, M., Wang, X., & Abbes, D. (2022). An IGDT-based day-ahead co-optimization of energy and reserve in a VPP considering multiple uncertainties. *IEEE Transactions on Industry Applications, 58*(3), 4037–4049.

18. Hasankhani, A., & Hakimi, S. M. (2021). Stochastic energy management of smart microgrid with intermittent renewable energy resources in electricity market. *Energy, 219*, 119668.
19. Sheidaei, F., & Ahmarinejad, A. (2020). Multi-stage stochastic framework for energy management of virtual power plants considering electric vehicles and demand response programs. *International Journal of Electrical Power & Energy Systems, 120*, 106047.
20. Shinde, P., Kouveliotis-Lysikatos, I., & Amelin, M. (2022). Multistage stochastic programming for VPP trading in continuous intraday electricity markets. *IEEE Transactions on Sustainable Energy, 13*(2), 1037–1048.
21. Shabanzadeh, M., Sheikh-El-Eslami, M. K., & Haghifam, M. R. (2015). The design of a risk-hedging tool for virtual power plants via robust optimization approach. *Applied Energy, 155*, 766–777.
22. Seijo, S., del Campo, I., Echanobe, J., & García-Sedano, J. (2016). Modeling and multi-objective optimization of a complex CHP process. *Applied Energy, 161*, 309–319.
23. Gazijahani, F. S., & Salehi, J. (2019). IGDT-based complementarity approach for dealing with strategic decision making of price-maker VPP considering demand flexibility. *IEEE Transactions on Industrial Informatics, 16*(4), 2212–2220.
24. Mohajerin Esfahani, P., & Kuhn, D. (2018). Data-driven distributionally robust optimization using the Wasserstein metric: Performance guarantees and tractable reformulations. *Mathematical Programming, 171*(1), 115–166.
25. Zhu, R., Wei, H., & Bai, X. (2019). Wasserstein metric based distributionally robust approximate framework for unit commitment. *IEEE Transactions on Power Systems, 34*(4), 2991–3001.
26. Guo, Y., Baker, K., Dall'Anese, E., Hu, Z., & Summers, T. H. (2018). Data-based distributionally robust stochastic optimal power flow—Part I: Methodologies. *IEEE Transactions on Power Systems, 34*(2), 1483–1492.
27. Guo, Y., Baker, K., Dall'Anese, E., Hu, Z., & Summers, T. H. (2018). Data-based distributionally robust stochastic optimal power flow—Part II: Case studies. *IEEE Transactions on Power Systems, 34*(2), 1493–1503.
28. He, C., Zhang, X., Liu, T., & Wu, L. (2019). Distributionally robust scheduling of integrated gas-electricity systems with demand response. *IEEE Transactions on Power Systems, 34*(5), 3791–3803.
29. Chang, X., Xu, Y., Sun, H., & Khan, I. (2021). A distributed robust optimization approach for the economic dispatch of flexible resources. *International Journal of Electrical Power & Energy Systems, 124*, 106360.
30. Bertsimas, D., Natarajan, K., & Teo, C. P. (2006). Persistence in discrete optimization under data uncertainty. *Mathematical Programming, 108*, 251–274.
31. Duan, C., Fang, W., Jiang, L., Yao, L., & Liu, J. (2018). Distributionally robust chance-constrained approximate AC-OPF with Wasserstein metric. *IEEE Transactions on Power Systems, 33*(5), 4924–4936.
32. Zhao, C., & Guan, Y. (2015). Data-driven stochastic unit commitment for integrating wind generation. *IEEE Transactions on Power Systems, 31*(4), 2587–2596.
33. Boyd, S., Parikh, N., Chu, E., Peleato, B., & Eckstein, J. (2011). Distributed optimization and statistical learning via the alternating direction method of multipliers. *Foundations and Trends® in Machine Learning, 3*(1), 1–122.
34. Yan, X., Abbes, D., & Francois, B. (2018). Development of a tool for urban microgrid optimal energy planning and management. *Simulation Modelling Practice and Theory, 89*, 64–81.

Part III
Local Energy Market Constrained by the Distribution Network

Chapter 7
Distribution Network-Constrained Local Peer-to-Peer Trading Among Multi-microgrids

Abstract This chapter introduces a two-layer network-constrained peer-to-peer (P2P) energy framework for multi-microgrids (MGs) systems. The framework aims to maintain the security of the distribution power network while enabling flexible energy trading between MGs. In the lower layer, MGs engage in energy trading through a P2P manner. The trading dynamics are emulated using a multi-leader multi-follower (MLMF) Stackelberg game model, while we prove the existence and uniqueness of the Stackelberg equilibrium (SE) and provide an explicit formula. A set of distributed algorithms is proposed to address privacy issues to determine the SE. In the upper layer, the distribution system operator (DSO) adjusts the network operation based on the outcomes of the P2P energy transactions, employing AC optimal power flow techniques and considering network reconfiguration. If network congestion arise, the DSO may request adjustments at the lower level to ensure network security. The DSO's operational problem is reformulated into a mixed-integer second-order cone programming (MISOCP) model to ensure precision in solutions. Results from simulations of a 4-MG setup, as well as modified IEEE 33-bus and 123-bus distribution systems, illustrate the practical benefits of the proposed trading model and its solving method.

7.1 Introduction

Multi-microgrids (MGs) function as controllable loads, enabling energy transactions with the main grid and other MGs, thereby facilitating the integration of distributed energy resources (DERs) [1, 2]. Local energy market (LEM), which promotes and incentivizes energy trading among MGs, is viewed as an advanced energy management strategy for accommodating the growth of DERs [3, 4]. This concept is increasingly attracting interest from utility operations, as evidenced by various pilot projects. References [5–7] offer a framework for the design, implementation, and deployment of LEM, showing the real-world applications in Melbourne, Australia, and Brooklyn, New York.

M. Song et al., *Local Energy Markets*, https://doi.org/10.1007/978-981-97-9750-9_7

Numerous studies have focused on the development of LEMs. References [3, 8] proposed a LEM for multiple MGs using centralized optimization methods, where a market operator gathers all information from MGs and optimizes the DERs within them. However, a significant drawback of this centralized approach is the lack of privacy, as market operators can access sensitive information such as generator costs and DER output limits. To address privacy concerns, Refs. [9, 10] introduced decentralized optimization methods for LEMs, in which MGs share only minimal information with the market operator [11]. Additionally, Refs. [12, 13] suggested an auction-based LEM where MGs submit bids and offers to a third party, or auctioneer, who then clears the market and communicates the results back to each MG.

Recent advancements in information and communication technologies, such as blockchain, have facilitated the development of reliable peer-to-peer (P2P) trading for participants [14, 15]. In the framework of LEM, MGs can directly communicate with each other to trade energy. For example, Ref. [16] designed a bilateral contract network for P2P trading, introducing an associated market mechanism. However, this study did not consider power network constraints, which are crucial for energy trading markets. Ignoring these constraints can lead to impractical results. To address this, the inclusion of power network constraints in P2P energy trading among MGs is essential. Additionally, gaming methods, which have been extensively used in P2P trading, can model and analyze the strategic behavior and preferences of each market participant [4].

Reference [17] explored the use of the Stackelberg game in a P2P trading for multiple MGs and demonstrated the existence of SE. However, this study overlooked MG operation constraints, such as DER output limits. In contrast, Refs. [18–20] incorporated renewable DERs (e.g., solar PV) into the Stackelberg game framework and accounted for the output limits of these renewable resources. Nonetheless, these studies did not consider other types of non-renewable DERs, such as microturbines (MTs).

Reference [21] expanded the analysis to include multiple types of DERs (both renewable and non-renewable) in MGs. However, it did not address the heterogeneity among different DER types. For example, renewable DERs have very low generation costs as they do not consume fuel, while non-renewable DERs like microturbines have quadratic generation cost functions. Ignoring these heterogeneous characteristics can lead to impractical market equilibria. Additionally, the gaming approaches in Refs. [17–21] neglected power network constraints and often required participants to share proprietary data to achieve Stackelberg equilibrium (SE), raising further privacy concerns.

The following issues will be further discussed in our chapter when applying the Stackelberg game to P2P trading: (1) Heterogeneous Characteristics of DERs: The diverse attributes of different DER types will be considered, such as the low generation cost of renewable DERs versus the quadratic generation cost function of non-renewable DERs like microturbines. (2) Efficient Distributed Algorithm: We will develop an efficient distributed algorithm to achieve SE without revealing private information, addressing privacy concerns. (3) Power Network Constraints: Power network constraints will be incorporated into the P2P trading, ensuring practical and feasible energy trading outcomes. (4) Distribution system operator (DSO) Operation and Network Reconfiguration: The operations of the DSO, such as network reconfiguration, are integrated into the energy trading process. Network reconfiguration involves adjusting the topology of the distribution power system by controlling line switches, which helps in maintaining line flow limits, reducing power losses, and addressing voltage violations in distribution systems [22–24]. This chapter introduces a two-level network-constrained P2P scheme. This scheme ensures network security and encourages MGs to participate in energy trading. The key differences between this proposed method and existing P2P energy approaches are summarized in Table 7.1.

Figure 7.1 illustrates the framework of the proposed dual-layer P2P energy trading system. At the foundational level, P2P energy transactions are facilitated, where MGs engage in energy exchanges, establishing trade prices and quantities independently of a centralized market operator. The dynamics among these MGs are represented through a multi-leader, multi-follower Stackelberg game strategy. Here, sellers, in their role as leaders, propose energy prices and surplus quantities, while buyers, as followers, respond by indicating their energy requirements. Once the P2P market equilibrium is established at this foundational tier, the DSO at the upper echelon adjusts the distribution network configuration, informed by the trading outcomes from below. The DSO's role is strictly limited to overseeing the operational integrity of the distribution network, without direct involvement in market activities. To address the privacy issues, MGs only report their aggregate load to the DSO. For instances, where the trading outcomes do not align with the network's security constraints, such as voltage levels and power flow limits, the DSO communicates with the MGs to initiate an iterative process of trading adjustments. This ensures the system remains compliant with operational standards.

The contributions of this chapter are summarized as follows: (1) At the lower level, we introduce a P2P energy trading system for MGs, leveraging a multi-leader multi-follower (MLMF) Stackelberg game to model the trading dynamics. The model is innovatively expanded to accommodate a mix of renewable and non-renewable DERs. We establish the existence and uniqueness of the SE and provide a formulaic representation for calculating it, factoring in the diversity of DERs. Additionally,

Table 7.1 Comparative features of previous P2P studies

Reference	Gaming approach	Privacy protection	Heterogeneous characteristics of DERs	Network constraints or reconfiguration
[3, 8]	✖	✖	✓	✓
[9, 10]	✖	✓	✖	✓
[12, 13]	✖	✓	✖	✖
[17, 18, 20, 21]	✓	✖	✖	✖
[19]	✓	✓	✖	✖
Proposed method	✓	✓	✓	✓

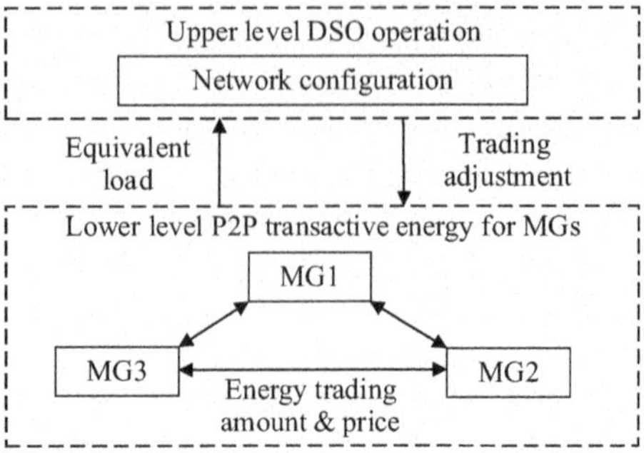

Fig. 7.1 Structure of the proposed two-level P2P trading

we introduce a suite of distributed algorithms designed to safeguard MG privacy, enabling them to converge to SE through an iterative process that only requires the disclosure of energy demand/surplus and pricing information. (2) At the upper level, the chapter details the DSO strategy for optimizing the distribution network. The traditional AC optimal power flow problem is reconceptualized as a mixed-integer second-order cone programming (MISOCP) model, streamlining the representation of network reconfiguration. The iterative sequential cone programming (SCP) technique is employed to enhance the precision of MISOCP solutions. The chapter showcases the effectiveness of this approach in both radial and meshed distribution network scenarios.

7.2 Lower-Level P2P Energy Trading

In this section, we delve into P2P trading at the lower level. The primary challenge in this domain is establishing the price and quantity for energy transactions between sellers and buyers. To address these challenges, we employ the MLMF Stackelberg game to model the decision-making interactions between sellers and buyers. Sellers, acting as leaders, present offers (referred to as optimal strategies) that include energy prices and surplus energy, while buyers, acting as followers, respond to these offers by specifying their energy demands. This process of P2P trading aims to achieve market equilibrium.

7.2.1 Game Method Analyses

We assume the P2P energy trading operates under perfect competition for several reasons: (1) Energy is a homogeneous product with no differentiation in terms of features and quality. The only distinction among energy provided by different sellers is the price. Consequently, buyers do not exhibit any preference for energy from different sellers. (2) The LEM comprises massive buyers and sellers within a distribution network, ensuring that no single buyer or seller can influence the market price. (3) Due to the limited capacity of DERs, no individual seller or buyer holds significant market power. Therefore, in a perfectly competitive market equilibrium, each seller's energy price will converge to the same level [25, 26]. Even if sellers initially set different prices, they will adjust their prices over time to reach equilibrium. Thus, the energy prices set by sellers in this market are:

$$\pi_{1,t} = \pi_{2,t} = \ldots = \pi_{s,t} = \pi_t \, \forall t \tag{7.1}$$

Constraint (7.1) is applied solely to analyze the behaviors of buyers and sellers when the trading market is at equilibrium. In Sect. 7.2.5, we introduce Algorithm 3, which demonstrates how sellers, starting with different energy prices, eventually converge to a uniform energy price. Moreover, the P2P trading market becomes imperfectly competitive when considering MG preferences, such as a preference for clean energy. In such an imperfectly competitive market, each participant's energy price may vary at equilibrium. The methodology for setting energy prices in an imperfectly competitive market is discussed in Sect. 7.4.1.

In this chapter, DER refers to MTs and renewable DERs, such as wind turbines and solar photovoltaics. Each MG is represented by its aggregated load, MTs, and renewable DERs. MGs make decisions on energy exchanges based on the prices offered by other MGs and the DSO. For MG i at hour t, the MG operation problem is formulated as follows:

$$\max \; U_{i,t}(PD_{i,t}) + \pi_t(PS_{i,t} - PB_{i,t}) - \pi_t^{\max} PB_{i,t}^{DSO} + \pi_t^{\min} PS_{i,t}^{DSO} - \sum_n C(PG_{i,n,t}) \tag{7.2}$$

$$U_{i,t}(PD_{i,t}) = \lambda_{i,t} PD_{i,t} - \beta_{i,t}(PD_{i,t})^2 \tag{7.3}$$

$$PD_{i,t} + PS_{i,t} + PS_{i,t}^{DSO} = \sum_n PG_{i,n,t} + \sum_m PW_{i,m,t} + PB_{i,t} + PB_{i,t}^{DSO} \tag{7.4}$$

$$PD_{i,t}^{\min} \le PD_{i,t} \le PD_{i,t}^{\max} \tag{7.5}$$

$$0 \le PG_{i,n,t} \le PG_{i,n,t}^{\max} \tag{7.6}$$

$$0 \le PW_{i,m,t} \le PW_{i,m,t}^{\max} \tag{7.7}$$

$$0 \le PS_{i,t} + PS_{i,t}^{DSO} \le k_{i,t}(PG_{i,t} - PD_{i,t}) \tag{7.8}$$

$$0 \le PB_{i,t} + PB_{i,t}^{DSO} \le (1 - k_{i,t})(PD_{i,t} - PG_{i,t}) \tag{7.9}$$

where $PB_{i,t}$ and $PS_{i,t}$ denote the power purchases from other MGs and sell to other MGs for MG i. $PB_{i,t}^{DSO}$ and $PS_{i,t}^{DSO}$ mean the power purchases from and sale to DSO for MG i at hour t, respectively. $PD_{i,t}$, $PG_{i,n,t}$ and $PW_{i,m,t}$ are the load, output of MT n, and renewable DER m in MG i at hour t, respectively. Let $k_{i,t}$ represent the status of MG if the MG is a seller and 0 if it is a buyer. Objective (7.2) is the MG's payoff, where the first term $U_{i,t}(PD_{i,t})$ measures the MG's revenue from energy consumption and its preference for consuming a certain amount of energy. The revenue function $U_{i,t}(PD_{i,t})$ should satisfy the following properties [25]: (1) The MG prefers to consume more energy until it reaches its target consumption. (2) The MG's revenue and preference for energy consumption become saturated as it reaches the target level. Therefore, $U_{i,t}(PD_{i,t})$ should be a strictly concave and continuously differentiable function, as shown in (7.3) [27]. Constraint (7.4) ensures power balance. Constraints (7.5)–(7.6) set limits on load supply and DER output, respectively. Constraints (7.8)–(7.9) state that MGs cannot simultaneously sell and buy power. The power production cost of renewable DERs is ignored. For Micro Turbines (MT), the power production cost is represented as [28]:

$$C(PG_{i,n,t}) = a_{i,n}(PG_{i,n,t})^2 + b_{i,n}PG_{i,n,t} + c_{i,n} \tag{7.10}$$

After solving the MG operation problem, the optimal solutions for $PS_{i,t}$ and $PB_{i,t}$ are obtained. MG i will be a seller at hour t if $PS_{i,t} > 0$ and a buyer if $PB_{i,t} > 0$. We assume each MG acts rationally. Thus, sellers will set their price within $[\pi_t^{\min}, \pi_t^{\max}]$ that ensures mutually beneficial transactions. If π_t is larger than $\pi_t^{\max}$, buyers will prefer to purchase energy from the DSO. If π_t is less than $\pi_t^{\min}$, sellers will choose to sell their surplus energy to the DSO rather than to buyers. For the following analyses, we assume that the surplus energy from sellers can meet the demand of buyers. If the sellers' surplus energy is insufficient, buyers will procure the shortfall from the DSO.

7.2.2 Buyer (Follower) Model

This section focuses on analyzing the buyer's behavior. By solving the MG operation problem (7.2)–(7.9), we determine the status of each MG as either a buyer or a seller. To specifically analyze buyer behavior, we simplify the MG operation problem to (7.11)–(7.15) for each buyer. These two sets of problems are equivalent when the MG is a buyer, meaning the optimal solutions (e.g., power demand) obtained from both formulations are identical.

$$\max\ WB_{b,t} = U_{b,t}(PD_{b,t}) - \pi_t PB_{b,t} - \sum_n C(PG_{b,n,t}) \tag{7.11}$$

$$PD_{b,t}^{\min} \le PD_{b,t} \le PD_{b,t}^{\max} \tag{7.12}$$

$$0 \leq PG_{b,n,t} \leq PG_{b,n,t}^{\max} \tag{7.13}$$

$$0 \leq PW_{b,m,t} \leq PW_{b,m,t}^{\max} \tag{7.14}$$

$$PB_{b,t} = PD_{b,t} - \sum_{n} PG_{b,n,t} - \sum_{m} PW_{b,m,t} \tag{7.15}$$

Let $WB_{b,t}$ represent the payoff function of buyer b and let $C(PG_{b,t})$ represent the power production cost. The buyer's objective, as shown in (7.11), is to maximize its payoff for a given price π_t. Constraints (7.12) to (7.14) limit the load and outputs of Micro Turbines (MTs) and renewable DERs, respectively. Constraint (7.15) represents the power purchase from other sellers. Initially, we analyze the scenario where the MG has one MT or one renewable DER. Subsequently, the analysis is expanded to MGs with multiple MTs and renewable DERs.

If buyer b is equipped with a renewable DER, it will buy its energy deficit, which could not be supplied by DER, from sellers. The buyer's problem (7.10)–(7.13) is restated as:

$$\max\ WB_{b,t} = U_{b,t}(PD_{b,t}) - \pi_t\left(PD_{b,t} - PW_{b,t}^{\max}\right) \tag{7.16}$$

$$PD_{b,t}^{\min} \leq PD_{b,t} \leq PD_{b,t}^{\max} \tag{7.17}$$

The first-order and second-order derivatives of (7.16) concerning are listed as follows:

$$\partial WB_{b,t} / \partial PD_{b,t} = \lambda_{b,t} - \pi_t - 2\beta_{b,t} PD_{b,t} \tag{7.18}$$

$$\partial^2 WB_{b,t} / \partial PD_{b,t}^2 = -2\beta_{b,t} \tag{7.19}$$

Since $\beta_{i,t}$ is a positive constant, the second-order derivative (7.19) is negative. Therefore, (7.16) is a strictly concave function and the optimal solution is calculated by $\partial WB_{b,t} / \partial PD_{b,t} = 0$, which is restated as (7.20). Thus, buyer b' best response is calculated as (7.21):

$$PD_{b,t}^{*} = \begin{cases} PD_{b,t}^{\max}\ \lambda_{b,t} - \pi_t > 2\beta_{b,t} PD_{b,t}^{\max} \\ \dfrac{\lambda_{b,t} - \pi_t}{2\beta_{b,t}}\ PD_{b,t}^{\min} \leq \dfrac{\lambda_{b,t} - \pi_t}{2\beta_{b,t}} \leq PD_{b,t}^{\max} \\ PD_{b,t}^{\min}\ \lambda_{b,t} - \pi_t < 2\beta_{b,t} PD_{b,t}^{\min} \end{cases} \tag{7.20}$$

$$PB_{b,t}^{*} = PD_{b,t}^{*} - PW_{b,t}^{\max} \tag{7.21}$$

Now we analyze the buyer with one MT. Using the MT's cost function (7.10), the buyer will increase the MT output to supply the load until the upper limit of MT is reached or the marginal cost of MT is equal to π_t. The marginal cost of MT is the first-order derivative of (7.10) concerning $PG_{b,t}$. By setting $\partial C(PG_{b,t})/\partial PG_{b,t} = \pi_t$, the optimal $PG^*_{b,t}$ is calculated as:

$$PG^*_{b,t} = \begin{cases} (\pi_t - b_b)/2a_b & \pi_t - b_b \le 2a_b PG^{\max}_{b,t} \\ PG^{\max}_{b,t} & \pi_t - b_b > 2a_b PG^{\max}_{b,t} \end{cases} \tag{7.22}$$

Accordingly, the best response $PB^*_{b,t}$ is calculated as:

$$PB^*_{b,t} = PD^*_{b,t} - PG^*_{b,t} \tag{7.23}$$

Now we analyze buyers with several renewable DERs and MTs. Similar to (7.16) and (7.22), outputs of each renewable DER and each MT are stated as (7.24) and (7.25), respectively:

$$PW^*_{b,m,t} = PW^{\max}_{b,m,t} \tag{7.24}$$

$$PG^*_{b,n,t} = \begin{cases} (\pi_t - b_{b,n})/2a_{b,n} & \pi_t - b_{b,n} \le 2a_{b,n} PG^{\max}_{b,n,t} \\ PG^{\max}_{b,n,t} & \pi_t - b_{b,n} > 2a_{b,n} PG^{\max}_{b,n,t} \end{cases} \tag{7.25}$$

Accordingly, the best response $PB^*_{b,t}$ is calculated as:

$$PB^*_{b,t} = PD^*_{b,t} - \sum_n PG^*_{b,n,t} - \sum_m PW^*_{b,m,t} \tag{7.26}$$

Theorem 1 *There exists a unique best response for any price π_t offered by sellers for each buyer.*

Proof Each buyer determines its best response using the closed-form expressions in (7.26) for a given price. Consequently, there is a unique best response for each buyer.

7.2.3 Seller (Leader) Model

This section is used to analyze the seller's behavior. Similarly, the MG operation problem (7.2)–(7.9) is simplified as (7.27)–(7.31) for each seller.

$$\max\ WS_{s,t} = U_{s,t}(PD_{s,t}) + \pi_t PS_{s,t} - \sum_n C(PG_{s,n,t}) \tag{7.27}$$

$$PD_{s,t}^{\min} \leq PD_{s,t} \leq PD_{s,t}^{\max} \tag{7.28}$$

$$0 \leq PG_{s,n,t} \leq PG_{s,n,t}^{\max} \tag{7.29}$$

$$0 \leq PW_{b,m,t} \leq PW_{b,m,t}^{\max} \tag{7.30}$$

$$PS_{s,t} = \sum_{n} PG_{s,n,t} - \sum_{m} PW_{s,m,t} - PD_{s,t} \tag{7.31}$$

where $WS_{s,t}$ represents the payoff function of seller s at hour t. Constraints (7.28)–(7.30) limit the load, and outputs of MT and renewable DER, respectively. Constraint (7.31) represents the power sale to other buyers. Here, we use the backward induction method to analyze the seller. First, we analyze the seller s to obtain the closed-form solution for load and DER output in which π_t is a parameter. Then, we set π_t to calculate the optimal load and DER out-put. Similar to (7.20), the optimal load for seller s at a given price π_t is stated as:

Let $WS_{s,t}$ represent the payoff function of seller s. Constraints (7.28)–(7.30) limit the load and outputs of MTs and renewable DERs, respectively. Constraint (7.31) represents the power sale to other buyers. To analyze the seller's behavior, we use the backward induction method. First, we derive the closed-form solution for load and DER output, with π_t as a parameter. Then, we set π_t to determine the optimal load and DER output. Similar to (7.20), the optimal load for seller s at a given price is expressed as:

$$PD_{s,t}^{*} = \begin{cases} PD_{s,t}^{\max} \; \lambda_{s,t} - \pi_t > 2\beta_{s,t} PD_{s,t}^{\max} \\ \dfrac{\lambda_{s,t} - \pi_t}{2\beta_{s,t}} \; PD_{s,t}^{\min} \leq \dfrac{\lambda_{s,t} - \pi_t}{2\beta_{s,t}} \leq PD_{s,t}^{\max} \\ PD_{s,t}^{\min} \; \lambda_{s,t} - \pi_t < 2\beta_{s,t} PD_{s,t}^{\min} \end{cases} \tag{7.32}$$

Similar to (7.24) and (7.25), the outputs of each renewable DER and MT with a given price are stated as (7.33) and (7.34), respectively:

$$PW_{s,m,t}^{*} = PW_{s,m,t}^{\max} \tag{7.33}$$

$$PG_{s,n,t}^{*} = \begin{cases} (\pi_t - b_{s,n}) / 2a_{s,n} \; \pi_t - b_{s,n} \leq 2a_{s,n} PG_{s,n,t}^{\max} \\ PG_{s,n,t}^{\max} \; \pi_t - b_{s,n} > 2a_{s,n} PG_{s,n,t}^{\max} \end{cases} \tag{7.34}$$

Accordingly, the surplus power of seller s is calculated as:

$$PS_{s,t}^{*} = \sum_{n} PG_{s,n,t}^{*} + \sum_{m} PW_{s,m,t}^{*} - PD_{s,t}^{*} \tag{7.35}$$

To meet the requirement of supply–demand equilibrium, sellers need to determine the electricity price π_t as

$$\sum_s PS_{s,t}^* = \sum_b PB_{b,t}^* \tag{7.36}$$

If min/max limits for loads and MTs are relaxed, both $\sum_s PS_{s,t}^*$ and $\sum_b PB_{b,t}^*$ are continuous linear functions concerning π_t. The value of π_t which satisfies (7.36) is at the intersection of the two linear functions. Therefore, π_t is obtained by substituting (7.26) and (7.35) into (7.36) as:

$$\pi_t = \frac{\sum_i \frac{\lambda_{i,t}}{2\beta_{i,t}} + \sum_i \sum_n \frac{b_{i,n}}{2a_{i,n}} - \sum_i \sum_m PW_{i,m,t}}{\sum_i \frac{1}{2\beta_{i,t}} + \sum_i \sum_n \frac{1}{2a_{i,n}}} \tag{7.37}$$

When considering the min/max limits, both $\sum_s PS_{s,t}^*$ and $\sum_b PB_{b,t}^*$ are piecewise linear functions concerning π_t. Therefore, we calculate each segment of piecewise functions to find the intersection of the two piecewise functions. The following approach is applied to calculate π_t:

(1) Calculate $\lambda_{i,t} - 2\beta_{i,t}PD_{i,t}^{\max}$, $\lambda_{i,t} - 2\beta_{i,t}PD_{i,t}^{\min}$, and $2a_{i,n}PG_{i,n,t}^{\max} + b_{i,n}$ for each MG and rank them in ascending order.

(2) Divide $[\pi_t^{\min}, \pi_t^{\max}]$ into n segments $\left[\pi_t^{q,\min}, \pi_t^{q,\max}\right] \forall q \in n$. The bounds $\pi_t^{q,\min}$ and $\pi_t^{q,\max}$ for each interval correspond to the values calculated in Step 1).

(3) For each interval $\left[\pi_t^{q,\min}, \pi_t^{q,\max}\right]$, calculate π_t^q by using (7.38). If $\pi_t^q \in \left[\pi_t^{q,\min}, \pi_t^{q,\max}\right]$, set $\pi_t = \pi_t^q$.

$$\pi_t^q = \left(\sum_{i \notin (\Omega_{da} \cup \Omega_{di})} \frac{\lambda_{i,t}}{2\beta_{i,t}} + \sum_{i \notin \Omega_g} \frac{b_{i,n}}{2a_{i,n}} - \sum_{m \in \Omega_r} PW_{i,m,t}^{\max} - \sum_{i \in \Omega_g} PG_{i,n,t}^{\max} + \sum_{i \in \Omega_{da}} PD_{i,t}^{\max} + \sum_{i \in \Omega_{di}} PD_{i,t}^{\min} \right) \Big/ \left(\sum_{n \notin \Omega_g} \frac{1}{2a_{i,n}} + \sum_{i \notin (\Omega_{da} \cup \Omega_{di})} \frac{1}{2\beta_{i,t}} \right) \tag{7.38}$$

where Ω_r represents the set of renewable DERs. Ω_g represents the set of MTs whose outputs reach the maximum limits at intervals $\left[\pi_t^{q,\min}, \pi_t^{q,\max}\right]$. Ω_{da} and Ω_{di} respectively represent the set of loads that reach the maximum and minimum limits at intervals $\left[\pi_t^{q,\min}, \pi_t^{q,\max}\right]$. The closed expression (7.38) is utilized to obtain the optimal solution π_t. However, such an expression requires all MGs' relevant parameters. For privacy concerns, we develop several distributed algorithms in Sect. 7.2.5 to obtain the energy price.

Theorem 2 *Uniqueness of Optimal Pricing Strategy for Sellers.*

Proof When the minimum and maximum constraints for loads and MTs are relaxed, there exists a unique price, which can be calculated using Eq. (7.37). Considering the constraints, from Eqs. (7.20) and (7.25), it can be seen that the total MG load is a piecewise function that decreases monotonically as the price π_t increases. Conversely, the total output of DERs is also a piecewise function but increases monotonically with the price π_t. The unique solution to Eq. (7.38) is found at the intersection of these two functions, ensuring a unique optimal pricing strategy.

7.2.4 Existence and Uniqueness of SE

In this section, we demonstrate the existence and uniqueness of the SE. Each seller determines an optimal price that maximizes its payoff, creating a non-cooperative game among the sellers. First, we establish the existence of a Nash equilibrium within this non-cooperative game. Subsequently, we show both the existence and uniqueness of the SE.

Lemma 1 *A Nash equilibrium exists in a non-cooperative game if the following conditions are met* [29]: (1) *The player set in the non-cooperative game is finite.* (2) *The strategy sets are closed, bounded, and convex.* (3) *The payoffs for all players are continuous and quasi-concave over the strategy space.*

Theorem 3 *Existence of Nash Equilibrium in the Non-Cooperative Game Among Sellers.*

Proof In the non-cooperative game, the players consist of all sellers, and their number is finite, denoted by SSS. Thus, Condition 1 is met. As analyzed before, each seller's strategy lies within a specific interval $\left[\pi_t^{\min}, \pi_t^{\max}\right]$, ensuring that the strategy sets are both bounded and convex, satisfying Condition 2. By substituting Eq. (7.1) into Eq. (7.27), we can derive the second-order derivative of Eq. (7.27) for each seller, which further confirms the existence of a Nash equilibrium in this non-cooperative game.

$$\frac{\partial^2 WS_{s,t}}{\partial \pi_{s,t}^2} = -\left(\frac{1}{2\beta_s} + \frac{1}{2a_s}\right) < 0 \tag{7.39}$$

Therefore, the MG's payoff is both continuous and quasi-concave, satisfying condition 3.

Theorem 4 *A unique SE exists in the P2P trading between sellers and buyers.*

Proof Based on Theorem 3, a Nash equilibrium is present in the non-cooperative game among sellers. As demonstrated in Theorem 1 and Theorem 2, each buyer's optimal response and each seller's optimal strategy are unique. Consequently, a unique SE, which includes Eqs. (7.26), (7.35), and (7.38), exists in the MLMF Stackelberg game.

7.2.5 Distributed Algorithm for SE

The closed-form expression (7.38) can be directly applied to find the optimal solution. However, this requires access to all relevant parameters from each MG, which may not be shared due to privacy concerns. Therefore, we have developed several distributed algorithms that respect different privacy requirements.

Algorithm 1 This central market approach at the lower level uses the closed-form expression (7.38), necessitating all relevant information (e.g., parameters and values) from each MG.

Algorithm 2 In this approach, MGs only reveal their best strategies and responses. For instance, each seller's surplus power and price are made public. Consequently, all sellers start the process at the same initial price. The steps for this algorithm are as follows:

(1) Initialize the iteration index and set a starting price.
(2) Solve the MG operation problem (7.2)–(7.9) with the given price to identify whether MGs are sellers or buyers. Each seller calculates its surplus power, and each buyer determines its power demand $PB^*_{b,t}$ by solving the MG operation problem.
(3) Distribute buyers' power demands to sellers using Eq. (7.40).
(4) Each seller collects the total power demand from buyers and updates its price using Eq. (7.41).
(5) If π_t^h and π_t^{h-1} in (7.42) are satisfied, terminate the algorithm. Otherwise, increment the iteration index and repeat steps 2–5.

$$PB^*_{b,s,t} = PB^*_{b,t} PS^*_{s,t} \Big/ \sum_s PS^*_{s,t} \tag{7.40}$$

$$\pi_t^h = \pi_t^h + \sigma_t \left(\sum_s PS^*_{s,t} - \sum_b PB^*_{b,t} \right) \tag{7.41}$$

$$\left| \pi_t^h - \pi_t^{h-1} \right| \le \varepsilon \tag{7.42}$$

where ε is a sufficiently small positive number, and σ_t is the speed adjustment parameter. The convergence criterion (7.42) indicates that Algorithm 2 will terminate when the energy prices from two consecutive iterations are nearly identical. The value of σ_t should be chosen carefully. A very small σ_t could lead to slow convergence, while a larger σ_t might cause the algorithm to fail to converge.

Algorithm 3 Here, MGs do not share any information. Sellers start the process at different prices, determining their energy prices and surplus energy independently. Buyers then respond to each seller. The steps for this algorithm are as follows:

(1) Set the iteration index $h = 1$. Each MG sets an initial price.
(2) Each MG collects prices from other MGs. Then, each MG solves the MG operation problem (7.2)–(7.9) based on the highest price π_t^h to determine whether it is a seller or a buyer. Each seller calculates its surplus power $PS_{s,t}^*$, and each buyer determines its power demand $PB_{b,t}^*$ by solving the MG operation problem.
(3) Each buyer calculates the average price and obtains a new variable. Then, each buyer allocates its demand to sellers using Eq. (7.43).
(4) Each seller updates the price using Eq. (7.44).
(5) Terminate the algorithm when seller prices satisfy Eq. (7.45). Otherwise, increment the iteration index and repeat steps 2–4.

$$PB_{b,s,t}^* = PB_{b,t}^* \eta_{s,t}^* \Big/ \sum_s \eta_{s,t}^* \tag{7.43}$$

$$\pi_{s,t}^h = \pi_{s,t}^h + \sigma_{s,t} \left(\sum_b PB_{b,s,t}^* - PS_{s,t}^* \right) \tag{7.44}$$

$$\left| \pi_{s,t}^h - \pi_{s,t}^{h-1} \right| \le \varepsilon \tag{7.45}$$

In this algorithm, if the seller s's energy price $\pi_{s,t}^h$ is higher than the average price $\overline{\pi}_t^h$, the power sold by the seller s will be less than its surplus power, as determined by Eq. (7.43). Consequently, sellers with higher prices will lower their prices using the updating strategy in Eq. (7.44) to sell more energy. When the algorithm converges to equilibrium, all sellers' prices will be equal.

7.3 Upper-Level DSO Operation

7.3.1 *Mathematical Model of DSO Operation*

At the upper level, the DSO reconfigures the network based on the energy trading outcomes of the MGs. To protect privacy, each MG is treated as an equivalent load. If the current energy trading by MGs breaches network security, the DSO will instruct MGs to adjust their trading by restricting their equivalent loads. Thus, the DSO's decision variables include trading adjustments $\Delta E_{i,t}$, line status $J_{ij,t}$, and the power purchased $PB_{i,t}^{DSO}$ from and sold to the DSO $PS_{i,t}^{DSO}$. The objective of the DSO is to minimize the total operation costs, which encompass buying, selling, and trading adjustment costs, by optimizing the network reconfiguration.

$$\min \pi_t^{\max} \sum_i PB_{i,t}^{DSO} - \pi_t^{\min} \sum_i PS_{i,t}^{DSO} + \sum_i \gamma \left| \Delta E_{i,t} \right| \tag{7.46}$$

By introducing auxiliary variables $U_{i,t} = \left(V_{i,t}\right)^2 / \sqrt{2}$, $R_{ij,t} = V_{i,t}V_{j,t}\cos\left(\theta_{i,t} - \theta_{j,t}\right)$, and $T_{ij} = V_{i,t}V_{j,t}\sin\left(\theta_{i,t} - \theta_{j,t}\right)$. The DSO's operation constraints are listed as follows [30]:

$$\left(V_i^{\min}\right)^2 / \sqrt{2} \le U_{i,t} \le \left(V_i^{\max}\right)^2 / \sqrt{2}\,\forall i, \forall t \tag{7.47}$$

$$\begin{aligned} &P_{i,t} + PB_{i,t}^{DSO} - PS_{i,t}^{DSO} = \sum_j PL_{ji,t} - \sum_i PL_{ij,t}\,\forall i \in ST, \forall t \\ &Q_{i,t} + QB_{i,t}^{DSO} = \sum_j QL_{ji,t} - \sum_i QL_{ij,t}\,\forall i \in ST, \forall t \end{aligned} \tag{7.48}$$

$$\begin{aligned} &P_{i,t} + \Delta E_{i,t} = \sum_j PL_{ji,t} - \sum_i PL_{ij,t}\,\forall i, \forall t \\ &Q_{i,t} = \sum_j QL_{ji,t} - \sum_i QL_{ij,t}\,\forall i, \forall t \end{aligned} \tag{7.49}$$

$$P_{i,t} = \begin{cases} PS_{i,t}^{*}\,\forall i \in S, \forall t \\ -PB_{i,t}^{*}\,\forall i \in B, \forall t \end{cases} \tag{7.50}$$

$$\begin{aligned} &-PS_{i,t}^{*} \le \Delta E_{i,t} \le 0\,\forall i \in S, \forall t \\ &0 \le \Delta E_{i,t} \le PB_{i,t}^{*}\,\forall i \in B, \forall t \end{aligned} \tag{7.51}$$

$$\left(PL_{ij,t}\right)^2 + \left(QL_{ij,t}\right)^2 \le \left(S_{ij}^{\max}\right)^2\,\forall ij, \forall t \tag{7.52}$$

$$\begin{aligned} &PL_{ij,t} = J_{ij,t}\left(\sqrt{2}g_{ij}U_{i,t} - g_{ij}R_{ij,t} - b_{ij}T_{ij,t}\right)\forall ij, \forall t \\ &QL_{ij,t} = J_{ij,t}\left(-\sqrt{2}U_{i,t}b_{ij} - g_{ij}T_{ij,t} + b_{ij}R_{ij,t}\right)\forall ij, \forall t \end{aligned} \tag{7.53}$$

$$\left(R_{ij,t}\right)^2 + \left(T_{ij,t}\right)^2 = 2U_{i,t}U_{j,t}\,\forall ij, \forall t \tag{7.54}$$

$$\tan\left(\theta_{i,t} - \theta_{j,t}\right) = R_{ij,t} / T_{ij,t}\,\forall ij, \forall t \tag{7.55}$$

where $PL_{ij,t}$ and $QL_{ij,t}$ represent the active and reactive power flow on line ij at hour t, respectively. Constraint (7.47) sets limits on bus voltage magnitudes. Constraints (7.48)–(7.49) define active and reactive bus power injections. Constraint (7.50) defines the equivalent load of an MG, where it equals the surplus power for sellers and the power demand for buyers. Constraint (7.51) sets limits on trading adjustments for both buyers and sellers. Constraints (7.52)–(7.53) impose limits on line flows, accounting for network reconfiguration. Constraint (7.54) establishes the relationship among auxiliary variables, while constraint (7.55) sets limits on bus angles.

7.3.2 Convex Reformulation of Model

Here, we present a convex reformulation of the primal model by linearizing the nonlinear constraints (7.53), (7.54), and (7.55).

(1) Linearization of Constraint (7.53):

Constraint (7.53) is linearized into constraints (7.56)–(7.59) using the big-M method:

$$\begin{aligned} PL_{ij,t} &= \sqrt{2}g_{ij}UL_{ij,t} - g_{ij}RL_{ij,t} - b_{ij}TL_{ij,t}\ \forall ij, \forall t \\ QL_{ij,t} &= -\sqrt{2}UL_{ij,t}b_{ij} - g_{ij}TL_{ij,t} + b_{ij}RL_{ij,t}\ \forall ij, \forall t \end{aligned} \tag{7.56}$$

$$\begin{aligned} PL_{ij,t} &= \sqrt{2}g_{ij}UL_{ij,t} - g_{ij}RL_{ij,t} - b_{ij}TL_{ij,t}\ \forall ij, \forall t \\ QL_{ij,t} &= -\sqrt{2}UL_{ij,t}b_{ij} - g_{ij}TL_{ij,t} + b_{ij}RL_{ij,t}\ \forall ij, \forall t \end{aligned} \tag{7.57}$$

$$\begin{aligned} &-J_{ij,t}M \le RL_{ij,t} \le J_{ij,t}M\ \forall ij, \forall t \\ &-(1 - J_{ij,t})M \le RL_{ij,t} - R_{ij,t} \le (1 - J_{ij,t})M\ \forall ij, \forall t \end{aligned} \tag{7.58}$$

$$\begin{aligned} &-J_{ij,t}M \le TL_{ij,t} \le J_{ij,t}M\ \forall ij, \forall t \\ &-(1 - J_{ij,t})M \le TL_{ij,t} - T_{ij,t} \le (1 - J_{ij,t})M\ \forall ij, \forall t \end{aligned} \tag{7.59}$$

where $UL_{ij,t}$, $RL_{ij,t}$, and $TL_{ij,t}$ are auxiliary variables.

(2) Conic Relaxation of Constraint (7.54):

Constraint (7.54) is relaxed into the following conic constraint:

$$\left(R_{ij,t}\right)^2 + \left(T_{ij,t}\right)^2 \le 2U_{i,t}U_{j,t}\ \forall ij, \forall t \tag{7.60}$$

(3) Linearization of Constraint (7.55):

Constraint (7.61) is employed to restrict θ_i instead of (7.55).

$$T_{ij,t} = V_{i,t}V_{j,t}\sin\left(\theta_{i,t} - \theta_{j,t}\right) \tag{7.61}$$

Assuming that $V_{i,t} \approx 1$, $V_{j,t} \approx 1$ and $\sin\left(\theta_i - \theta_j\right) \approx \theta_i - \theta_j$, (7.61) is linearized as:

$$T_{ij,t} \approx \theta_{i,t} - \theta_{j,t} \tag{7.62}$$

7.3.3 Iterative SCP

Since (7.60) is a relaxation of the primal constraint (7.54), iterative Successive Convex Programming (SCP) [31] is applied to ensure the accuracy of this relaxation. By adding the following constraint to the original problem, the relaxed constraint is made exact:

$$\left(R_{ij,t}\right)^2 + \left(T_{ij,t}\right)^2 \geq 2U_{i,t}U_{j,t}\,\forall ij, \forall t \tag{7.63}$$

However, adding (7.63) to the original problem results in a non-convex and nonlinear problem. Therefore, constraint (7.63) is linearized as (7.64) using the first-order Taylor expansion:

$$\begin{aligned} & w_{ij,t}^{u} + 2R_{ij,t}^{u-1}R_{ij,t}^{u} - \left(R_{ij,t}^{u-1}\right)^2 + 2T_{ij,t}^{u-1}T_{ij,t}^{u} - \left(T_{ij,t}^{u-1}\right)^2 \geq \\ & 2U_{i,t}^{u-1}U_{j,t}^{u} + 2U_{i,t}^{u}U_{j,t}^{u-1} - 2U_{i,t}^{u-1}U_{j,t}^{u-1}\,\forall ij, \forall t \end{aligned} \tag{7.64}$$

where u represents the current iteration. $R_{ij,t}^{u-1}$, $T_{ij,t}^{u-1}$, $U_{i,t}^{u-1}$, and $U_{j,t}^{u-1}$ are the solutions obtained from the previous iteration. $w_{ij,t}^{u}$ is a slack variable. With the introduction of SCP, the MISOCP problem is revised as follows:

$$\begin{aligned} & \min\ z_t^h = \pi_t^{\max} P_{i,t}^{DSO} - \pi_t^{\min} P_{i,t}^{DSO} + \sum_i \gamma\left|\Delta E_{i,t}\right| + \sum_{ij} \psi_{ij,t}^{u} w_{ij,t}^{u} \\ & \text{s.t. Constraints } (47)-(52), (56)-(60), (62) \end{aligned} \tag{7.65}$$

where γ is a large number and $\psi_{ij,t}^{u}$ is a penalty factor. The initial point for the iterative SCP is found by solving the primal MISOCP problem, which includes the objective (7.46) and constraints (7.47)–(7.52), (7.60), and (7.62).

7.3.4 Procedure of Upper-Level DSO Operation

The upper-level DSO operation is solved through the following iterative process:

(1) Collect each MG's equivalent load $P_{i,t}$.
(2) Initialize the penalty factors $\psi_{ij,t}^{u}$ and $v_{ij,t}$. Set the iteration index $u = 1$ and residual tolerances ε^z and ε^w.
(3) Replace nonlinear constraints (7.53), (7.54), and (7.55) with convex constraints (7.56)–(7.60) and (7.62). Then solve the MISOCP problem to obtain the initial point.
(4) Solve (7.65) using the initial point to calculate $gap^z = z_t^u - z_t^{u-1}$ and $gap^w = \sum_{ij} \psi_{ij,t}^{u} w_{ij,t}^{u}$.

(5) for $gap^w > \varepsilon^w$, updating the iteration index $u = u + 1$ and the penalty factor $\psi_{ij,t}^u = v_{ij,t}\psi_{ij,t}^u$. Repeat steps 1–5.
(6) If $gap^z \le \varepsilon^z$ and $gap^w \le \varepsilon^w$, the trading adjustment $\Delta E_{i,t}^*$ is obtained. If (7.66) is satisfied, the current energy trading results meet network security requirements. Terminate the loop and obtain the optimal network configuration. Otherwise, add constraint (7.67) to the MG operation problem and update each MG's equivalent load at the lower level.

$$\Delta E_{i,t}^* = 0\,\forall i, \forall t \tag{7.66}$$

$$\begin{aligned} &PD_{i,t} + PS_{i,t} + PS_{i,t}^{DSO} - \sum_n PG_{i,n,t} - \sum_m PW_{i,m,t} \\ &- PB_{i,t} - PB_{i,t}^{DSO} = PE_{i,t}^* \end{aligned} \tag{7.67}$$

$$PE_{i,t}^* = P_{i,t} - \delta \Delta E_{i,t}^* \tag{7.68}$$

where $\psi_{ij,t}^u$ and $v_{ij,t}$ are penalty factors, and δ is the convergence adjustment parameter. ε^z and ε^w are sufficiently small numbers to ensure accuracy. $v_{ij,t}$ is used to ensure the precision of the relaxed constraint (7.60) by limiting the values of $U_{i,t}$. Increasing the penalty factor $v_{ij,t}$ reduces the number of upper-level iterations, but a large $v_{ij,t}$ can limit $U_{i,t}$ and result in a more conservative solution. The number of iterations is also reduced as residual tolerances ε^z and ε^w increase, though setting these too high may compromise the accuracy of the relaxed constraint (7.60). Constraint (7.67) limits each MG's equivalent load, affecting its load and generation. For example, if a buyer needs to reduce its power demand DDD, it could decrease its load and increase its generation, such as by increasing total MT output. The closed-form expression for the load and MG outputs, considering $PE_{i,t}^*$, is provided in the Appendix.

7.4 Case Studies

The proposed model and solution method are tested on a 4-MG system, as well as modified IEEE 33-bus and 123-bus distribution power systems. All computations were performed on a personal computer using Matlab 2018a with Gurobi, running on an Intel Core i7-7700 CPU (3.6 GHz) and 16 GB of memory. The parameter $\beta_{i,t}$ is set to 0.24. The value of $\lambda_{i,t}$, which varies for each MG at different hours, ranges from 35 to 45.

7.4.1 Microgrid System

Figure 7.2 shows the trading energy price with the DSO using data from [32]. Figure 7.3 illustrates the topology of the 4-MG system, which includes 4 MGs and 6 lines, each with a switch. G1 and G2 are MTs, while W1 and W2 are renewable DERs. Table 7.2 presents the cost curve coefficients for the MTs, with G2 being cheaper than G1. Algorithm 2 is used for this test system. Three cases are proposed to compare the results with P2P trading.

Case I: P2P transactions are not considered; MGs only trade with the DSO as shown in Fig. 7.1.

Case II: A central market is considered at the lower level, where the central operator clears the market to maximize all MG payoffs based on marginal DER costs before the upper-level DSO reconfigures the network, as illustrated in Fig. 7.1 [9].

Case III: Both P2P trading at the lower level and DSO operation at the upper level are considered, as proposed in Fig. 7.1.

Fig. 7.2 Price for buying energy from DSO and selling energy to DSO

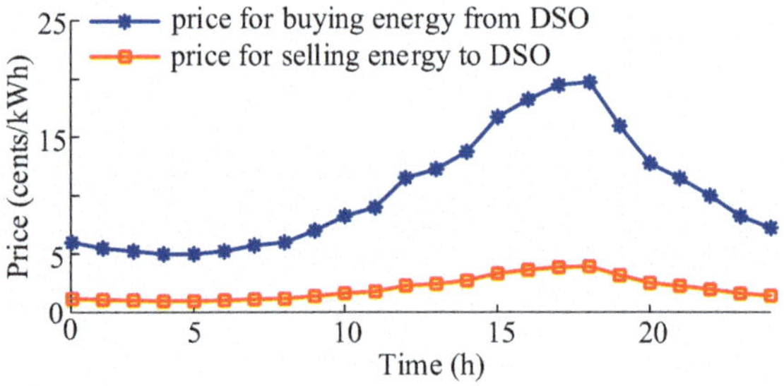

Fig. 7.3 Topology of the 4-MG system

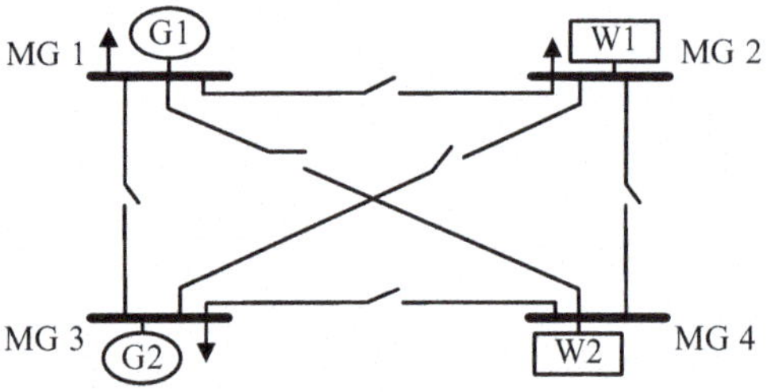

Table 7.2 Cost curve coefficients of MTs G1 and G2

MT	a_i (\$/kW^2)	b_i (\$/kW)	c_i (\$)
G1	0.00021	0.079	2.01
G2	0.00017	0.068	2.01

Table 7.3 shows the allocated costs for different scenarios. In Case I, MGs trade energy solely with the DSO, buying at high prices and selling at low prices, resulting in the highest trading cost and the lowest total MG payoff. In Case II, MGs trade energy through a central market operator at the lower level, reducing trading costs with the DSO and increasing total MG payoffs due to the benefits of energy trading. However, this approach compromises the privacy of each MG, as they must share proprietary information with the central market operator. Here, each MG is compensated based on its DER marginal cost, leading to minimal participation incentives for those with renewable DERs due to their low generation costs.

Achieving equilibrium in such markets is challenging, especially during energy shortages when sellers' surplus power doesn't meet buyers' demand. Market equilibrium means no participant can gain higher benefits by altering the current trading result. During shortages, MGs can earn more by setting prices above marginal costs since trading continues if prices remain below the DSO's rate. Thus, equilibrium is unattainable in shortages as MGs seek better trading prices.

In Case III, MGs find optimal prices to maintain supply–demand equilibrium, incentivizing all MGs to trade energy. At equilibrium, the total MG payoff is maximized, matching the payoff in Case II. No participant can gain higher benefits by disrupting the market equilibrium, and MG privacy is preserved as there's no central operator at the lower level.

Table 7.4 compares individual MG payoffs in Cases II and III. Although total MG payoffs are the same in both cases, individual payoffs differ. MG2 and MG4, equipped with renewable DERs, balance local loads with low-cost onsite generation and sell surplus energy to other MGs, resulting in higher payoffs than MG1 and MG3, which use MTs. In Case II, MGs are paid based on marginal costs, while in Case III, they receive trading prices. The higher trading prices in Case III lead to higher payoffs for MG2 and MG4 compared to Case II. Conversely, MG1 and MG3 have lower payoffs in Case III as they purchase energy from MG2 and MG4 at higher prices. MGs with renewable DERs are thus incentivized to trade energy, achieving higher payoffs at equilibrium.

Figure 7.4a–d illustrates the energy trading outcomes for MG1-MG4 in Case III. MG2 and MG4, which have low-cost renewable DERs, generate energy throughout the entire operation period. They sell surplus energy (e.g., during hours 0–6) and purchase energy when deficient (e.g., during hours 15–16). Since MT energy is more expensive, MG1 and MG3 opt to buy energy from MG2 and MG4, shutting down their MTs. When the surplus energy from MG2 and MG4 is insufficient to

Table 7.3 Cost allocations of 4-MG system in different cases

Case	Total MG payoff ($)	Total MG cost for buying energy from DSO ($)	Total MG revenues for selling energy to DSO ($)
I	1278	103	35
II	1379	24	0
III	1379	24	0

Table 7.4 MG's payoffs in cases II and III

Case	Each MG payoff ($)			
	MG1	MG2	MG3	MG4
II	244	412	287	436
III	225	435	247	472

meet MG1 and MG3's demands, they either purchase energy from the DSO (e.g., during hours 0–8) or activate their MTs (e.g., during hours 14–20), depending on the DSO's price.

Table 7.5 details the surplus energy allocation for MGs at hours 15 and 16. At hour 15, MG2 and MG4 act as sellers, while MG1 and MG3 are buyers. MG2 and MG4's surplus energy is fully allocated to MG1 and MG3. Specifically, MG1 buys 47.07 kW from MG2, and MG3 purchases the remaining 49.67 kW. At hour 16, MG1 and MG3 become sellers, and MG2 and MG4 are buyers. MG2 acquires 18.62 kW from MG1, while MG4 purchases the remaining 16.82 kW.

7.4.2 *Modified IEEE 33-Bus Distribution Power System*

A modified IEEE 33-bus distribution power system, illustrated in Fig. 7.5, is used to test the proposed model. This system includes 10 MGs with MTs, 10 MGs with renewable DERs, 13 load buses, and 37 lines, all equipped with switches. The distribution load buses can also purchase energy from MGs. The trading price with the DSO remains consistent with Fig. 7.2. Three scenarios are considered to evaluate the proposed strategy:

P2P-I: The proposed P2P trading is considered without network constraints.
P2P-II: The proposed network-constrained P2P trading is considered without applying any network reconfiguration (i.e., all distribution switches remain stationary).
P2P-III: The proposed P2P transactive strategy is considered with network reconfiguration at the DSO level.

Figure 7.6 shows the network reconfiguration at hour 16 in P2P-III, where switches at lines 3–23, 15–16, and 26–27 are opened, while other switches remain closed. This hourly network reconfiguration ensures the DSO's network security and provides an optimal network topology for MGs to trade energy.

Figure 7.7 compares the apparent power flows $\sqrt{(PL_{ij})^2+(QL_{ij})^2}$ at hour 16 in P2P-I and P2P-III under the network topology illustrated in Fig. 7.6. In P2P-I, MGs trade energy without considering network constraints, which can lead to violations of line flow limits. However, in P2P-III, the consideration of network constraints restricts line flows, ensuring compliance with distribution grid security.

Fig. 7.4 Trading for **a** MG1, **b** MG2, **c** MG3, and **d** MG4 in Case III

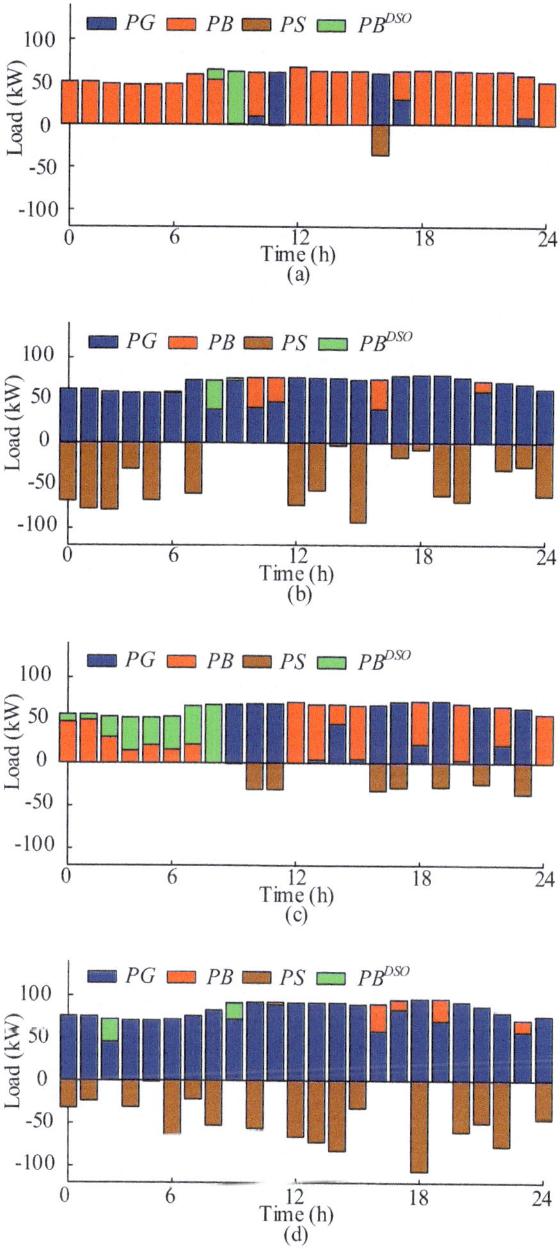

Table 7.5 Surplus power allocations at hours 15 and 16

Hour	Seller	Traded power provided by sellers (kW)				Surplus power (kW)
		MG1	MG2	MG3	MG4	
15	MG2	47.07	0	49.67	0	96.67
	MG4	15.88	0	16.73	0	32.61
16	MG1	0	18.62	0	16.82	35.44
	MG3	0	16.98	0	15.34	32.32

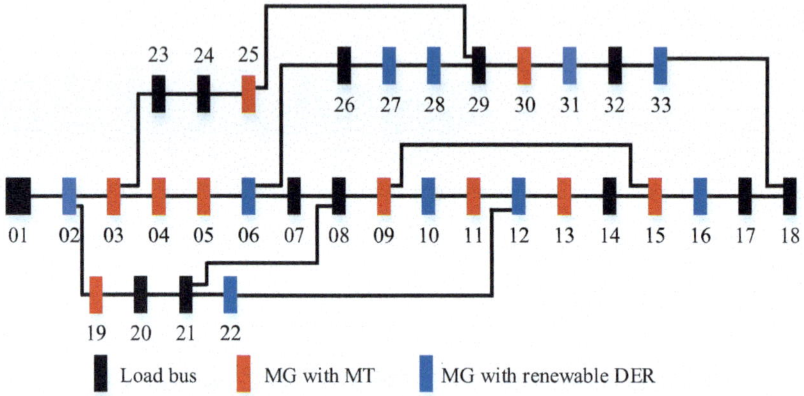

Fig. 7.5 Topology of the 33-bus system

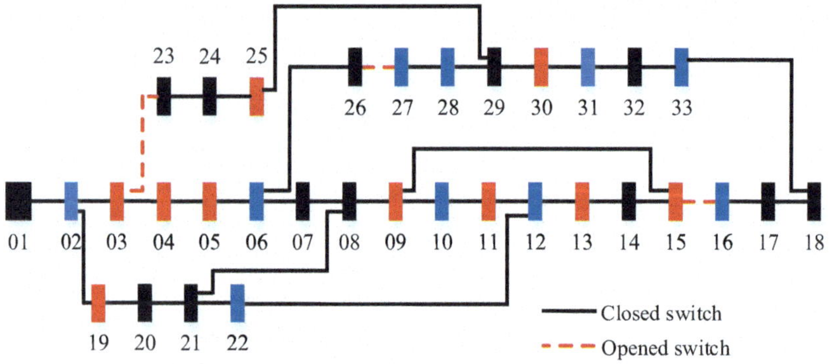

Fig. 7.6 Network reconfiguration at hour 16 in P2P-III

Figure 7.8 compares the MG energy trading during the entire operation period across three cases. Energy trading is highest in P2P-I since network constraints are relaxed, allowing MGs to sell all surplus energy, as shown in Fig. 7.7, which may lead to line congestion. In P2P-II, energy trading is constrained by the network. The total surplus energy that MGs can sell is limited not only by DER capacity but also

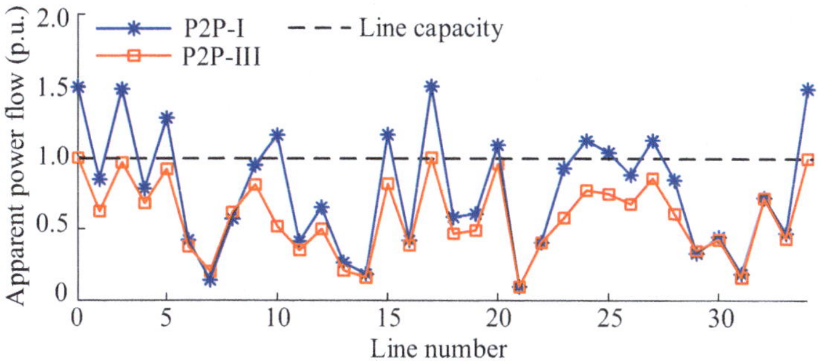

Fig. 7.7 Line flows at hour 16 in P2P-I and P2P-III

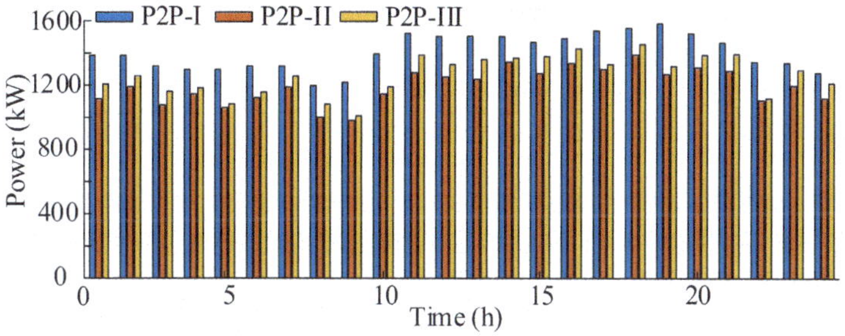

Fig. 7.8 Traded power among MGs in three cases

by network constraints, resulting in less trading than in P2P-I. However, network security is ensured in this case. In P2P-III, network reconfiguration helps alleviate line congestion, allowing for more energy trading compared to P2P-II.

Figure 7.9 shows the trading prices obtained by the three proposed algorithms. All three algorithms converge to the same prices, demonstrating the effectiveness of the proposed distributed methods for calculating prices while protecting MGs' privacy. Compared to the prices in Fig. 7.2, the energy price in P2P trading is lower than the DSO price, especially when the DSO price is high (e.g., hours 15–20). This indicates that the proposed P2P trading method effectively incentivizes MGs to trade energy at the lower level, offering them higher earnings compared to trading with the DSO.

Table 7.6 presents the computational performance of the three proposed algorithms, with iteration numbers averaged over 24 time steps. Upper-level iterations occur between upper and lower levels, while lower-level iterations reflect the average number of iterations at the lower level to obtain the trading price. As seen in Fig. 7.6, all three algorithms yield the same results for lower-level P2P trading. Consequently, the upper-level iterations are identical for all three algorithms. Algorithm 1 involves no lower-level iterations, resulting in the shortest solution time, but it compromises each MG's privacy. Algorithm 2 protects MGs' privacy to a certain extent by not disclosing their payoff functions, loads, and DER outputs. However, it requires

7.5.2 *Applications to the Real-World*

This section outlines the implementation of the proposed P2P energy market in practical scenarios. As trading evolves, existing market and grid operations may need to adapt to accommodate these new trading models [34]. In a real-world application, electricity utilities would act as DSOs in this framework. MGs would report their equivalent loads to the utility, which would then reconfigure the distribution network and provide trading adjustments to each MG. Utilities could also participate in the LEM if they manage their MGs.

The P2P energy market could be integrated with existing wholesale markets. For instance, in the day-ahead market, utilities submit bids to an independent system operator (ISO) to balance the distribution system's load. Utilities would forecast electricity prices and share these forecasts with MGs. MGs would then use the P2P market to trade energy based on these prices. After trading, MGs submit their energy needs to the DSO, which consolidates the data and places bids with the ISO to procure the necessary electricity.

7.6 Conclusions

In this chapter, we introduce a bi-level distribution network-constrained local P2P trading approach among multi-microgrids. At the lower level, a P2P trading system is implemented among MGs, which is modelled as a multi-leader multi-follower Stackelberg game. We prove the uniqueness of the SE and provide a closed-form expression for it. To address the privacy concerns, we develop several distributed algorithms to compute the SE. At the upper level, a MISOCP model is proposed to determine the AC power flow for network reconfiguration, utilizing iterative SCP to maintain model accuracy. Simulation results validate the effectiveness of our model and approach in ensuring network security and encouraging MGs to engage in energy trading.

In practical applications, ignoring forecast errors in renewable DER generation could lead to infeasible trading outcomes. Future work will focus on integrating the uncertainty of renewable DERs and incorporating various types of DERs, such as batteries, into the P2P trading framework.

Appendix

To provide closed-form expressions for the optimal solutions of MGs considering trading adjustments, we focus on the buyers. For simplicity, the analysis is restricted to buyers only.

The buyer's operation problem, incorporating trading adjustments, is outlined below:

$$\max\ WB_{b,t} = U_{b,t}(PD_{b,t}) - \pi_t PB_{b,t} - \sum_n C(PG_{b,n,t}) \tag{7.81}$$

$$\text{Constraints (7.12)–(7.15)} \tag{7.82}$$

$$PB_{b,t} \le PE^*_{i,t} \tag{7.83}$$

If (7.83) is relaxed, the optimal solutions for $PG^*_{b,n,t}$, $PD^*_{b,t}$, $PB^*_{b,t}$ and $PW^*_{b,m,t}$ can be obtained using (7.20) and (7.24)–(7.26). The trading adjustment $\Delta E^*_{b,t}$ of buyer b at hour t is calculated as follows:

$$\Delta E^*_{b,t} = PB^*_{b,t} - PE^*_{i,t} \tag{7.84}$$

Constraint (7.84) presents that buyer b must restrict $PB^*_{b,t}$ to $PE^*_{i,t}$ by adjusting $PD_{b,t}$ and $PG_{b,n,t}$. Also, the total adjustments in $PD_{b,t}$ and $PG_{b,n,t}$ should be equal to $\Delta E^*_{b,t}$. More specifically, compared with $PG^*_{b,n,t}$ and $PD^*_{b,t}$, $PD_{b,t}$ is decreased and $PG_{b,n,t}$ is increased [renewable DER output cannot be increased since the output is equal to its forecast value by using (7.24)]. We assume the reduction in $PD_{b,t}$ is stated as $\eta^L_t \Delta E^*_{b,t}$ while the increase in $PG_{b,n,t}$ is stated as $\eta^G_{b,n,t}\Delta E^*_{b,t}$. The buyer's payoff is reduced when the trading adjustment is considered. Accordingly, if min/max limits of $PD_{b,t}$ and $PG_{b,n,t}$ are relaxed, $PD_{b,t}$ and $PG_{b,n,t}$ are stated as:

$$PD_{b,t} = PD^*_{b,t} - \eta^L_t \Delta E^*_{b,t} \tag{7.85}$$

$$PG_{b,n,t} = PG^*_{b,n,t} + \eta^G_{n,t}\Delta E^*_{b,t}\ \forall n \tag{7.86}$$

$$\eta^L_t + \sum_n \eta^G_{n,t} = 1 \tag{7.87}$$

Therefore, we find the optimal values of η^L_t and

$$\begin{aligned} \min\ FB_{b,t} = {} & UB_{b,t}(PD^*_{b,t})^2 - \sum_n C(PG^*_{b,n,t}) \\ & -\left(U_{b,t}(PD_{b,t})^2 - \sum_n C(PG_{b,n,t})\right) \end{aligned} \tag{7.88}$$

$$\frac{\partial C(PG_{b,1,t})}{\partial PG_{b,1,t}} = \frac{\partial C(PG_{b,2,t})}{\partial PG_{b,2,t}} = \dots = \frac{\partial C(PG_{b,n,t})}{\partial PG_{b,n,t}} \tag{7.89}$$

$$\text{Constraints (85)–(87)} \tag{7.90}$$

By adding (7.25) and (7.86) into (7.89), we can gain Eq. (7.91) as follows:

$$\eta_{n,t}^{G} = \eta_{1,t}^{G} a_{b,1} / a_{b,n} \, \forall n \tag{7.91}$$

Constraint (7.92) is calculated by substituting (7.20), (7.25), and (7.85)–(7.87) into (7.88):

$$\begin{aligned} FB_{b,t} = & \left(a_{b,n} + \beta_{b,t}\right)\left(\Delta E_{b,t}^{*}\right)^{2}\left(\sum_{n}\left(a_{b,1}/a_{b,n}\right)\right)^{2}\left(\eta_{1,t}^{G}\right)^{2} \\ & - 2\beta_{b,t}\left(\Delta E_{b,t}^{*}\right)^{2}\left(\sum_{n}\left(a_{b,1}/a_{b,n}\right)\right)\eta_{1,t}^{G} + \Delta E_{b,t}^{*}\left(\pi_{t} + \beta_{b,t}\Delta E_{b,t}^{*}\right) \end{aligned} \tag{7.92}$$

By setting $\partial FB_{b,t} / \partial \eta_{1,t}^{G} = 0$, we obtain the closed-form expressions for η_{t}^{L} and $\eta_{n,t}^{G}$. Accordingly, the closed-form expressions of load and each MT output are stated as (7.93)–(7.94), respectively:

$$PD_{b,t} = \frac{\lambda_{b,t} - \pi_{t}}{2\beta_{b,t}} - \Delta E_{b,t}^{*} + \frac{2\beta_{b,t} a_{b,1} \Delta E_{b,t}^{*}}{\left(a_{b,n} + \beta_{b,t}\right)\left(\sum_{n}\left(a_{b,1}/a_{b,n}\right)\right) b_{b,n}} \tag{7.93}$$

$$PG_{b,n,t} = \frac{\pi_{t} - b_{b,n}}{2a_{b,n}} + \frac{2\beta_{b,t} a_{b,1} \Delta E_{b,t}^{*}}{\left(a_{b,n} + \beta_{b,t}\right)\left(\sum_{n}\left(a_{b,1}/a_{b,n}\right)\right) b_{b,n}} \, \forall n \tag{7.94}$$

If considering the min/max limits for $PD_{b,t}$ and $PG_{b,n,t}$, an iterative method is used to calculate $PD_{b,t}$ and $PG_{b,n,t}$:

(1) Calculate $PD_{b,t}$ and use (7.93)–(7.94).
(2) If $PD_{b,t}$ and $PG_{b,n,t}$ beyond the min/max constraints, set $PD_{b,t}$ and $PG_{b,n,t}$ to their min/max scopes. Thereafter, we can obtain energy adjustments $\Delta E_{b,t}^{*}$ like:

$$\Delta E_{b,t}^{*} = \Delta E_{b,t}^{*} - \left(PD_{b,t}^{*} - PD_{b,t}^{\min}\right) - \sum_{n \in \Omega_{g}}\left(PG_{b,n,t}^{\max} - PG_{b,n,t}^{*}\right) \tag{7.95}$$

(3) For $PD_{b,t}$ and that do not reach the min/max limits, calculate $PD_{b,t}$ and $PG_{b,n,t}$ again using (7.93)–(7.94) with the updated trading adjustment.
(4) Repeat steps 2–4 until $PD_{b,t}$ and $PG_{b,n,t}$ do not exceed the min/max limits.

References

1. Shahidehpour, M., Li, Z., Bahramirad, S., Li, Z., & Tian, W. (2017). Networked microgrids: Exploring the possibilities of the IIT-Bronzeville grid. *IEEE Power and Energy Magazine, 15*(4), 63–71.
2. Zhou, Q., Shahidehpour, M., Alabdulwahab, A., & Abusorrah, A. (2019). Flexible division and unification control strategies for resilience enhancement in networked microgrids. *IEEE Transactions on Power Systems, 35*(1), 474–486.
3. Renani, Y. K., Ehsan, M., & Shahidehpour, M. (2017). Optimal transactive market operations with distribution system operators. *IEEE Transactions on Smart Grid, 9*(6), 6692–6701.
4. Khorasany, M., Mishra, Y., & Ledwich, G. (2018). Market framework for local energy trading: A review of potential designs and market clearing approaches. *IET Generation, Transmission & Distribution, 12*(22), 5899–5908.
5. Council, T. (2015). Gridwise transactive energy framework version 1.0. *The GridWise Architecture Council*, Tech. Rep.
6. Khorasany, M., Azuatalam, D., Glasgow, R., Liebman, A., & Razzaghi, R. (2020). Transactive energy market for energy management in microgrids: The monash microgrid case study. *Energies, 13*(8), 2010.
7. Mengelkamp, E., Gärttner, J., Rock, K., Kessler, S., Orsini, L., & Weinhardt, C. (2018). Designing microgrid energy markets: A case study: The Brooklyn microgrid. *Applied Energy, 210*, 870–880.
8. Wang, Z., Chen, B., Wang, J., Begovic, M. M., & Chen, C. (2014). Coordinated energy management of networked microgrids in distribution systems. *IEEE Transactions on Smart Grid, 6*(1), 45–53.
9. Wang, Y., Huang, Z., Shahidehpour, M., Lai, L. L., Wang, Z., & Zhu, Q. (2019). Reconfigurable distribution network for managing transactive energy in a multi-microgrid system. *IEEE Transactions on Smart Grid, 11*(2), 1286–1295.
10. Liu, W., Zhan, J., & Chung, C. Y. (2018). A novel transactive energy control mechanism for collaborative networked microgrids. *IEEE Transactions on Power Systems, 34*(3), 2048–2060.
11. Khorasany, M., Mishra, Y., & Ledwich, G. (2019). A decentralized bilateral energy trading system for peer-to-peer electricity markets. *IEEE Transactions on Industrial Electronics, 67*(6), 4646–4657.
12. Vakili, R., Afsharnia, S., & Golshannavaz, S. (2018). Interconnected microgrids: Optimal energy scheduling based on a game-theoretic approach. *International Transactions on Electrical Energy Systems, 28*(10), 2603.
13. Zaidi, B. H., & Hong, S. H. (2018). Combinatorial double auctions for multiple microgrid trading. *Electrical Engineering, 100*, 1069–1083.
14. Li, Z., Bahramirad, S., Paaso, A., Yan, M., & Shahidehpour, M. (2019). Blockchain for decentralized transactive energy management system in networked microgrids. *The Electricity Journal, 32*(4), 58–72.
15. Tushar, W., Saha, T. K., Yuen, C., Smith, D., & Poor, H. V. (2020). Peer-to-peer trading in electricity networks: An overview. *IEEE Transactions on Smart Grid, 11*(4), 3185–3200.
16. Morstyn, T., Teytelboym, A., & McCulloch, M. D. (2018). Bilateral contract networks for peer-to-peer energy trading. *IEEE Transactions on Smart Grid, 10*(2), 2026–2035.
17. Lee, J., Guo, J., Choi, J. K., & Zukerman, M. (2015). Distributed energy trading in microgrids: A game-theoretic model and its equilibrium analysis. *IEEE Transactions on Industrial Electronics, 62*(6), 3524–3533.
18. Jadhav, A. M., Patne, N. R., & Guerrero, J. M. (2018). A novel approach to neighborhood fair energy trading in a distribution network of multiple microgrid clusters. *IEEE Transactions on Industrial Electronics, 66*(2), 1520–1531.
19. Liu, N., Yu, X., Wang, C., Li, C., Ma, L., & Lei, J. (2017). Energy-sharing model with price-based demand response for microgrids of peer-to-peer prosumers. *IEEE Transactions on Power Systems, 32*(5), 3569–3583.

20. Jadhav, A. M., & Patne, N. R. (2017). Priority-based energy scheduling in a smart distributed network with multiple microgrids. *IEEE Transactions on Industrial Informatics, 13*(6), 3134–3143.
21. Anoh, K., Maharjan, S., Ikpehai, A., Zhang, Y., & Adebisi, B. (2019). Energy peer-to-peer trading in virtual microgrids in smart grids: A game-theoretic approach. *IEEE Transactions on Smart Grid, 11*(2), 1264–1275.
22. Sultana, B., Mustafa, M. W., Sultana, U., & Bhatti, A. R. (2016). Review on reliability improvement and power loss reduction in distribution system via network reconfiguration. *Renewable and Sustainable Energy Reviews, 66*, 297–310.
23. Dorostkar-Ghamsari, M. R., Fotuhi-Firuzabad, M., Lehtonen, M., & Safdarian, A. (2015). Value of distribution network reconfiguration in presence of renewable energy resources. *IEEE Transactions on Power Systems, 31*(3), 1879–1888.
24. Liu, Y., Li, J., & Wu, L. (2018). Coordinated optimal network reconfiguration and voltage regulator/DER control for unbalanced distribution systems. *IEEE Transactions on Smart Grid, 10*(3), 2912–2922.
25. Maharjan, S., Zhu, Q., Zhang, Y., Gjessing, S., & Başar, T. (2016). Demand response management in the smart grid in a large population regime. *IEEE Transactions on Smart Grid, 7*(1), 189–199.
26. Wei, F., Jing, Z. X., Wu, P. Z., & Wu, Q. H. (2017). A Stackelberg game approach for multiple energies trading in integrated energy systems. *Applied Energy, 200*, 315–329.
27. Chai, B., Chen, J., Yang, Z., & Zhang, Y. (2014). Demand response management with multiple utility companies: A two-level game approach. *IEEE Transactions on Smart Grid, 5*(2), 722–731.
28. Yan, M., Zhang, N., Ai, X., Shahidehpour, M., Kang, C., & Wen, J. (2018). Robust two-stage regional-district scheduling of multi-carrier energy systems with a large penetration of wind power. *IEEE Transactions on Sustainable Energy, 10*(3), 1227–1239.
29. Neel, J. O., Reed, J. H., & Gilles, R. P. (2004). Convergence of cognitive radio networks. In *Proceedings of IEEE WCNC* (pp. 2250–2255).
30. Jabr, R. A. (2008). Optimal power flow using an extended conic quadratic formulation. *IEEE Transactions on Power Systems, 23*(3), 1000–1008.
31. He, Y., Yan, M., Shahidehpour, M., Li, Z., Guo, C., Wu, L., & Ding, Y. (2017). Decentralized optimization of multi-area electricity-natural gas flows based on cone reformulation. *IEEE Transactions on Power Systems, 33*(4), 4531–4542.
32. [online] Available: https://www.pjm.com/markets-and-operations/energy.aspx
33. Wang, X., Li, Z., Shahidehpour, M., & Jiang, C. (2017). Robust line hardening strategies for improving the resilience of distribution systems with variable renewable resources. *IEEE Transactions on Sustainable Energy, 10*(1), 386–395.
34. EC–European Commission. (2016). *Proposal for a Regulation of the European Parliament and of the Council on the internal market for electricity.*

Chapter 8
Fast Sequential Clearing of Congestion-Aware Local Energy and Flexibility Markets

Abstract To address the uncertainty of distributed renewable energy (RE), real-time and efficient operation of the local energy market (LEM) at the distribution level is imperative. Traditional security-constrained economic dispatch methods prove inefficient and incompatible with the high-frequency operational needs of LEM in the presence of large-scale prosumers. This study proposes an optimal operation mechanism based on the sequential clearing of energy and flexibility markets, where energy trading and congestion management are decoupled. Prosumers are offered two bidding modes to cater to their trading needs under different risk preferences. The Guaranteed Price Mode (GPM) provides a secure and predictable trading environment, while the Energy Resale Mode (ERM) allows prosumers to capitalize on market fluctuations. A continuous double auction method is proposed to clear the LEM. Dynamic calculation of power loss and network fees is implemented to incentivize prosumer trading within neighborhoods. Furthermore, a flexibility market for congestion management is established to encourage the active exploration of prosumers' flexibility and to provide services to the distribution network. Case studies demonstrate the effectiveness of the proposed mechanism in enhancing transaction efficiency and ensuring the operational security of distribution networks.

8.1 Introduction

The rapid growth of renewable energy (RE) sources presents significant challenges for maintaining the power system's supply and demand balance. However, the local energy market (LEM) at the distribution system level offers a promising solution. By adjusting the profit margins between the time of use (TOU) and feed-in tariff (FiT) prices for prosumers [1] and promoting direct transactions, the local market can enhance RE consumption and balance power supply and demand [2, 3]. Unlike wholesale markets, demand-side resources (DRs) have diverse characteristics and greater uncertainty, with prosumers differing in decision-making behaviors and preferences [4, 5]. Hence, a tailored mechanism is needed for the LEM.

M. Song et al., *Local Energy Markets*, https://doi.org/10.1007/978-981-97-9750-9_8

To address these challenges, existing research has proposed market frameworks based on auction mechanisms [6, 7] and game theory [8, 9]. Distributed solution techniques have also been used to protect prosumer privacy [10, 11].

Compared to auction-based methods, transaction models based on game theory are more complex and often rely on iterative solutions that require longer computation times [6]. As the number of prosumers increases, market clearing time increases, reducing market efficiency. Thus, achieving rapid and frequent calculations for safe and efficient market clearing is crucial. Unlike large market players, prosumers are smaller-scale with limited profit margins and less professional expertise, making it difficult to handle complex market rules or strategies. Consequently, game theory-based models are more suited for scenarios with fewer, larger market players, such as microgrids [8].

Auction theory-based market mechanisms offer a simpler auction process, which is convenient for market players. Although a central authority is needed to facilitate the matching process, it only needs to collect bidding information, thus minimizing privacy concerns. A straightforward double-sided auction mechanism can clear the market efficiently by sorting bids without complex optimization or iterative processes, making it suitable for large numbers of market players.

Besides market clearing efficiency, transaction results must ensure the safe operation of the distribution network. Therefore, power flow constraints are a critical factor. Similar to wholesale markets, individual market players cannot access real-time network structure and operating conditions. Thus, the market architecture must involve the distribution system operator (DSO) to clear the market considering power flow constraints. This approach ensures safe network operation in a market environment [12]. However, most studies overlook security constraints during market mechanism design [7, 9, 13]. Checking transmission constraints post-market clearing can result in transaction cancellations, reducing scalability. The radial structure of distribution networks with high *r/x* ratios further complicates market models, increasing computational burdens.

To tackle these challenges posed by both the local market architecture and the physical transmission constraints of the distribution network, existing research can be broadly categorized into two approaches:

(1) Extensive research on Distribution Locational Marginal Price (DLMP) [14–16] combines the Locational Marginal Price (LMP) method from wholesale markets with the physical architecture of distribution networks. Clearing prices are formed using unit commitment and economic dispatch with security constraints, and duality theory derives DLMP components.
(2) Continuous Double Auction (CDA) involves rapid market clearing by temporarily ignoring security constraints and matching transactions using simple double auction rules. Security issues are addressed by reducing or canceling transactions based on designed principles [6]. Some studies use a verification approach until no further transactions can be made [17].

While DLMP ensures network safety [18, 19], it involves solving complex optimization models with power flow constraints, which is less efficient. CDA compensates for DLMP's efficiency issues, but post-hoc security verification can cancel many transactions, hindering supply–demand balance and RE consumption. Encouraging transactions while ensuring fairness remains unresolved.

LEM transforms DSOs into service providers. DSOs must calculate network losses from transactions and collect service fees to cover network investments and operating costs [20]. Transmission fees and network loss fees vary with transaction orders, and energy transactions between distant prosumers may incur higher fees [21]. A proposed method considers network losses and fees but simplifies power flow to DC flow, ignoring reactive power and voltage impacts [22]. Transmission fees modeled as a Nash bargaining problem achieve global optimization but increase complexity as market players grow [21]. CDA for rapid market clearing and line impedance-based post-settlement of fees and costs has been proposed, but price-only matching causes higher network costs [23]. Some studies incorporate network costs in CDA by solving DLMP, but solving speed challenges remains [24]. Using CDA to improve efficiency while reducing or canceling transactions through post-hoc verification raises fairness concerns [6].

This chapter proposes a sequential clearing mechanism for local energy and flexibility markets with congestion management. Key contributions include:

(1) Proposing a sequential clearing market architecture for rapid transaction matching among large-scale prosumers, enhancing efficiency while considering distribution network constraints.
(2) Introducing a CDA mechanism that accurately calculates network losses and fees based on AC distribution power flow sensitivity factors. This method dynamically incorporates network costs into market bids, reducing total transaction costs.
(3) Designing a local flexibility market for congestion management to avoid transaction cancellations and ensure fairness. Using the Fictitious Nodal Demand (FND) model and approximations for linear constraints improves clearing efficiency, incentivizing prosumers to provide flexible services and manage potential power flow issues while increasing revenue.

8.2 Local Energy and Flexibility Market Operating Framework

It is assumed that the DSO is responsible for organizing and clearing the LEM. Prosumers across various distribution network nodes can engage directly in electricity transactions. Each prosumer uses smart metering devices connected to the Home Energy Management System (HEMS), enabling automatic market bidding [25].

This mechanism includes energy and flexibility markets, which operate sequentially, as illustrated in Fig. 8.1. In the energy market, each prosumer's HEMS selects different bidding modes and submits transaction demands in real-time. After bidding,

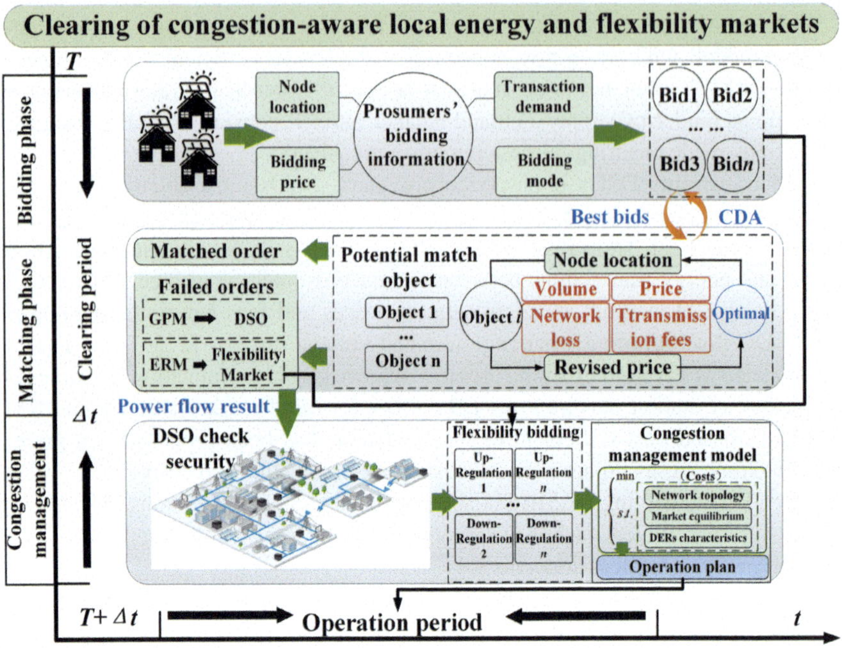

Fig. 8.1 Operation framework of the LEM

the DSO revises the prices based on network losses and transmission fees. The DSO then matches all trading orders using a CDA method to determine the trading results. Based on the chosen bidding mode, untraded bids are settled or transferred to the flexibility market.

Following the energy market clearing, the DSO performs a security assessment of the transaction results. All matched orders are executed to maintain market fairness and encourage prosumer participation. If the transaction results from the energy market violate power flow constraints, a flexibility market is introduced for congestion management. Prosumers with excess generation, consumption capacity, or regulatory capabilities can place additional bids. The DSO manages congestion by minimizing resource allocation costs.

As depicted in Fig. 8.1, the LEM provides two bidding modes for prosumers:

(1) Guaranteed price mode (GPM): All unsuccessful bids in the energy market are settled directly with the DSO at the TOU or FiT prices.
(2) Energy Resale Mode (ERM): successful bids in the energy market are transferred to the flexibility market with new flexibility bidding prices to participate in congestion management managed by the DSO.

The energy resale bidding mode increases the chances of prosumers winning a bid. However, it also means they may ultimately forfeit the chance to generate or consume electricity if the new bidding prices are less favorable. Prosumers' bidding strategies are influenced by their operating characteristics and risk preferences, which are beyond the scope of this chapter. Related research on these aspects can be found in [26, 27].

8.3 LEM Mechanism with Network Losses and Transmission Fees Correction

8.3.1 Calculation Method of Transmission Fees and Network Losses Based on Sensitivity Factors

We assumed that there are N_{DC} DCs operated by an operator in the LEM. Due to DCs' higher power consumption, market participants are categorized into DC-type and conventional-type prosumers. The total number of prosumers in the ξ-th LEM is denoted by K_{ξ}.

Transactions over longer electrical distances use more network resources and incur higher transmission fees. Consequently, transmission fees can serve as price signals to encourage prosumers to engage in local energy trading. Many existing studies utilize the Power Transfer Distribution Factor (PTDF) to characterize the impact of power transmission on distribution lines [22]. However, PTDFs derived from the DC power flow model overlook the effects of reactive power and voltage on power flow in distribution systems, leading to accuracy issues. To address these limitations, this chapter employs a linearized AC power flow model, proven effective in radial distribution networks with high *r*/*x* ratios [28]. Specifically, it can be expressed as:

$$P_i = \sum_j G_{ij} V_j - \sum_j B'_{ij} \theta_j \tag{8.1}$$

$$Q_i = -\sum_j B_{ij} V_j - \sum_j G'_{ij} \theta_j \tag{8.2}$$

$$P_{ij} = g_{ij}(V_i - V_j) - b_{ij}(\theta_i - \theta_j) \tag{8.3}$$

$$Q_{ij} = -b_{ij}(V_i - V_j) - g_{ij}(\theta_i - \theta_j) \tag{8.4}$$

where i and j refer to the start and end node of distribution lines; P_i and Q_i mean the net injection at node i; P_{ij} and Q_{ij} refer to the active and reactive power of line $i-j$; V_i, θ_i mean the voltage magnitude and phase angle of node i; $Y_{ij} = G_{ij} + jB_{ij}$ is the element of the node admittance matrix; the superscript' means that the node admittance of the parallel branch to ground is ignored. $g_{ij} + jb_{ij}$ denotes the admittance of distribution line *i*-*j*.

Based on (8.1)–(8.4), the power flow sensitivity factors of the distribution network can be derived as detailed in Appendix8.1 [29].

The transmission fees based on the PTDF of transmission lines can be expressed as:

$$C_{ab}^{\mathrm{T}} = \sigma_{\mathrm{p}} u_{ab}^{\mathrm{P}} P_{ab}^{\mathrm{T}} + \sigma_{\mathrm{q}} u_{ab}^{\mathrm{Q}} Q_{ab}^{\mathrm{T}} \tag{8.5}$$

where *CT ab* is the transmission fee that should be paid for the transaction between the prosumers located at nodes *a* and *b*, respectively. σ_{p}, σ_{q} mean transmission fee rates per active and reactive power unit, respectively. *uP ab, uQ ab* are the occupation rates of transmission line capacity per unit of active and reactive power in the transaction between nodes *a* and *b*, respectively. *PT ab* and *QT ab* are the active and reactive power sold by the prosumer at node *a* to the prosumer at node *b*, respectively.

uP ab and *uQ ab* are calculated as:

$$u_{ab}^{\mathrm{P}} = \sum_{i} \sum_{j} \left[\left| S_{\mathrm{P}_a}^{\mathrm{P}_{ij}} - S_{\mathrm{P}_b}^{\mathrm{P}_{ij}} \right| + \left| S_{\mathrm{Q}_a}^{\mathrm{P}_{ij}} - S_{\mathrm{Q}_b}^{\mathrm{P}_{ij}} \right| \right] \tag{8.6}$$

$$u_{ab}^{\mathrm{Q}} = \sum_{i} \sum_{j} \left[\left| S_{\mathrm{P}_a}^{\mathrm{Q}_{ij}} - S_{\mathrm{P}_b}^{\mathrm{Q}_{ij}} \right| + \left| S_{\mathrm{Q}_a}^{\mathrm{Q}_{ij}} - S_{\mathrm{Q}_b}^{\mathrm{Q}_{ij}} \right| \right] \tag{8.7}$$

Similarly, a longer electrical distance may lead to higher network losses. Including the cost of network losses in transactions further incentivizes prosumers to trade locally, thereby reducing overall network loss costs. The network loss cost resulting from prosumer transactions is given by:

$$C_{ab}^{\mathrm{L}} = \pi_{\mathrm{p}}^{\mathrm{TOU}} w_{ab}^{\mathrm{P}} P_{ab}^{\mathrm{T}} + \pi_{\mathrm{q}}^{\mathrm{TOU}} w_{ab}^{\mathrm{Q}} Q_{ab}^{\mathrm{T}} \tag{8.8}$$

where, *CL ab* denotes the network loss cost that should be paid for the transaction between the prosumers located at nodes *a* and *b*, respectively. πTOU p, πTOU q are the compensation costs for network losses per unit of active and reactive power, respectively *wP ab* and *wQ ab* mean the sensitivities of system network losses per unit of active and reactive power in the transaction between nodes *a* and *b*.

wP ab and *wQ ab* are stated as:

$$w_{ab}^{\mathrm{P}} = 0.5 * \left(\left| S_{\mathrm{P}_a}^{\mathrm{P}_{\mathrm{loss}}} - S_{\mathrm{P}_b}^{\mathrm{P}_{\mathrm{loss}}} \right| + \left| S_{\mathrm{Q}_a}^{\mathrm{P}_{\mathrm{loss}}} - S_{\mathrm{Q}_b}^{\mathrm{P}_{\mathrm{loss}}} \right| \right) \tag{8.9}$$

$$w_{ab}^{\mathrm{Q}} = 0.5 * \left(\left| S_{\mathrm{P}_a}^{\mathrm{Q}_{\mathrm{loss}}} - S_{\mathrm{P}_b}^{\mathrm{Q}_{\mathrm{loss}}} \right| + \left| S_{\mathrm{Q}_a}^{\mathrm{Q}_{\mathrm{loss}}} - S_{\mathrm{Q}_b}^{\mathrm{Q}_{\mathrm{loss}}} \right| \right) \tag{8.10}$$

The network loss is calculated using the approximate method in the Appendix. Because the Eqs. (8.44)–(8.51) is twice the actual network loss [29], the coefficients in (8.9) and (8.10) is set as 0.5.

8.3.2 CDA Mechanism of LEM

Considering transmission fees and network losses, a CDA-based trading mechanism is proposed for the LEM. The mechanism operates as follows:

(1) Bidding Phase: Prosumers provide transaction details, including energy demand and pricing, and select a bidding mode. In the traditional model, prosumers buy and sell electricity to the DSO at TOU or FiT prices. The profit margin between TOU and FiT prices determines the local market's price range, so all bids should fall within this range.

After collecting all bids, the DSO selects winners based on the provided information. The criteria for choosing the optimal bids are as follows:

$$\pi^* = \min_{m,n}\left(\pi_{\mathrm{p}}^{\mathrm{TOU}} - \pi_m, \pi_n - \pi_{\mathrm{p}}^{\mathrm{FiT}}\right) \tag{8.11}$$

where, $m \in \{1,2,3,\ldots,M\}$ and $n \in \{1,2,3,\ldots,N\}$ mean the prosumers purchasing and selling electricity, respectively. πTOU p and πFIT p mean TOU and FiT prices, respectively. πm and π_n are the bidding prices of prosumers. The closer their offer/bid is to the contract prices, the more prior they are to reach a deal. π^* denotes the optimal bid in the current round of trading.

(2) Matching Phase: Once the winning bid is determined, the DSO calculates the network losses and transmission fees associated with all potential transactions involving the winning prosumer. The DSO then adjusts all bidding prices to account for these network costs to find the optimal matching partner. For example, if the winning prosumer is a buyer, the DSO will compute the network losses and transmission fees for transactions between this buyer and each potential seller. The seller with the lowest revised price will be selected as the optimal match for the buyer. If the revised prices for both parties still fall within the TOU and FiT price range, the transaction will be confirmed.

It is assumed that the power factor of prosumers is fixed, so reactive power is treated as a by-product of energy transactions. Therefore, prosumers do not bid for reactive power. The cost per kWh of transmission fees *cT ab* and network losses *cL ab* can be expressed as:

$$c_{ab}^{\mathrm{T}} = 0.5C_{ab}^{\mathrm{T}}/P_{ab}^{\mathrm{T}} \tag{8.12}$$

$$c_{ab}^{\mathrm{L}} = 0.5C_{ab}^{\mathrm{L}}/P_{ab}^{\mathrm{T}} \tag{8.13}$$

where 0.5 denotes that the seller and buyer share the additional cost equally.

However, appendix (8.44)–(8.47) shows that the loss sensitivity factor is proportional to the power injection and inversely proportional to the voltage square. Before

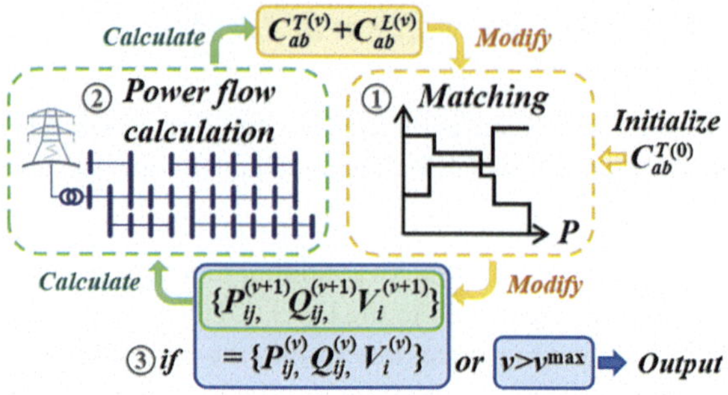

Fig. 8.2 LEM clearing process

the energy market is cleared, each node's power flow and voltage are unknown. Therefore, it is difficult to accurately measure the impact of network losses on transactions during the clearing process.

To tackle the issues, this chapter proposes an iterative correction method for network loss sensitivity, analogous to the forward–backward power flow calculation shown in Fig. 8.2. Fix the power flow P_{ij} and Q_{ij} and the voltage V_i in each iteration. Then, the additional cost of network losses is calculated for transaction matching. Subsequently, P_{ij}, Q_{ij}, and V_i are corrected based on the results of the transaction matching, and this process is repeated until convergence is achieved. To reduce the algorithm error and improve the convergence performance, initial values of $P(0)$ ij, $Q(0)$ ij, and $V(0)$ j are calculated by the transaction matching result, which only considers the transmission fee.

It should be noted that as the number of prosumers increases, the scale of the model will also become larger. The iterative method may have difficulty in convergence within the required time. Considering that network losses have a relatively small impact on power flow results, this chapter also sets a maximum iteration time $v^{\max}$. It can be proved that the transaction results after $v^{\max}$ iterations can meet practical needs.

(3) Terminal Condition: When prosumer mmm and prosumer n agree on a transaction, the traded volume is determined by the smaller of the two values: surplus from prosumer mmm or demand from prosumer n. Any remaining volume is carried over to the next auction round at its original price. The bids that have been matched are removed from the bidding list. The matching process continues until the highest bid price of the buyer is less than the lowest offer price of the seller.
(4) Failed Orders: Offers or bids that remain unmatched are handled by the DSO in two ways: the DSO either fulfills the prosumer's transaction demand at the TOU or FiT price or transfers them to the flexibility market for further bidding.

The following algorithm can outline the transaction matching procedure: v denotes the current round of the matching process. ε represents the acceptable error margin for convergence in the matching process.

Algorithm: Transaction Matching Method Based on Correction of Network Losses and Transmission Fee
Input: π_m, π_n, πTOU p, πFIT p, $P(v)\ ij$, $Q(v)\ ij$, $V(v)\ i$, σ_p, $v = 0$
Sort buyers in descending order and sellers in ascending order based on their bid/offer prices, forming the pending transaction list
while $\max(\pi_m) \geq \min(\pi_n)$
The winner * is determined (* is assumed to be a buyer) based on (11)
Based on (12) and (13), the network cost for * and every seller $n \in N$ is calculated as $cT\ {*n} + cL\ {*n}$, and the offer price $\pi_n + cT\ {*n} + cL\ {*n}$ is revised accordingly
Sellers are sorted in ascending order based on the revised prices, forming an awaiting selling list
for $n = 1,2,\ldots,N$
if $0.5(\pi^* + \pi_n) + cT\ {*n} + cL\ {*n} \leq \pi$TOU p and $0.5(\pi^* + \pi_n) - cT\ {*n} - cL\ {*n} \geq \pi$FIT p and $\pi^* \geq \pi_n$
The buyer * matching with seller n
Record the transaction results and update the pending transaction list. Break
Otherwise, remove seller n from the awaiting selling list
end if
end for
if no sellers are available to match with buyer *
the bidding of * fails, and * is moved from the pending transaction list to either (1) the guaranteed price list, or (2) the energy resale list, depending on *'s bidding mode
end if
end while
Based on the transaction results and (1)–(4), the power flow in the distribution network is calculated
if $\max_i(V(v+1)\ I - V(v)\ i) \geq \varepsilon$ or $\max_i(P(v+1)\ ij - P(v)\ ij) \geq \varepsilon$ or $\max_i(Q(v+1)\ ij - Q(v)\ ij) \geq \varepsilon$ or $v \leq v^{\max}$
$v = v + 1$. Turn to the Input step for the next iteration
end if
Output: Transaction results

To ensure the equitable distribution of trading profits and maintain market balance, the clearing price for matched orders is determined as the average of the bid price and the offer price. If buyer m and seller n are positioned at nodes a and b, respectively, the transaction cost or profit can be expressed as:

$$\pi_{mn} = 0.5 * (\pi_m + \pi_n) \tag{8.14}$$

$$C_m = 0.5 * (C_{ab}^{\mathrm{T}} + C_{ab}^{\mathrm{L}}) + \pi_{mn} P_{mn} \tag{8.15}$$

$$C_n = 0.5 * (C_{ab}^{\mathrm{T}} + C_{ab}^{\mathrm{L}}) - \pi_{mn} P_{mn} \tag{8.16}$$

$$C_{\mathrm{DSO}} = -C_{ab}^{\mathrm{T}} - C_{ab}^{\mathrm{L}} \tag{8.17}$$

where π_{mn} and P_{mn} mean the transaction prices and volume of the buyer m and the seller n, respectively; C_m, $C_{n,}$ and C_{DSO} mean the transaction costs of m, n, and the DSO, respectively. $C_m + C_n + C_{\mathrm{DSO}} = 0$ indicates the LEM satisfies the market equilibrium.

8.4 Congestion-Aware Local Flexibility Market

After completing the energy transactions, the distribution network's operation plan is established. However, transaction outcomes that do not account for physical constraints may result in line capacity breaches and voltage issues in the network.

A local flexibility market is introduced for congestion management based on the energy trading results to address these concerns. The DSO assesses the security of the energy transaction outcomes. If any violations are detected, prosumers can place additional bids based on their ability to adjust their energy profiles. Once all flexibility bids are collected, the DSO adjusts the power flow to minimize the cost of acquiring flexibility and restore the distribution network to its normal operating state.

In the flexibility market, each prosumer can bid simultaneously based on their expected revenue and bid for upward and downward capacities. Let $g \in \{1,2,3,\ldots, G\}$ mean the prosumers participating in congestion management. The optimization model to minimize the DSO's congestion management cost can be expressed as:

$$\min\ C_{\mathrm{M}} = \sum_g \mu_g P_g + \sum_h \mu_h Q_h + \sum_i (\pi_{\mathrm{p}}^{\mathrm{TOU}} P_i + \pi_{\mathrm{q}}^{\mathrm{TOU}} Q_i) \tag{8.18}$$

where C_M denotes the DSO's congestion management cost; μ_g is the compensation price per unit of electricity adjusted by the prosumer g; P_g denotes the adjustment capacity called upon by the prosumer g, and Q_h and μ_h are the power generated by the reactive power compensation devices and its unit cost, respectively. P_i and Q_i are the active and reactive power at node i, respectively. The third term denotes the cost of distribution network losses.

P_g and Q_h should satisfy the following constraints:

$$0 \leq P_g \leq P_g^{\max} \tag{8.19}$$

$$0 \leq Q_h \leq Q_h^{\max} \tag{8.20}$$

where Pmax g denotes the maximum flexibility of prosumers; Qmax h is the output limit of the reactive power compensation devices owned by the DSO.

In the flexibility market, network losses must also be addressed. The Fictitious Nodal Demand (FND) model incorporates a load power equal to 0.5 times the line loss at both ends of the line, effectively treating the line as lossless [30]. The line losses for each segment are calculated as follows:

$$P_{ij}^{\text{loss}} = 0.5 \sum_{\delta} \left(S_{\mathrm{P}_\delta}^{\mathrm{P}_{ij}^{\text{loss}}} P_\delta + S_{\mathrm{Q}_\delta}^{\mathrm{P}_{ij}^{\text{loss}}} Q_\delta \right) \tag{8.21}$$

$$Q_{ij}^{\text{loss}} = 0.5 \sum_{\delta} \left(S_{\mathrm{P}_\delta}^{\mathrm{Q}_{ij}^{\text{loss}}} P_\delta + S_{\mathrm{Q}_\delta}^{\mathrm{Q}_{ij}^{\text{loss}}} Q_\delta \right) \tag{8.22}$$

Accordingly, the corrected power injection and line flow for each node based on the FND are stated as:

$$P_i' = P_i - 0.5 \sum_{j} P_{ij}^{\text{loss}} \tag{8.23}$$

$$Q_i' = Q_i - 0.5 \sum_{j} Q_{ij}^{\text{loss}} \tag{8.24}$$

$$P_{ij}' = \sum_{\delta} \left(S_{\mathrm{P}_\delta}^{\mathrm{P}_{ij}} P_\delta' + S_{\mathrm{Q}_\delta}^{\mathrm{P}_{ij}} Q_\delta' \right) \tag{8.25}$$

$$Q_{ij}' = \sum_{\delta} \left(S_{\mathrm{P}_\delta}^{\mathrm{Q}_{ij}} P_\delta' + S_{\mathrm{Q}_\delta}^{\mathrm{Q}_{ij}} Q_\delta' \right) \tag{8.26}$$

where P'_i and Q'_i mean the modified active and reactive power injections at each node; P'_{ij} and Q'_{ij} mean the modified active and reactive line flow with the compensation by FND.

The power flow constraints that need to be satisfied can be expressed as follows:

$$\sum_{i} P_i' = 0 \tag{8.27}$$

$$\sum_{i} Q_i' = 0 \tag{8.28}$$

$$V_i^{\min} \leq \sum_{i} \mathbf{1} + S_{\mathrm{P}_i}^{\mathrm{V}} P_i' + S_{\mathrm{Q}_i}^{\mathrm{V}} Q_i' \leq V_i^{\max} \tag{8.29}$$

$$\sqrt{P_{ij}'^2 + Q_{ij}'^2} \leq S_{ij}^{\max} \tag{8.30}$$

where $\mathbf{1}$ denotes a unit vector with a dimension consistent with the number of system nodes.

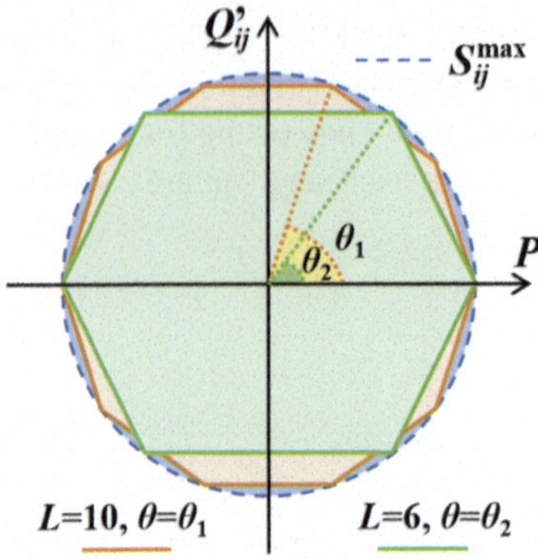

Fig. 8.3 Power circle inner polygon approximation

To improve the solution efficiency, the quadratic constraint (8.30) for line transmission capacity can be linearized by approximating it with an inner polygon, shown in Fig. 8.3.

As shown in Fig. 8.3, assuming that the power circle is approximated by an inscribed regular L-sided polygon (where L is an even number) with a corresponding central angle of θ, (8.30) can be further rewritten as:

$$\begin{cases} Q'_{ij} \le k_l P'_{ij} + b_l, \ \forall 1 \le l \le L/2 \\ Q'_{ij} \ge k_l P'_{ij} + b_l, \ \forall L/2 + 1 \le l \le L \end{cases} \tag{8.31}$$

where k_l and b_l mean the slope and y-intercept of the lth side of the inscribed regular L-sided polygon, respectively. They can be calculated as:

$$k_l = \frac{\sin l\theta - \sin(l-1)\theta}{\cos l\theta - \cos(l-1)\theta} \tag{8.32}$$

$$b_l = S_{ij}^{\max} \sin l\theta - k_l S_{ij}^{\max} \cos l\theta \tag{8.33}$$

An iterative algorithm addresses the nonlinearity introduced by line losses in power flow. The process begins by solving the congestion management model without accounting for line losses. The resulting dispatch is then used to calculate the system loss sensitivity. Adjustments are made based on the power flow equations. This iterative process continues until the solution meets the desired accuracy or reaches the maximum number of iterations. The final transaction results of the flexibility market are derived from this model. An overview of the rapid sequential clearing of the local energy and flexibility markets is shown in Fig. 8.4.

In addition, this chapter assumes that the DSO pays the costs arising from congestion management, and these costs can be included in the transmission fee rates σp based on cost recovery principles. The determination of σp should be jointly decided by the DSO's operation costs and surcharges, which is not the focus of this chapter. Therefore, σp is considered a constant parameter in this chapter.

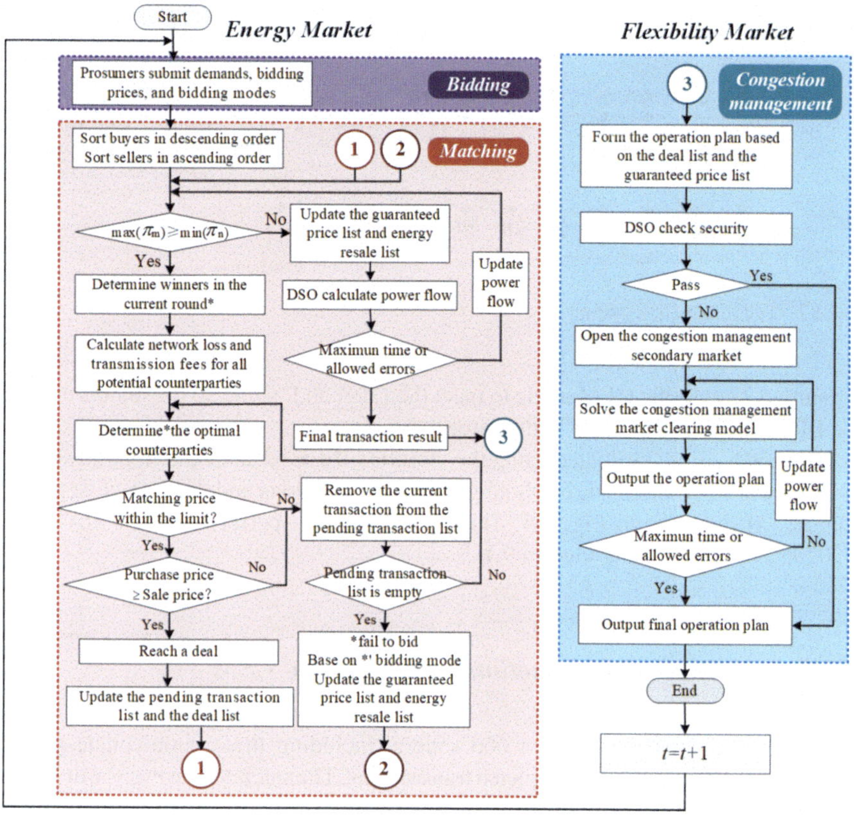

Fig. 8.4 Fast sequential clearing process of local energy and flexibility markets

8.5 Case Studies

8.5.1 *Parameter Setting*

This section demonstrates that the proposed mechanism can accommodate the high-frequency operational needs of LEM. With a clearing period set at $\Delta t = 5$ min, the process is divided as follows: from 0 to 2 min is allocated for collecting all bidding information during the bidding phase. The matching phase occurs between 2 and 4 min. The remaining time is dedicated to congestion management.

The bidding period is from 9:00 to 10:00 a day. πTOU p = 1.49 CNY/kWh and πFIT p = 0.27 CNY/kWh. The transmission fee rate $\sigma_p = 0.005$ CNY/kWh, and the bidding information is detailed in Table 8.7. Use a 36-sided inscribed polygon to approximate the power circle, $\varepsilon = 10^{-5}$ and $\nu^{max} = 5$. Each node is assumed to have 100 prosumers participating in the LEM, with 20 acting as sellers and 80 as buyers. Half of these prosumers submit flexibility bids simultaneously. It is assumed that 70

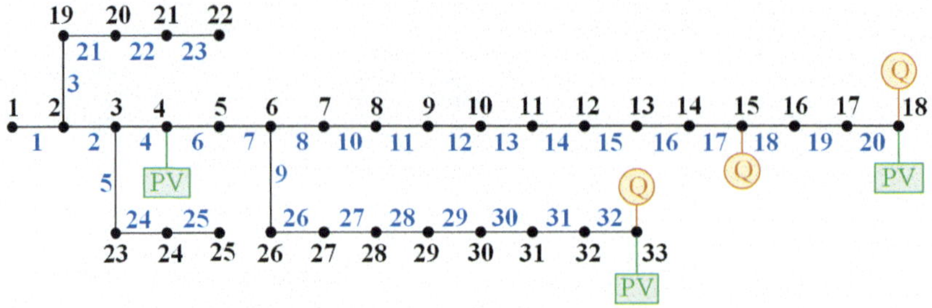

Fig. 8.5 Test distribution system

prosumers choose the GPM mode to meet their demand, while 30 prosumers choose the ERM mode in pursuit of higher profits.

All scenarios are evaluated using the modified IEEE 33-node distribution system [31], which incorporates three photovoltaic power stations and three reactive power sources, as depicted in Fig. 8.5. The simulations are performed with MATLAB R2020a and solved using Gurobi 9.1.

8.5.2 Analysis of Transaction Results in the LEM

The LEM has 2640 buyers and 660 sellers (including three photovoltaic power stations), resulting in 1,333 completed transactions. The stack area in Fig. 8.6 displays the transaction costs/income (green area) and network costs (red area) under the proposed mechanism. This is compared with the costs/income of the traditional contract-based model for buying and selling electricity (blue area).

Figure 8.6a illustrates that, for each transaction, the blue area representing the TOU contract price is consistently larger than the combined green and red areas, which reflect the total costs for buyers. This confirms that all electricity buyers have lowered their purchase costs through transactions in the LEM.

Network costs average 2.96% of the total electricity purchase cost. The highest proportion of network loss costs and transmission fees occurred in the 1164th transaction, where the buyer and seller were at nodes 16 and 1, respectively. As seen in the enlarged section of Fig. 8.6a, network costs accounted for 16.58% of the total expenditure in this case. The transaction matching process considers both the bidding price and network costs. Although the buyer at node 16 offered the highest bid, the higher network costs due to the location resulted in fewer suitable matching options. The pattern for sellers, shown in Fig. 8.6b, is similar, with all sellers benefiting from increased profits in the proposed market.

Table 8.1 compares the costs and revenues between the traditional contract model and the proposed energy market. In the LEM, prosumers incur only minimal network costs and see a significant increase in surplus. If transactions fail due to suboptimal bidding strategies, failed orders can be settled with the DSO at the contract price or

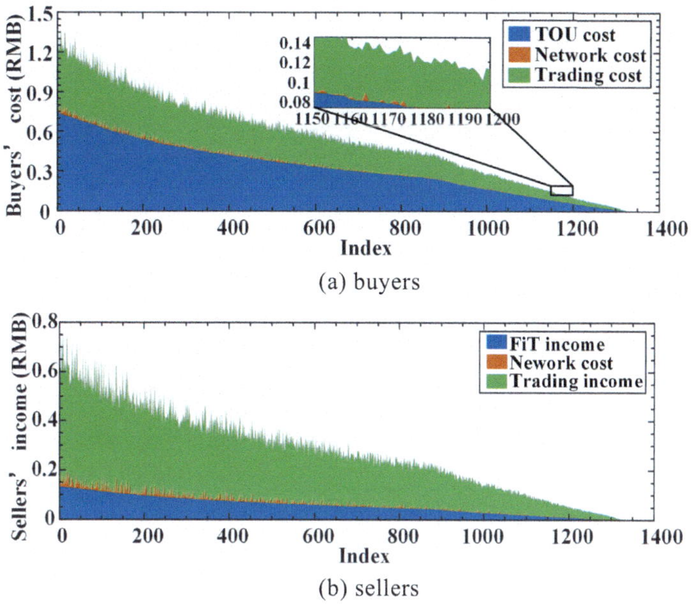

Fig. 8.6 Costs and revenue of prosumers

Table 8.1 Trading results of the LEM

	Costs (CNY)	Revenues (CNY)	Network costs (CNY)
Traditional contract model	434.0	78.6	0
LEM	291.0	284.7	20.1

moved to the flexibility market for further bidding. Thus, the proposed mechanism offers a strong incentive for prosumers.

It's important to note that reactive power may not be fully balanced through market transactions due to varying power factors among prosumers. Consequently, the difference in costs and revenues reflects additional reactive power costs that buyers might need to cover for the upper-level grid. Nevertheless, the overall expenses remain balanced within the market framework.

8.5.3 Analysis of the Transaction Results of Prosumers

To illustrate the effects of network losses and transmission fees, four scenarios are compared: (1) no network costs, (2) no network loss costs, (3) no transmission fees, and (4) the proposed mechanism. The analysis focuses on the first five matched orders from the transaction results, as shown in Table 8.2.

Table 8.2 Trading results of the energy market considering different types of network costs (first 5 transactions)

Scenario 1		Scenario 2		Scenario 3		Scenario 4	
S1/14	B1/8	S2/6	B1/8	S2/6	B1/8	S2/6	B1/8
S1/14	B2/25	S4/5	B1/8	S7/28	B1/8	S7/28	B1/8
S2/6	B2/25	S4/5	B2/25	S3/22	B2/25	S13/25	B2/25
S2/6	B3/12	S1/14	B3/12	S4/5	B2/25	S1/14	B3/12
S3/22	B3/12	S1/14	B7/13	S1/14	B3/12	S1/14	B7/13

Table 8.2 lists buyers and sellers by their bidding prices, with B1 being the highest bidder and S1 the lowest-priced seller. When network costs are excluded, transactions are based solely on bidding prices, disregarding locations and potentially resulting in higher network costs. With network loss and transmission fees incorporated, transactions generally favor nearby matches. For instance, S1 at node 14 would be the ideal seller for B1 at node 8 without additional costs, but this pair would incur high network costs after adjustments. Instead, the first matched transaction is between S2 at node 6 and B1 at node 8.

When only transmission fees are considered, matched transactions are closer in node locations because these fees are based on online flow sensitivity factors linked to the network's electrical layout. Network loss costs, related to line impedance and geographical distance, provide more detailed results. Therefore, the proposed mechanism allows the DSO to effectively manage network operating costs, achieve goals of localized trading, and reduce system network loss through the combined consideration of transmission fees and network losses.

Table 8.3 shows the trading volumes across the four scenarios. The total trading volume remains approximately 291.3 kWh in all scenarios, indicating that accounting for network loss and transmission fees does not impact the traded volume. Furthermore, the minimal total network costs confirm the relationship between network loss and transmission fees. Scenario 3 shows significantly higher network loss costs compared to Scenario 4, while Scenario 2 has slightly higher transmission fees than Scenario 4. This illustrates the combined impact of considering both network loss and transmission fees.

Table 8.3 Trading volume and network fees in different scenarios

Scenario	Network loss costs (CNY)	Transmission fees (CNY)	Traded volume (kWh)
1	124.27	24.90	291.26
2	7.95	15.49	291.26
3	81.60	12.87	291.26
4	7.32	12.80	291.26

8.5.4 Analysis of Transaction Results in the Local Flexibility Market

After the LEM is cleared, all confirmed transactions are recorded as physical contracts for execution. Bids that fail are managed according to their bidding modes. Bids in the GPM are incorporated into the distribution network's operational plan and settled at the TOU or FiT contract price. In contrast, bids in the ERM are moved to the flexibility market with revised bidding prices.

The DSO then performs a power flow security assessment based on the trading outcomes. If there are violations, all flexibility bids are used for congestion management. The flexibility market is cleared by minimizing dispatch costs. Figure 8.7 shows the voltage profiles at various nodes in the distribution network before and after congestion management. The resulting trades might not ensure operational safety since physical constraints are not considered in the LEM. Some nodes in the network experience voltage violations, with end nodes, such as node 18, experiencing significant voltage drops, reaching 0.87 p.u.

The DSO will initiate congestion management and acquire downward flexibility capacity from prosumers as a unilateral buyer. This approach involves injecting power at each node, which reduces the power flow through the lines and alleviates voltage sag. Additionally, the DSO adjusts the reactive power distribution using its compensation equipment, helping to return the distribution network to a stable operating state.

Similarly, line power flow issues arise with congestion. With its single-ended power supply, the radial network structure complicates load transfer. Figure 8.8 illustrates the changes in active power flow before and after congestion management. The flexibility market helps reduce power flow at each node, addressing line congestion.

Demand-side Distributed Energy Resources (DERs) can significantly ease distribution line congestion. For example, a photovoltaic station connected to node 33 might contribute to power flow issues on branch 32. Traditionally, this would reduce or cancel the offer from node 33, leading to curtailment. With the flexibility market, a high bid for upward capacity at node 33 enables the surplus energy to be consumed

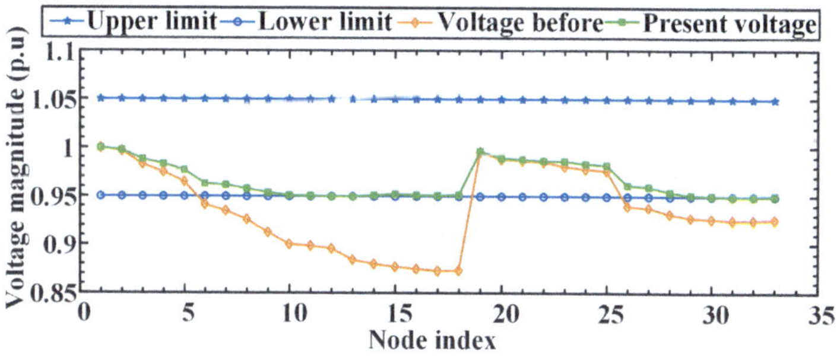

Fig. 8.7 Node voltage magnitude before and after congestion management

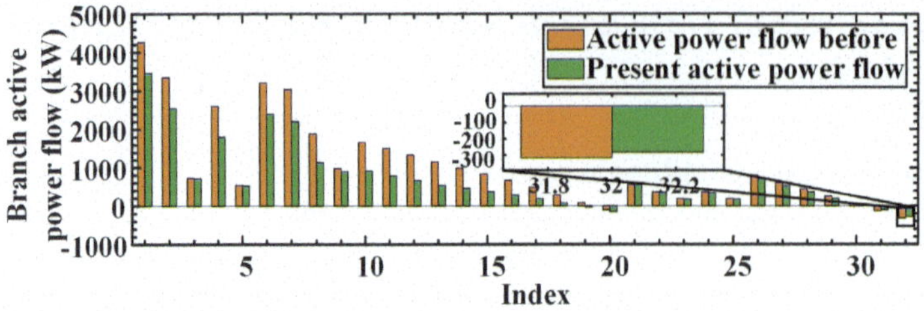

Fig. 8.8 Active power flow before and after congestion management

locally, alleviating congestion on branch 32. The proposed method allows for better utilization of DER flexibility and distributed generation as the network grows.

To fairly distribute congestion management costs, the DSO can adjust transmission fees in the energy market monthly according to these costs and make updates as needed. Additionally, market access conditions for prosumers could require continuous participation in transactions for at least a quarter.

8.5.5 *Analysis of Model Efficiency and Scalability*

A total of 581 bids are moved to the flexibility market for energy resale. Table 8.3 illustrates the efficiency of each stage in the rapid sequential clearing process. Table 8.4

The bidding pattern from prosumers roughly follows a Poisson distribution [27]. The linearized congestion management model is efficiently solvable, with each phase converging quickly. Compared to the total time window of $\Delta t = 5$ min, this approach supports high-frequency operation in the real-time local market, even with a large number of prosumers.

To assess the proposed method's efficiency and scalability, it was compared with the Security Constrained Economic Dispatch (SCED) model. As shown in Tables 8.5 and 8.6, the SCED method struggles with computational demands when there are 140 prosumers per node. However, the proposed method maintains high computational efficiency even with 1600 prosumers per node. This capability is essential for managing large-scale prosumer participation and ensuring efficient energy transactions.

Table 8.4 Trading efficiency of 3300 prosumers

Market phase	Iteration	Time (s)	Window limit (s)
Bidding	–	–	120
Matching	2	0.35	120
Congestion Management	3	4.82	60

Table 8.5 Solution efficiency of the SCED with different scales of prosumers

Prosumer number per node	20	40	60	80	140
Solving time (s)	6.73	21.53	48.60	83.41	313.47

Table 8.6 Solution efficiency of the proposed method with different scales of prosumers

Prosumer number per node	200	400	800	1600	3200
Solving time of energy market (s)	2.19	12.86	38.67	110.54	519.21
Solving time of flexibility market (s)	7.80	11.02	13.89	39.48	138.97
Total (s)	9.99	23.88	52.56	150.02	658.18

In the SCED model, with m buyers and n sellers, the complexity arises from the need to introduce $m \times n$ decision variables to accurately account for network losses and transmission fees in peer-to-peer transactions. In contrast, the congestion management model proposed in this chapter requires only $m \times n$ decision variables, focusing on the injected power at each node. This results in a considerably shorter optimization time compared to the SCED model.

All test cases consistently converged within 2–4 iterations. Importantly, results that neglect network losses are close to actual values. Typically, 2–3 iterations suffice to mitigate system network losses. The iteration termination criteria can be adjusted according to the trading cycle's needs, making the method adaptable to real-time operations with numerous prosumers.

Unlike the SCED model, which may restrict trading opportunities for loads at the network's end during peak periods due to voltage and power flow limitations, the proposed LEM model allows for greater trading volume by initially ignoring power flow constraints. This approach is followed by organizing a flexible market to utilize the adjusted capacities of market players. Transactions are executed physically, and failed bids in the GPM are settled with the DSO, while those in the ERM are moved to the flexibility market for further profit potential. This method accommodates various risk preferences among prosumers, enhancing both system reliability and economic efficiency.

8.6 Conclusions

This chapter introduces a sequential market mechanism designed to handle local energy and flexibility markets, ensuring reliability and high-frequency transactions even with a large number of prosumers. The mechanism employs the CDA approach, which adjusts prices dynamically based on network losses and transmission fees. Additionally, the DSO manages a flexible market to enhance local transactions and maintain operational security. Key findings include:

(1) By adjusting bidding prices for network losses and transmission fees, the proposed model encourages prosumers to trade locally and effectively recover service costs, unlike conventional CDA methods.
(2) Compared to traditional contract models, this approach guarantees minimum revenue for prosumers, motivating them to utilize their flexibility and contribute to the distribution network.
(3) The proposed mechanism outperforms the SCED model by increasing overall market profits and improving local transaction matching. It also offers lower complexity, effectively meeting the real-time demands of large-scale local markets.

Future studies should explore prosumer self-management and market decision-making processes and how the DSO can set transmission fee standards based on congestion management costs.

Appendix 8.1: Sensitivity Factors of Power Flow in Distribution Networks

Ignoring the line-to-ground conductance, the matrix form of Eqs. (8.34) and (8.35) can be expressed as follows:

$$\begin{bmatrix} \mathbf{P} \\ \mathbf{Q} \end{bmatrix}_{2N\times 1} = \begin{bmatrix} \mathbf{G} & -\mathbf{B}' \\ -\mathbf{B} & -\mathbf{G} \end{bmatrix}_{2N\times 2N} \begin{bmatrix} \mathbf{V} \\ \boldsymbol{\theta} \end{bmatrix}_{2N\times 1} \tag{8.34}$$

Delete the row and column of the slack node in the coefficient matrix of equation (8.34), and calculate the inverse matrix of equation (8.34),

$$\begin{bmatrix} \mathbf{V} \\ \boldsymbol{\theta} \end{bmatrix}_{2(N-1)\times 1} = \begin{bmatrix} \mathbf{K} & \mathbf{L} \\ \mathbf{M} & \mathbf{N} \end{bmatrix}_{2(N-1)\times 2(N-1)} \begin{bmatrix} \mathbf{P} \\ \mathbf{Q} \end{bmatrix}_{2(N-1)\times 1} \tag{8.35}$$

In the formula, **K**, **L**, **M,** and **N** are the $(N - 1)$ order matrix, and 0 elements are added in the row and column where the slack node is located. Then **K**, **L**, **M,** and **N** are the Nst order matrix, where N denotes the number of nodes in the distribution system.

Energy trading among prosumers influences the power injected at each node in the distribution network, which subsequently impacts voltage levels and line power flows. To determine how these changes affect power flow, the sensitivity factor based on the injected power can be derived as follows:

(1) Sensitivity factor of voltage

Voltage sensitivity is the partial derivative of the voltage V_i at each node to the injected power P_δ and Q_δ. Specifically, it can be expressed as follows:

$$S_{\mathrm{P}_\delta}^{\mathrm{V}} = \frac{\partial V_i}{\partial P_\delta} = \frac{\partial\left(\sum_j \mathbf{K}_{ij} P_j + \mathbf{L}_{ij} Q_j\right)}{\partial P_\delta} = \mathbf{K}_{i\delta} \tag{8.36}$$

$$S_{\mathrm{Q}_\delta}^{\mathrm{V}} = \frac{\partial V_i}{\partial Q_\delta} = \frac{\partial\left(\sum_j \mathbf{K}_{ij} P_j + \mathbf{L}_{ij} Q_j\right)}{\partial Q_\delta} = \mathbf{L}_{i\delta} \tag{8.37}$$

$$S_{\mathrm{P}_\delta}^{\theta} = \frac{\partial \theta_i}{\partial P_\delta} = \frac{\partial\left(\sum_j \mathbf{M}_{ij} P_j + \mathbf{N}_{ij} Q_j\right)}{\partial P_\delta} = \mathbf{M}_{i\delta} \tag{8.38}$$

$$S_{\mathrm{Q}_\delta}^{\theta} = \frac{\partial \theta_i}{\partial Q_\delta} = \frac{\partial\left(\sum_j \mathbf{M}_{ij} P_j + \mathbf{N}_{ij} Q_j\right)}{\partial Q_\delta} = \mathbf{N}_{i\delta} \tag{8.39}$$

where, $S_{\mathrm{P}_\delta}^{\mathrm{V}}$, $S_{\mathrm{Q}_\delta}^{\mathrm{V}}$, $S_{\mathrm{P}_\delta}^{\theta}$ and $S_{\mathrm{Q}_\delta}^{\theta}$ mean the sensitivity of node voltage amplitude and phase angle to injected active and reactive power, respectively.

(2) Sensitivity factor of line flow.

The sensitivity of line flow is defined as the partial derivative of the line flow P_{ij} and Q_{ij} to the node injected power P_δ and Q_δ. Combining Eqs. (8.3), (8.36), and (8.38), the sensitivity of the active power flow of the line to the injected active power of the node can be expressed as follows:

$$\begin{aligned} S_{\mathrm{P}_\delta}^{\mathrm{P}_{ij}} &= \frac{\partial P_{ij}}{\partial P_\delta} = \frac{\partial\left[g_{ij}\left(V_i - V_j\right) - b_{ij}\left(\theta_i - \theta_j\right)\right]}{\partial P_\delta} \\ &= g_{ij}\left(\mathbf{K}_{i\delta} - \mathbf{K}_{j\delta}\right) - b_{ij}\left(\mathbf{M}_{i\delta} - \mathbf{M}_{j\delta}\right) \end{aligned} \tag{8.40}$$

Similarly, the sensitivity factors of other power flow lines can be expressed as follows:

$$S_{\mathrm{Q}_\delta}^{\mathrm{P}_{ij}} = \frac{\partial P_{ij}}{\partial Q_\delta} = g_{ij}\left(\mathbf{L}_{i\delta} - \mathbf{L}_{j\delta}\right) - b_{ij}\left(\mathbf{N}_{i\delta} - \mathbf{N}_{j\delta}\right) \tag{8.41}$$

$$S_{\mathrm{P}_\delta}^{\mathrm{Q}_{ij}} = \frac{\partial Q_{ij}}{\partial P_\delta} = -b_{ij}\left(\mathbf{K}_{i\delta} - \mathbf{K}_{j\delta}\right) - g_{ij}\left(\mathbf{M}_{i\delta} - \mathbf{M}_{j\delta}\right) \tag{8.42}$$

$$S_{\mathrm{Q}_\delta}^{\mathrm{Q}_{ij}} = \frac{\partial Q_{ij}}{\partial Q_\delta} = -b_{ij}\left(\mathbf{L}_{i\delta} - \mathbf{L}_{j\delta}\right) - g_{ij}\left(\mathbf{N}_{i\delta} - \mathbf{N}_{j\delta}\right) \tag{8.43}$$

where $S_{\mathrm{P}_\delta}^{\mathrm{P}_{ij}}$, $S_{\mathrm{Q}_\delta}^{\mathrm{P}_{ij}}$, $S_{\mathrm{P}_\delta}^{\mathrm{Q}_{ij}}$ and$S_{\mathrm{Q}_\delta}^{\mathrm{Q}_{ij}}$ mean the sensitivity of line flow P_{ij} and Q_{ij} to node injection power P_δ and Q_δ.

(3) Sensitivity factor of network loss.

The sensitivity of network loss is defined as the partial derivative of system network loss P_{loss} and Q_{loss} to the node injected power P_δ and Q_δ. Specifically, it can be expressed as follows:

$$
\begin{aligned}
S_{\mathrm{P}_\delta}^{\mathrm{P}_{\mathrm{loss}}} = \frac{\partial P_{\mathrm{loss}}}{\partial P_\delta} &= \frac{\partial\left[\sum_i \sum_j r_{ij}\left(P_{ij}^2 + Q_{ij}^2\right)/V_i^2\right]}{\partial P_\delta} \\
&\approx \sum_i \sum_j \left[\frac{r_{ij}}{V_i^2}\left(2P_{ij}\frac{\partial P_{ij}}{\partial P_\delta} + 2Q_{ij}\frac{\partial Q_{ij}}{\partial P_\delta}\right)\right] \\
&= \sum_i \sum_j \left[\frac{2r_{ij}}{V_i^2}\left(P_{ij}S_{\mathrm{P}_\delta}^{\mathrm{P}_{ij}} + Q_{ij}S_{\mathrm{P}_\delta}^{\mathrm{Q}_{ij}}\right)\right]
\end{aligned} \tag{8.44}
$$

Similarly, the sensitivity factors for other network losses can be expressed as follows:

$$
S_{\mathrm{Q}_\delta}^{\mathrm{P}_{\mathrm{loss}}} = \frac{\partial P_{\mathrm{loss}}}{\partial Q_\delta} = \sum_i \sum_j \left[\frac{2r_{ij}}{V_i^2}\left(P_{ij}S_{\mathrm{Q}_\delta}^{\mathrm{P}_{ij}} + Q_{ij}S_{\mathrm{Q}_\delta}^{\mathrm{Q}_{ij}}\right)\right] \tag{8.45}
$$

$$
S_{\mathrm{P}_\delta}^{\mathrm{Q}_{\mathrm{loss}}} = \frac{\partial Q_{\mathrm{loss}}}{\partial P_\delta} = \sum_i \sum_j \left[\frac{2x_{ij}}{V_i^2}\left(P_{ij}S_{\mathrm{P}_\delta}^{\mathrm{P}_{ij}} + Q_{ij}S_{\mathrm{P}_\delta}^{\mathrm{Q}_{ij}}\right)\right] \tag{8.46}
$$

$$
S_{\mathrm{Q}_\delta}^{\mathrm{Q}_{\mathrm{loss}}} = \frac{\partial Q_{\mathrm{loss}}}{\partial Q_\delta} = \sum_i \sum_j \left[\frac{2x_{ij}}{V_i^2}\left(P_{ij}S_{\mathrm{Q}_\delta}^{\mathrm{P}_{ij}} + Q_{ij}S_{\mathrm{Q}_\delta}^{\mathrm{Q}_{ij}}\right)\right] \tag{8.47}
$$

In the formula, $S_{\mathrm{P}_\delta}^{\mathrm{P}_{\mathrm{loss}}}$, $S_{\mathrm{Q}_\delta}^{\mathrm{P}_{\mathrm{loss}}}$, $S_{\mathrm{P}_\delta}^{\mathrm{Q}_{\mathrm{loss}}}$, and $S_{\mathrm{Q}_\delta}^{\mathrm{Q}_{\mathrm{loss}}}$ mean the sensitivity of active and reactive losses P_{loss} and Q_{loss} to the node injected power P_δ and Q_δ in the entire distribution system, respectively. In contrast, r_{ij} and x_{ij} mean the resistance and reactance of the line *i-j*.

Similarly, the sensitivity factors of each line's network loss to the injected power at each node can be expressed as follows:

$$
S_{\mathrm{P}_\delta}^{\mathrm{P}_{ij}^{\mathrm{loss}}} = \frac{\partial P_{ij}^{\mathrm{loss}}}{\partial P_\delta} == \left[\frac{2r_{ij}}{V_i^2}\left(P_{ij}S_{\mathrm{P}_\delta}^{\mathrm{P}_{ij}} + Q_{ij}S_{\mathrm{P}_\delta}^{\mathrm{Q}_{ij}}\right)\right] \tag{8.48}
$$

$$
S_{\mathrm{Q}_\delta}^{\mathrm{P}_{ij}^{\mathrm{loss}}} = \frac{\partial P_{ij}^{\mathrm{loss}}}{\partial Q_\delta} = \left[\frac{2r_{ij}}{V_i^2}\left(P_{ij}S_{\mathrm{Q}_\delta}^{\mathrm{P}_{ij}} + Q_{ij}S_{\mathrm{Q}_\delta}^{\mathrm{Q}_{ij}}\right)\right] \tag{8.49}
$$

$$
S_{\mathrm{P}_\delta}^{\mathrm{Q}_{ij}^{\mathrm{loss}}} = \frac{\partial Q_{ij}^{\mathrm{loss}}}{\partial P_\delta} = \left[\frac{2x_{ij}}{V_i^2}\left(P_{ij}S_{\mathrm{P}_\delta}^{\mathrm{P}_{ij}} + Q_{ij}S_{\mathrm{P}_\delta}^{\mathrm{Q}_{ij}}\right)\right] \tag{8.50}
$$

$$
S_{\mathrm{Q}_\delta}^{\mathrm{Q}_{ij}^{\mathrm{loss}}} = \frac{\partial Q_{ij}^{\mathrm{loss}}}{\partial Q_\delta} = \left[\frac{2x_{ij}}{V_i^2}\left(P_{ij}S_{\mathrm{Q}_\delta}^{\mathrm{P}_{ij}} + Q_{ij}S_{\mathrm{Q}_\delta}^{\mathrm{Q}_{ij}}\right)\right] \tag{8.51}
$$

Table 8.7 Bidding parameters of prosumers

Bidding phase	Bidding price (CNY/kWh)	Bidding power (kW)	Power factor (purchase/sale of electricity)
Energy Market	U(0.27,1.49)	U(2,6)	U(0.85,0.95)/0.95
Flexibility market (Initial bidding)	U(0.88,2.1)	U(1,3)	U(0.85,0.95)/0.95
Flexibility market (Energy resale)	U(0.27,1.49)	-	U(0.85,0.95)/0.95

In the formula, $S_{\mathrm{P}_\delta}^{\mathrm{P}_{ij}^{\mathrm{loss}}}$, $S_{\mathrm{Q}_\delta}^{\mathrm{P}_{ij}^{\mathrm{loss}}}$, $S_{\mathrm{P}_\delta}^{\mathrm{Q}_{ij}^{\mathrm{loss}}}$, and $S_{\mathrm{Q}_\delta}^{\mathrm{Q}_{ij}^{\mathrm{loss}}}$ mean the sensitivity of the active and reactive losses and Q_{ij}^{loss} of line *i-j* to the injected power P_δ and Q_δ at each node, respectively.

Appendix 8.2: Bidding Parameters for Prosumers

See Table 8.7.

References

1. Wu, Z., Wang, J., Zhong, H., Gao, F., Pu, T., Tan, C. W., & Xia, Q. (2022). Sharing economy in local energy markets. *Journal of Modern Power Systems and Clean Energy, 11*(3), 714–726.
2. Jogunola, O., Ajagun, A. S., Tushar, W., Olatunji, F. O., Yuen, C., Morley, C., & Shongwe, T. (2024). Peer-to-Peer Local Energy Market: Opportunities, Barriers, Security and Implementation Options. IEEE Access.
3. Chen, Y., Wang, J., Zhang, Y., & Wang, X. (2021). Consensus-based approach to day-ahead joint local energy-reserve market design considering uncertainties of renewable energy. *CSEE Journal of Power and Energy Systems, 9*(6), 2076–2089.
4. Song, M., Cai, Y., Gao, C., Chen, T., Yao, Y., & Ming, H. (2022). Transactive energy in power distribution systems: Paving the path towards cyber-physical-social system. *International Journal of Electrical Power & Energy Systems, 142*, 108289.
5. Khorasany, M., Paudel, A., Razzaghi, R., & Siano, P. (2020). A new method for peer matching and negotiation of prosumers in peer-to-peer energy markets. *IEEE Transactions on Smart Grid, 12*(3), 2472–2483.
6. Haggi, H., & Sun, W. (2021). Multi-round double auction-enabled peer-to-peer energy exchange in active distribution networks. *IEEE Transactions on Smart Grid, 12*(5), 4403–4414.
7. Xu, J., & He, X. (2020). Design of double auction mechanism based on social network. *IEEE Access, 8*, 8324–8335.
8. Yan, M., Shahidehpour, M., Alabdulwahab, A., Abusorrah, A., Gurung, N., Zheng, H., & Paaso, E. A. (2021). Blockchain for transacting energy and carbon allowance in networked microgrids. *IEEE Transactions on Smart Grid, 12*(6), 4702–4714.
9. Xu, X., Xu, Y., Wang, M. H., Li, J., Xu, Z., Chai, S., & He, Y. (2020). Data-driven game-based pricing for sharing rooftop photovoltaic generation and energy storage in the residential building cluster under uncertainties. *IEEE Transactions on Industrial Informatics, 17*(7), 4480–4491.

10. Cai, Y., Song, M., Gao, C., Chen, T., & Deng, F. (2022). Energy sharing strategy in green communities considering carbon credit and power traceability. In *Proceedings of the CSEE* (pp. 1–16).
11. Morstyn, T., & McCulloch, M. D. (2018). Multiclass energy management for peer-to-peer energy trading driven by prosumer preferences. *IEEE Transactions on Power Systems, 34*(5), 4005–4014.
12. Paredes, Á., Aguado, J. A., & Rodríguez, P. (2023). Uncertainty-aware trading of congestion and imbalance mitigation services for multi-DSO local flexibility markets. *IEEE Transactions on Sustainable Energy, 14*(4), 2133–2146.
13. Hoseinpour, M., Hoseinpour, M., Haghifam, M., & Haghifam, M. R. (2023). Privacy-preserving and approximately truthful local electricity markets: A differentially private VCG mechanism. *IEEE Transactions on Smart Grid.*
14. Papavasiliou, A. (2017). Analysis of distribution locational marginal prices. *IEEE Transactions on Smart Grid, 9*(5), 4872–4882.
15. Lezama, F., Soares, J., Hernandez-Leal, P., Kaisers, M., Pinto, T., & Vale, Z. (2018). Local energy markets: Paving the path toward fully transactive energy systems. *IEEE Transactions on Power Systems, 34*(5), 4081–4088.
16. Faqiry, M. N., Edmonds, L., Wu, H., & Pahwa, A. (2020). Distribution locational marginal price-based transactive day-ahead market with variable renewable generation. *Applied Energy, 259*, 114103.
17. Wang, B., Wu, W., Zhang, T., & Chen, T. (2022). Deduction of distributed power trading and operational interaction of distribution network based on dynamic network tariffs. In *Automation of electric power systems* (pp. 1–14).
18. Li, R., Wu, Q., & Oren, S. S. (2013). Distribution locational marginal pricing for optimal electric vehicle charging management. *IEEE Transactions on Power Systems, 29*(1), 203–211.
19. Liu, Z., Wu, Q., Oren, S. S., Huang, S., Li, R., & Cheng, L. (2016). Distribution locational marginal pricing for optimal electric vehicle charging through chance constrained mixed-integer programming. *IEEE Transactions on Smart Grid, 9*(2), 644–654.
20. Baroche, T., Pinson, P., Latimier, R. L. G., & Ahmed, H. B. (2019). Exogenous cost allocation in peer-to-peer electricity markets. *IEEE Transactions on Power Systems, 34*(4), 2553–2564.
21. Zhong, W., Xie, S., Xie, K., Yang, Q., & Xie, L. (2020). Cooperative P2P energy trading in active distribution networks: An MILP-based Nash bargaining solution. *IEEE Transactions on Smart Grid, 12*(2), 1264–1276.
22. Paudel, A., Sampath, L. P. M. I., Yang, J., & Gooi, H. B. (2020). Peer-to-peer energy trading in smart grid considering power losses and network fees. *IEEE Transactions on Smart Grid, 11*(6), 4727–4737.
23. Liaquat, S., Hussain, T., Celik, B., Fourney, R., & Hansen, T. M. (2023). Day-ahead continuous double auction-based peer-to-peer energy trading platform incorporating trading losses and network utilisation fee. *IET Smart Grid, 6*(3), 312–329.
24. Yao, H., Xiang, Y., Hu, S., Wu, G., & Liu, J. (2021). Optimal prosumers' peer-to-peer energy trading and scheduling in distribution networks. *IEEE Transactions on Industry Applications, 58*(2), 1466–1477.
25. Wang, J., Liu, J., Li, C., Zhou, Y., & Wu, J. (2020). Optimal scheduling of gas and electricity consumption in a smart home with a hybrid gas boiler and electric heating system. *Energy, 204*, 117951.
26. Zhou, Y., Wu, J., Song, G., & Long, C. (2020). Framework design and optimal bidding strategy for ancillary service provision from a peer-to-peer energy trading community. *Applied Energy, 278*, 115671.
27. Guerrero, J., Chapman, A. C., & Verbič, G. (2018). Decentralized P2P energy trading under network constraints in a low-voltage network. *IEEE Transactions on Smart Grid, 10*(5), 5163–5173.
28. Yang, J., Zhang, N., Kang, C., & Xia, Q. (2016). A state-independent linear power flow model with accurate estimation of voltage magnitude. *IEEE Transactions on Power Systems, 32*(5), 3607–3617.

29. Li, Z., Lai, C. S., Xu, X., Zhao, Z., & Lai, L. L. (2021). Electricity trading based on distribution locational marginal price. *International Journal of Electrical Power & Energy Systems, 124*, 106322.
30. Li, F., & Bo, R. (2007). DCOPF-based LMP simulation: Algorithm, comparison with ACOPF, and sensitivity. *IEEE Transactions on Power Systems, 22*(4), 1475–1485.
31. Baran, M. E., & Wu, F. F. (1989). Network reconfiguration in distribution systems for loss reduction and load balancing. *IEEE Transactions on Power Delivery, 4*(2), 1401–1407.

Part IV
Green Local Energy Market Considering the Carbon Emission

Chapter 9
Local Energy Market Mechanism Design with the Carbon Credit and Power Traceability

Abstract Under the environment of energy sharing between prosumers in power distribution systems, unified pricing, and transactions bring challenges to clarify the additional environmental value of heterogeneous electricity. This is not conducive to a low-carbon transition of end-prosumers and the sustainable development of distributed renewable energy. Therefore, an energy-sharing strategy for a green local energy market (LEM), which considers carbon credit and power traceability, is proposed. A carbon credit is used to monetize the environmental cost of heterogeneous electricity and circulate it together with it in the LEM. To minimize the energy and carbon credit cost of all prosumers/, an electricity-carbon credit joint clearing model is established, and a traceability-based pricing method for heterogeneous electricity is proposed to encourage prosumers to conduct their energy management in a low-carbon way. Using the alternating direction method of multipliers (ADMM), the proposed model is solved in a distributed manner via limited information exchange, thus realizing a distributed energy management of the green LEM. The case studies' results verify the proposed strategy's effectiveness in reducing the total energy cost of the green LEM and promoting green substitution in terminal energy consumption.

9.1 Introduction

Under the dual carbon goals, developing large-scale distributed energy resources (DERs) and advanced smart metering and information communication technologies enables prosumers to directly share energy on the distribution network side [1]. In this context, the concept of a local energy market (LEM) has emerged [2]. Prosumers with power generation capabilities can directly supply surplus energy to other prosumers through market-based transactions [3]. This helps to incentivize prosumers to explore their flexibility and actively participate in system operations, thereby promoting the local consumption of renewable energy and matching local supply and demand [3].

However, unlike the wholesale market on the transmission side, the distribution side has more prosumers, with more widely distributed and diverse DERs, making it challenging for distribution system operators (DSOs) to effectively monitor and

M. Song et al., *Local Energy Markets*, https://doi.org/10.1007/978-981-97-9750-9_9

dispatch them. Furthermore, in the LEM, electricity is no longer a homogeneous product due to the different production sources. Prosumers may be willing to pay a premium for green energy [4, 5], affordable energy [6], or local energy [7]. For example, prosumers with strong environmental awareness tend to purchase local renewable energy, while those lacking expertise and time may prefer to engage in traditional transactions with the distribution network at stable contract prices. Therefore, in the context of energy sharing among small-scale terminal prosumers, the traditional "rational actor" assumption of wholesale markets may no longer be applicable. It is necessary to consider the heterogeneous energy preferences of prosumers and model their bounded rational decision-making behavior in market transactions [8], assessing its impact on transaction outcomes.

Moreover, the increasing prominence of environmental issues is driving a rapid transformation of power systems and giving rise to market mechanisms to promote low-carbon and environmentally friendly practices. These mechanisms add environmental value to renewable energy. Currently, the most widely used include the renewable energy certificate (REC) market [9], the Chinese certified emission reduction (CCER), and the emissions trading system (ETS) [10]. The REC market primarily involves buyers on the sales side, while CCER and ETS are mainly used to impose carbon emission limits on the generation side, operating relatively independently. Due to their similar design goals in promoting low-carbon practices and the continuous development of renewable energy [11], the coordinated operation of these mechanisms can further promote the fair distribution of social costs [12]. The academic community has also conducted related research on these issues. For instance, Reference [13] proposes a coordinated operation model for REC and ETS, which helps reduce carbon emissions while promoting renewable energy development. Reference [14] achieves joint trading of green certificates and carbon emission rights using a blockchain system. Reference [15] proposes an optimal decision-making method for power generators to combine green certificates and carbon quotas in a market environment.

However, research on the coordinated operation of REC and ETS has primarily focused on the wholesale side. As market mechanisms incentivizing renewable energy development, REC and ETS currently do not evaluate or subsidize end-use prosumers in LEM. Most studies still treat renewable and fossil energy generation as homogeneous products in the context of energy sharing among distribution-side prosumers. Research on incentive mechanisms for low-carbon energy management for prosumers is relatively sparse, often overlooking the environmental value of green energy production and consumption, which is detrimental to the sustainable development of distributed renewable energy. For instance, Reference [16] assesses the additional value of renewable energy from the perspective of prosumers' initiative but lacks objective market incentives. Reference [17], based on cooperative game theory, proposes an energy-sharing strategy among multiple microgrids in a carbon market environment but does not consider the environmental value of renewable energy generation. References [18, 19] consider carbon trading in unit commitment and economic dispatch problems, yet they focus solely on the wholesale market on the transmission side.

This chapter proposes an energy-sharing strategy for distribution-side prosumers considering carbon credits and electricity traceability to address the aforementioned issues. Focusing on a green LEM composed of multiple building clusters, we design a carbon credit mechanism based on the traceability of electricity production methods and establish a trading model for the joint clearing of energy and carbon credits. This enables the traceable pricing and trading of heterogeneous electricity to incentivize prosumers to engage in low-carbon energy management. Additionally, to ensure fair and secure market operations while respecting the privacy of prosumers' trading preferences and equipment parameters, we solve the model using the alternating direction method of multipliers (ADMM). Simulation results validate the effectiveness of the proposed model. The main innovations and practical significance of this chapter are as follows:

(1) We propose a market-based pricing method based on different production methods and sources of electricity. This method monetizes the additional value of heterogeneous electricity's carbon and bounded rational behavior labels.
(2) We design a mechanism for acquiring and circulating carbon credits in green LEM and establish a combined energy-carbon credit-sharing strategy. This incentivizes prosumers to engage in low-carbon energy management and promotes the sustainable development of distributed renewable energy through market means.
(3) We implement distributed energy management among multiple prosumers in a green LEM through limited information exchange based on the ADMM. Quantitative calculations validate the strategy's effectiveness in reducing prosumers' energy costs and promoting the substitution of terminal green energy consumption.

9.2 Energy Sharing Framework of Green LEM

9.2.1 Carbon Credit and Energy Sharing Framework

In this section, a green LEM is used as the research subject. The LEM mainly consists of building prosumers and a community manager. Apart from inelastic rigid loads, the building prosumers have DERs, such as photovoltaic (PV) systems, micro turbines (MT), flexible loads represented by thermostatically controlled loads (TCL), and energy storage (ES) devices. The LEM manager, acting as a third-party operator, is responsible for relaying supply and demand signals within the internal market, purchasing carbon credits to cover deficits from external markets, and collecting operation and maintenance fees.

In the LEM, prosumers with surplus energy production or carbon credits share their excess with other prosumers in exchange for benefits. It is assumed that each building is equipped with a building energy management system (BEMS) [20] with metering, communication, and decision-making functions. Each prosumer forecasts

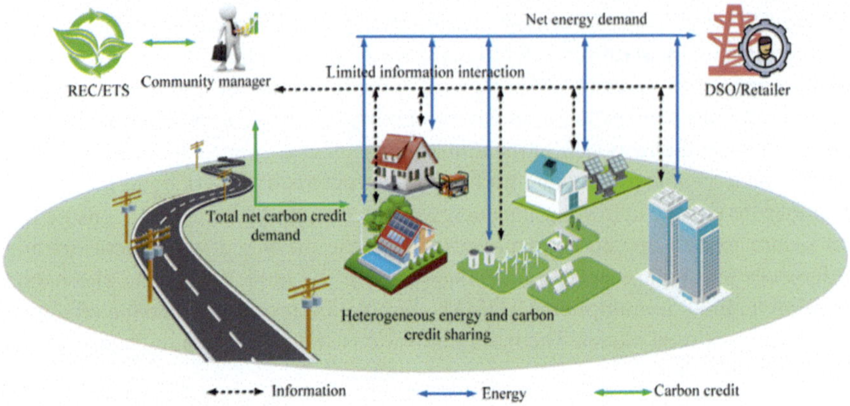

Fig. 9.1 Framework of carbon credit and energy sharing

distributed PV generation and load a day ahead, performing day-ahead energy and carbon credit joint management. The market clearing results then form the operational plan for the next day. The carbon credit and energy sharing operational framework is illustrated in Fig. 9.1.

9.2.2 *Market Operation Rules*

In the green LEM, prosumers with surplus energy production or carbon credits share their excess with other prosumers in exchange for benefits. Each building is assumed to have a building energy management system (BEMS) [20] with metering, communication, and decision-making functions. Each prosumer forecasts distributed PV generation and load a day ahead, performing day-ahead energy and carbon credit joint management. The market clearing results then form the operational plan for the next day. The carbon credit and energy sharing operational framework is illustrated in Fig. 9.1.

(1) Electricity generated through different production methods possesses distinct carbon attributes and personal preference attributes. For example, green electricity has carbon reduction attributes, while electricity produced by LEM-based Distributed Energy Resources (DERs) has local attributes. A joint energy-carbon credit clearing model is employed to evaluate the environmental added value of heterogeneous electricity. Prosumer decisions can be made autonomously based on their preference attributes.
(2) Carbon credits are used solely to measure prosumer contributions to the environment. In addition to being obtained or consumed through various electricity production methods, carbon credits are transferred among prosumers

during electricity transactions. Prosumers can circulate carbon credits internally through sharing mechanisms or purchase deficits directly from external markets. Carbon credits are settled and cleared daily to aid in achieving the zero-carbon goal of green LEM.

(3) In the absence of specific transaction preferences, all prosumers aim to clear the market to minimize the total energy and carbon credit costs for all participants. Local electricity is prioritized for nearby transactions to promote local renewable energy consumption.

(4) The introduction of energy sharing helps reduce energy costs for prosumers, while the carbon credit mechanism serves as a predefined market rule for building green LEMs. Its primary purpose is to improve the electricity production and consumption behaviors of prosumers, thereby promoting the construction of green LEMs. This mechanism is an obligation that prosumers must fulfill when participating in energy sharing within the green LEM. Ultimately, it achieves incentive compatibility between LEM energy costs and environmental benefits.

It should be noted that the carbon credit and energy sharing strategy for green LEMs is merely a business model. Prosumers can choose to participate voluntarily or continue to purchase electricity directly from the Distribution System Operator (DSO) or electricity retailers under traditional models.

9.3 Prosumer Model

9.3.1 Microturbine Model

Taking MT as a typical carbon-emitting power source, we assumed that some building prosumers are equipped with MT as backup power sources or for generating electricity during peak price periods. The operational characteristics of MT can be modeled as follows:

$$0 \leq P_i^{\mathrm{MT}}(t) \leq P_{i,\max}^{\mathrm{MT}} \tag{9.1}$$

where $I = \{1, 2, ..., I\}$ $i \in I$ denotes the set of prosumers, $T = \{1, 2, ..., T\}$ $t \in T$ denotes the set of time intervals, $P_i^{\mathrm{MT}}(t)$ denotes the active power output of the MT for prosumer i at time t interval, $P_{i,\max}^{\mathrm{MT}}$ denotes the maximum active power output limit of the MT for prosumer i.

9.3.2 Distributed Solar Generation Model

Take PV as a typical green power source, prosumers with PV systems can use the generated electricity for self-consumption, share energy within the LEM, or sell it directly to the external grid. Assuming that PV generation incurs no additional costs, the operational characteristics of PV can be modeled as follows:

$$0 \leq P_i^{\mathrm{PV}}(t) \leq P_{i,\max}^{\mathrm{PV}}(t) \tag{9.2}$$

where $P_i^{\mathrm{PV}}(t)$ and $P_{i,\max}^{\mathrm{PV}}(t)$ denotes the active power output and the forecasted maximum active power output of the PV and for prosumer i at time interval t.

9.3.3 Temperature-Controlled Load Model

TCLs are treated as typical flexible load resources for prosumers. Assuming that, under baseline operating conditions, TCLs operate at the most comfortable temperature set by the prosumer [21]. Taking the cooling mode as an example, the operational characteristics of TCLs can be modeled as follows:

$$T_i^{\mathrm{in}}(t+1) = T_i^{\mathrm{out}}(t+1) - \left[T_i^{\mathrm{out}}(t+1) - T_i^{\mathrm{in}}(t)\right]\mathrm{e}^{-1/(RC)} - \eta P_i^{\mathrm{TCL}}(t) \tag{9.3}$$

$$T_{i,\min}^{\mathrm{in}}(t) \leq T_i^{\mathrm{in}}(t) \leq T_{i,\max}^{\mathrm{in}}(t) \tag{9.4}$$

where $T_i^{\mathrm{in}}(t)$, $T_i^{\mathrm{out}}(t)$, $P_i^{\mathrm{TCL}}(t)$ represent the internal and external environmental temperatures and operating power of the TCL time interval t for prosumer. $T_{i,\min}^{\mathrm{in}}(t)$ and $T_{i,\max}^{\mathrm{in}}(t)$ Represent the internal temperature limits of the TCL, R, C, η represent the equivalent thermal resistance, equivalent thermal capacitance, and coefficient of performance of the TCL, respectively.

9.3.4 Energy Storage Model

Prosumers can manage their energy based on the operational characteristics of their Energy Storage (ES) systems, providing greater flexibility for market transactions and electricity production and consumption. The operation of ES equipment should meet the following constraints:

$$0 \leq P_{i,\mathrm{ch}}^{\mathrm{ES}}(t) \leq P_{i,\max}^{\mathrm{ES}} \tag{9.5}$$

$$0 \leq P_{i,\mathrm{dis}}^{\mathrm{ES}}(t) \leq P_{i,\max}^{\mathrm{ES}} \tag{9.6}$$

$$x_{i,\min}^{\mathrm{ES}} \leq x_i^{\mathrm{ES}}(t) \leq x_{i,\max}^{\mathrm{ES}} \tag{9.7}$$

$$x_i^{\mathrm{ES}}(t+1) = x_i^{\mathrm{ES}}(t) + \frac{P_{i,\mathrm{ch}}^{\mathrm{ES}}(t)\Delta t}{B_i^{\mathrm{ES}}} - \frac{P_{i,\mathrm{dis}}^{\mathrm{ES}}(t)\Delta t}{B_i^{\mathrm{ES}}} \tag{9.8}$$

where $P_{i,\mathrm{ch}}^{\mathrm{ES}}(t)$, $P_{i,\mathrm{dis}}^{\mathrm{ES}}(t)$, $x_i^{\mathrm{ES}}(t)$ represent the charging power, discharging power, and state of charge of the ES for prosumer, $P_{i,\max}^{\mathrm{ES}}$, $x_{i,\min}^{\mathrm{ES}}$, $x_{i,\max}^{\mathrm{ES}}$, B_i^{ES} represent the upper limits of charging and discharging power, state of charge limits, and capacity of the ES for the prosumers, respectively.

9.4 Green LEM Energy and Carbon Credit Sharing Mechanism

9.4.1 Green LEM Energy Sharing Model

Prosumers can manage their energy based on the operational characteristics of their ES systems, providing greater flexibility for market transactions and electricity production and consumption. The operation of ES equipment should meet the following constraints:

This chapter categorizes electrical energy based on different sources of electricity generation for product traceability, thereby achieving differentiated pricing for heterogeneous electrical energy. Specifically, $K = \{\mathrm{PV},\ \mathrm{MT},\ \mathrm{grid}\}$ $k \in K$ represent three different methods of electricity generation. The electrical energy from these different generation methods may possess distinct carbon attributes and prosumer preference characteristics, which confer additional value in the market. The power balance constraints based on different energy production sources can be expressed as follows:

$$\sum_{k \in K} P_{i,k}^{\mathrm{load}}(t) = P_i^{\mathrm{load}}(t) \tag{9.9}$$

$$\sum_{k \in K} P_{i,k}^{\mathrm{TCL}}(t) = P_i^{\mathrm{TCL}}(t) \tag{9.10}$$

$$\sum_{k \in K} P_{i,\mathrm{ch},k}^{\mathrm{ES}}(t) = P_{i,\mathrm{ch}}^{\mathrm{ES}}(t) \tag{9.11}$$

$$\sum_{k \in K} P_{i,\mathrm{dis},k}^{\mathrm{ES}}(t) = P_{i,\mathrm{dis}}^{\mathrm{ES}}(t) \tag{9.12}$$

where, $P_i^{\mathrm{load}}(t)$ represents the forecasted rigid load of prosumer i during the time interval t; $P_{i,k}^{*}(t)$ denotes the amount of electrical energy used by the production method i for route * during the time interval t. The above constraints indicate that for any given energy route, the electrical energy can originate only from three sources: PV generation, MT generation, and electricity purchased from the grid.

To ensure the balance of heterogeneous electrical energy within the ES, Eq. (9.8) should be further modified as follows:

$$x_{i,k}^{\mathrm{ES}}(t+1) = x_{i,k}^{\mathrm{ES}}(t) + \frac{P_{i,\mathrm{ch},k}^{\mathrm{ES}}(t)\Delta t}{B_i^{\mathrm{ES}}} - \frac{P_{i,\mathrm{dis},k}^{\mathrm{ES}}(t)\Delta t}{B_i^{\mathrm{ES}}} \tag{9.13}$$

$$\sum_{k \in K} x_{i,k}^{\mathrm{ES}}(t) = x_i^{\mathrm{ES}}(t) \tag{9.14}$$

where $x_{i,k}^{\mathrm{ES}}(t)$ indicates the state of charge of ES of the prosumer to the electric energy from the production mode at the time interval.

Based on the above analysis, the cost and benefits of each prosumer participating in LEM energy trading are described below.

(1) MT power generation cost.

The generation cost model of MT is constructed with the quadratic curve of the active power output of MT, then the generation cost of MT can be expressed as

$$C_i^{\mathrm{MT}}(t) = a_i\left[P_i^{\mathrm{MT}}(t)\Delta t\right]^2 + b_i P_i^{\mathrm{MT}}(t)\Delta t + c_i \tag{9.15}$$

where $C_i^{\mathrm{MT}}(t)$ represents the operating cost of the producer's MT at the time interval, a_i, b_i, c_i represents the cost factor of MT.

(2) ES wear and tear cost.

Energy and battery loss will be generated when ES is charged and discharged. Assuming a linear relationship between the loss of ES and the operating power, the cost model of ES equipment loss can be expressed as follows:

$$C_{i,\mathrm{ES}}^{\mathrm{loss}}(t) = \eta_i^{\mathrm{ES}}\left[P_{i,\mathrm{ch}}^{\mathrm{ES}}(t) + P_{i,\mathrm{dis}}^{\mathrm{ES}}(t)\right]\Delta t \tag{9.16}$$

where $C_{i,\mathrm{ES}}^{\mathrm{loss}}(t)$ represents the loss cost of the producer's ES at the time interval t; η_i^{ES} represents the loss coefficient of ES.

(3) The cost of purchasing electricity.

Prosumers purchase the energy deficit through market transactions, and the corresponding transaction cost can be expressed as

$$C_{i,k}^{\mathrm{p}}(t) = \lambda_k^{\mathrm{p}}(t) P_{i,k}^{\mathrm{net}}(t)\Delta t \tag{9.17}$$

where $P_{i,\mathrm{out}}^k$ indicates the power that the prosumer i sells electricity directly to the external grid from the electric energy derived from the k production mode at a time interval t.

Specifically, when $k = \text{grid}$, $P_{i,\text{grid}}^{\text{net}}$ can be rewritten as $P_{i,\text{grid}}^{\text{netbuy}}$ and $P_{i,\text{grid}}^{\text{netsell}}$, respectively, to represent the net purchase and sale of power to the external grid by prosumer. In this chapter, as for $\lambda_{\text{grid}}^{\text{p}}(t)$, the contract electricity price composed of real time price (RTP) and feed-in-tariff (FiT) is taken as the external purchasing and selling interface of each prosumer respectively, and the difference between the two constitutes the profit space of energy sharing within the LEM, which can be expressed as follows

$$\lambda_{\text{grid}}^{\text{p}}(t) = \begin{cases} \lambda_{\text{RTP}}^{\text{p}}(t), P_{i,\text{grid}}^{\text{netbuy}} \geq 0 \\ \lambda_{\text{FiT}}^{\text{p}}(t), P_{i,\text{grid}}^{\text{netsell}} \geq 0 \end{cases} \tag{9.18}$$

where $\lambda_{\text{RTP}}^{\text{p}}(t)$ and $\lambda_{\text{FiT}}^{\text{p}}(t)$ respectively represents the real-time electricity price and the on-grid electricity price of the external power grid.

In addition, considering that the producer's self-supplied power supply can sell excess electric energy to the grid, then if $k = \text{grid}$, Eq. (9.17) can be further rewritten as,

$$C_{i,\text{grid}}^{\text{p}}(t) = \left\{ \lambda_{\text{RTP}}^{\text{p}}(t) P_{i,\text{grid}}^{\text{netbuy}}(t) - \lambda_{\text{FiT}}^{\text{p}}(t) \cdot \left[P_{i,\text{grid}}^{\text{netsell}}(t) + \sum_{k \in \boldsymbol{K} \setminus \{\text{grid}\}} P_{i,\text{out}}^{k}(t) \right] \right\} \Delta t \tag{9.19}$$

where $P_{i,\text{out}}^{k}$ indicates the power that prosumer i sells electricity directly to the external grid from the electric energy derived from the k production mode at a time interval.

(4) Transaction utility of prosumer.

Each prosumer may have different degree of decision preferences [6], and make limited rational trading decisions. The corresponding relationship between different decision preferences and heterogeneous electric energy is shown in Table 9.1. In Table 9.1, the PV of prosumers $P_{i,\text{ins}}^{\text{PV}}$ is used for spontaneous self-use and internal circulation of the LEM.

Using marginal utility as a constant linear function [16], the finite rational behavior of green and local prosumers is described to quantify the additional value that prosumers are willing to pay for specific electricity. The prosumer transaction utility model based on Table 9.1 can be expressed as follows:

Table 9.1 Bounded rational behavior and trading preference of prosumers

Electric energy demand	Limited rational behavior label	Decision preferences
$P_{i,\text{PV}}^{\text{net}} + P_{i,\text{ins}}^{\text{PV}}$	Green environment type	Use more green electricity
$P_{i,\text{grid}}^{\text{netsell}} - P_{i,\text{grid}}^{\text{netbuy}}$	Local consumption type	Use more local power
P_{i}^{TCL}	Enjoyment type	Pursue a comfortable experience

$$U_i^{\mathrm{PV}}(t) = \beta_i^{\mathrm{PV}}\left[P_{i,\mathrm{PV}}^{\mathrm{net}}(t) + P_{i,\mathrm{ins}}^{\mathrm{PV}}(t)\right]\Delta t \tag{9.20}$$

$$U_i^{\mathrm{grid}}(t) = \beta_i^{\mathrm{grid}}\left[P_{i,\mathrm{grid}}^{\mathrm{netsell}}(t) - P_{i,\mathrm{grid}}^{\mathrm{netbuy}}(t)\right]\Delta t \tag{9.21}$$

Since TCL does not directly participate in the transaction of electric energy, and for those who enjoy consumption, the greater the deviation from the most comfortable setting temperature, the higher the degree of dissatisfaction [21]. Therefore, the quadratic function of increasing marginal utility is used to express the degree of dissatisfaction experienced by energy prosumers and prosumers, which can be expressed as follows:

$$U_i^{\mathrm{TCL}}(t) = \beta_i^{\mathrm{TCL}}\left[T_i^{\mathrm{in}}(t) - T_i^{\mathrm{set}}\right]^2 \tag{9.22}$$

where $U_i^{\mathrm{TCL}}(t)$ indicates the degree of dissatisfaction of the producer with TCL deviation from the set temperature within the time interval. β_i^{TCL} is a non-negative utility coefficient, indicating the sensitivity of the producer to TCL temperature; T_i^{set} represents the most comfortable TCL temperature set by the manufacturer.

Based on Eq. (9.15)–(9.22), the energy trading model of prosumers can be expressed as

$$\min C_i^{\mathrm{p}} = \sum_{t \in \boldsymbol{T}} [C_i^{\mathrm{MT}}(t) + C_{i,\mathrm{ES}}^{\mathrm{loss}}(t) + \sum_{k \in \boldsymbol{K}} C_{i,k}^{\mathrm{p}}(t) + U_i^{\mathrm{TCL}}(t) - U_i^{\mathrm{PV}}(t) - U_i^{\mathrm{grid}}(t)] \tag{9.23}$$

where C_i^{p} represents the energy transaction cost of the prosumer in a trading cycle.

In addition to the DERs operation characteristic constraint, the model should also meet the power balance constraint of prosumer, which can be expressed as:

$$P_{i,k}^{\mathrm{net}}(t) = P_{i,k}^{\mathrm{load}}(t) + P_{i,\mathrm{ch},k}^{\mathrm{ES}}(t) - P_{i,\mathrm{dis},k}^{\mathrm{ES}}(t) + P_{i,k}^{\mathrm{TCL}}(t) - P_{i,\mathrm{ins}}^{k}(t),\ k \in \{\mathrm{PV}, \mathrm{MT}\} \tag{9.24}$$

$$P_{i,\mathrm{grid}}^{\mathrm{netbuy}}(t) - P_{i,\mathrm{grid}}^{\mathrm{netsell}}(t) = P_{i,\mathrm{grid}}^{\mathrm{load}}(t) + P_{i,\mathrm{ch,grid}}^{\mathrm{ES}}(t) - P_{i,\mathrm{dis,grid}}^{\mathrm{ES}}(t) + P_{i,\mathrm{grid}}^{\mathrm{TCL}}(t) \tag{9.25}$$

where $P_{i,\mathrm{ins}}^{k}$ denotes the power generation power used by the k power source of the type of prosumer for self-use and internal circulation in the LEM.

Taking minimizing the cost of electricity purchase of all market members as the objective function, the green LEM clearing model can be expressed as follows:

$$\min C_{\mathrm{sum}}^{\mathrm{p}} = \sum_{i \in \boldsymbol{I}} \sum_{t \in \boldsymbol{T}} [C_i^{\mathrm{MT}}(t) + C_{i,\mathrm{ES}}^{\mathrm{loss}}(t) + \sum_{k \in \boldsymbol{K}} C_{i,k}^{\mathrm{p}}(t) + U_i^{\mathrm{TCL}}(t) - U_i^{\mathrm{PV}}(t) - U_i^{\mathrm{grid}}(t)] \tag{9.26}$$

where $C_{\text{sum}}^{\text{p}}$ represents the energy transaction cost of all prosumers in the LEM during a trading cycle.

In addition to Eqs. (9.1)–(9.7), (9.9)–(9.14), and (9.24)–(9.25), for $k \in \{\text{PV}, \text{MT}\}$, the following constraints must also be met:

$$P_{i,\text{ins}}^{k} + P_{i,\text{out}}^{k} = P_i^k \tag{9.27}$$

$$\sum_{i \in I} P_{i,k}^{\text{net}}(t) = 0 \tag{9.28}$$

9.4.2 Green LEM Carbon Credits Sharing Model

To reflect the carbon properties of electric energy through different production routes, and then add environmental value to low-carbon electric energy, this chapter establishes a green LEM carbon credits trading model as an incentive mechanism for low-carbon energy management for prosumers and prosumers. Define carbon credits as an indicator to validate environmentally friendly ways of generating electricity, as a tradable product that is only bought, sold and circulated within the LEM. Under the carbon credits mechanism, prosumers who actively carry out low-carbon energy management can obtain corresponding carbon credits, which can then be resold to other prosumers who have carbon credits deficit in exchange for profits.

It is stipulated that at the end of each trading cycle, the carbon credits account of each prosumer will be settled and cleared. The prosumer can obtain or consume different amounts of carbon credits based on different electric energy production and consumption methods. In addition to direct trading, carbon credits are also circulated in the market with energy trading. For $k \in \{\text{PV}, \text{MT}\}$, the carbon integral circulation rules are shown in Table 9.2.

When $k = \text{grid}$, the electricity considered in this chapter, whether from coal-fired power plants or new energy plants, has participated in the production side of the REC or ETS to fulfill the environmental protection rights and responsibilities. Therefore, as a prosumer of electric power from the grid, this chapter stipulates that prosumers do not need to bear additional environmental responsibilities for the net purchase of

Table 9.2 Circulating rules of carbon credit

Electric energy type	Electric energy flow	Carbon integral change rules
P_i^{PV}	Net production	Capture carbon credits
$P_{i,\text{PV}}^{\text{net}}$	Net buy/net sell	Earn/consume carbon credits
P_i^{MT}	Net production	Carbon consumption integral
$P_{i,\text{MT}}^{\text{net}}$	Net buy/net sell	Consume/earn carbon credits

electric power from the grid, that is, there is no change in carbon credits, to avoid repeated accounting of environmental value of different types of electric power at the production side. Based on the different types of electric energy in Table 9.2, the corresponding carbon integral calculation method can be expressed as:

$$\Delta S_{i,\text{gen}}^{\text{PV}}(t) = \omega^{\text{PV}} P_i^{\text{PV}}(t)\Delta t \tag{9.29}$$

$$\Delta S_{i,\text{trans}}^{\text{PV}}(t) = \omega^{\text{PV}} P_{i,\text{PV}}^{\text{net}}(t)\Delta t \tag{9.30}$$

$$\Delta S_{i,\text{gen}}^{\text{MT}}(t) = -\omega^{\text{MT}} P_i^{\text{MT}}(t)\Delta t \tag{9.31}$$

$$\Delta S_{i,\text{trans}}^{\text{MT}}(t) = -\frac{\omega^{\text{MT}}}{2} P_{i,\text{MT}}^{\text{net}}(t)\Delta t \tag{9.32}$$

where $\Delta S_{i,\text{gen}}^{*}$ and $\Delta S_{i,\text{trans}}^{*}$ respectively represent the change of carbon integral generated by producer producing or trading different types of electric energy at time intervals; ω^{PV} and ω^{MT} respectively represent the carbon integral conversion coefficient of a unit of heterogeneous electric energy. As shown in Table 9.2, to prevent the double accounting of environmental value brought by carbon credits of prosumers on the production side and the consumption side, this chapter defines the circulation rules of carbon credits based on the circulation direction of electric energy in the LEM, that is, the difference between positive and negative. Taking a pure green electricity transaction as an example, Fig. 9.2 shows that in a certain period of time, the PV of producer 1 produces 2 kW·h green electricity, of which 1 kW·h is used by itself and 1 kW·h is sold to producer 2. According to Eq. (9.29), prosumer1 can obtain carbon credits corresponding to 2 kW·h photovoltaic power generation due to the production of green electricity. Since producer 1 sells 1 kW·h green electricity to producer 2 at the same time, according to Eq. (9.30), producer 1 loses the carbon credits corresponding to 1 kW·h green electricity, and producer 2 gains the carbon credits corresponding to 1 kW·h photovoltaic power generation due to net purchase of green electricity, and only pays for 1 kW·h green electricity. At this time, all prosumers obtain carbon credits corresponding to 2 kW·h green electricity, which is balanced with the environmental benefits brought by the total green electricity production of 2 kW·h. Similarly, Eqs. (9.31)–(9.32) represents the carbon integral accounting method of MT power generation, and the negative sign indicates the opposite direction of photovoltaic power generation. The 0.5 coefficient of Eq. (9.32) indicates that the carbon emission cost of circulating MT power generation is shared by the buyer and the seller.

The conversion coefficient of the change of carbon credits generated by different types of electric energy production and trading is calculated as follows:

$$\omega^{\text{PV}} = \frac{\lambda^{\text{REC}}}{\lambda^{\text{s}}} \tag{9.33}$$

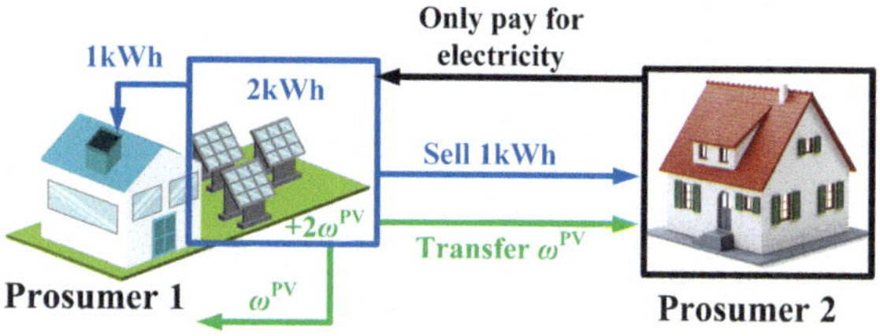

Fig. 9.2 Calculating method of carbon credit

$$\omega^{MT} = \frac{\sigma^{MT}\lambda^{ETS}}{\lambda^{s}} \tag{9.34}$$

where λ^{REC} represents the green certificate predicted value of the external market, yuan /kW; λ^{ETS} represents the predicted carbon price value of the external market, yuan /t; λ^{s} represents the price of carbon credits within the LEM, a parameter directly specified by the LEM manager, yuan/minute; σ^{MT} represents the carbon emission coefficient of MT unit, t/kW.

With the shortage of LEM carbon credits, the LEM manager buys green certificates or carbon emission rights in the external market and converts them into carbon credits for circulation in the internal market based on Eq. (9.33)–(9.34). In a trading cycle, the carbon credits transaction cost model of each prosumer can be expressed as

$$C_i^{s}(t) = \lambda^{s}\Delta S_i^{net}(t) \tag{9.35}$$

where $C_i^{s}(t)$ and $\Delta S_i^{net}(t)$, respectively, represent the carbon credits transaction cost and net transaction demand of prosumer i at time intervals.

If the carbon credits are settled daily and cleared to zero, and the carbon credits cannot be negative at settlement, then the carbon credits balance constraint of the prosumer can be expressed as:

$$\sum_{t\in T}\Delta S_i^{net}(t) = -\sum_{t\in T}\left[\Delta S_{i,gen}^{PV}(t) + \Delta S_{i,trans}^{PV}(t) + \Delta S_{i,trans}^{MT}(t) + \Delta S_{i,gen}^{MT}(t)\right] \tag{9.36}$$

$$\sum_{t\in \boldsymbol{T}}\Delta S_{grid}^{net}(t) = \sum_{t\in \boldsymbol{T}}\sum_{i\in \boldsymbol{I}}\Delta S_i^{net}(t) \tag{9.37}$$

where $\Delta S_{grid}^{net}(t)$ represents the carbon credits deficit of all prosumers in the LEM at t time interval.

The same as the energy market, with the objective function of minimizing the carbon credit purchase cost of all market members, the carbon credit market clearing model of the green LEM can be expressed as

$$\begin{cases} \min C_{sum}^{s} = \sum_{t\in T}\lambda^{s}\Delta S_{grid}^{net}(t) \\ \text{s.t. 式(36)— (37)} \end{cases} \tag{9.38}$$

where $C_{\text{sum}}^{\text{s}}$ represents the carbon credits trading cost of all prosumers in the LEM during a trading cycle.

9.4.3 Green LEM Energy-Carbon Integration Combined Clearing Model

Based on the above analysis, the energy-carbon integral model considering carbon integral can be expressed as

$$\begin{cases} \min C_{\text{sum}}^{\text{p}} + C_{\text{sum}}^{\text{s}} \\ \text{s.t.} \quad (9.1)\text{--}(9.7), (9.9)\text{--}(9.14), \\ (9.24)\text{--}(9.25), (9.27)\text{--}(9.28), (9.36)\text{--}(9.37) \end{cases} \tag{9.39}$$

In the model, carbon credits are not only directly resold, but also transferred between prosumers along with the production and trading of electric energy. Therefore, carbon credits and electric energy sharing are coupled together, and the coupling constraints are embodied in Eqs. (9.36)–(9.37). The optimization results of the model (Eq. (9.39)) include not only the market clearing results of heterogeneous electric energy, carbon integral trading volume and heterogeneous electric energy trading price, but also the operation scheme of each prosumer DERs, so as to realize the collaborative optimization of green LEM trading and operation.

It should be pointed out that the model (Eq. (9.39)) has good scalability. For any diverse and heterogeneous DERs, it can realize the fine management of energy consumption by prosumers. For example, a prosumer wants to use stable adjustable power supply and grid power to ensure basic power supply, and use flexible resources to smooth the randomness of renewable energy, just modify the heterogeneous power balance constraint (Eqs. (9.24)–(9.25)), and change the energy use mode under different energy traceability paths. For any other distributed power supply or load type DERs, just add or rewrite the constraint (Eqs. (9.9)–(9.12), Eqs. (9.24)–(9.25)), which will not be repeated here.

In the incentive mechanism of the model, energy sharing transfers the profit space composed of RTP and FiT price differences to prosumers. On this basis, carbon credits further monetize the environmental value of green power generation behavior, thus encouraging prosumers to take the initiative to carry out low-carbon energy management, and finally forming a green LEM coordinated management plan that considers energy use cost and environmental benefits.

In addition, the model (Eq. (9.39)) is centralized, and prosumers need to report all power generation information and equipment parameters, as well as limited rational behavior preference parameters that directly affect market prices, to the market organizer, which violates the privacy protection needs of prosumers. The risk of data breaches is also detrimental to the fair and stable operation of the market. In addition, with the further allocation of diversified and heterogeneous DERs and the further expansion of the scale of market players, higher demands are put forward for central computing resources. To make full use of prosumer computing resources and take into account the solving efficiency by parallel computing, this chapter adopts ADMM solving model (Eq. (9.39)) to take into account the scalability of the model and the privacy and security of prosumer.

9.4.4 Green LEM Energy-Sharing Price Mechanism

Energy sharing among prosumers within a LEM can give full play to the complementary characteristics of multiple DERs, promote the local and nearby consumption of new energy, and reduce energy transmission costs. Considering the limited professional knowledge and energy of prosumers, it is necessary to establish a fair and easy price mechanism to encourage prosumers to participate actively. The differentiated pricing method of heterogeneous energy proposed in this chapter is described below.

(1) PV electricity price

Considering the principle of fair distribution of benefits, the price of green electricity is set as the average of RTP and FiT in principle. Combined with the transaction preference parameters submitted by each prosumers, the green electricity price is calculated as follows:

$$\lambda_{i,\text{buy}}^{\text{PV}}(t) = \left[\lambda_{\text{RTP}}^{\text{P}}(t) + \lambda_{\text{FiT}}^{\text{P}}(t)\right]/2 + \beta_i^{\text{PV}} + \beta_i^{\text{grid}} \tag{9.40}$$

$$\lambda_{i,\text{sell}}^{\text{PV}}(t) = \frac{\sum_{i\in I}\lambda_{i,\text{buy}}^{\text{PV}}(t)\left[P_{i,\text{PV}}^{\text{net}}(t)\right]^{+}}{\sum_{i\in I}\left[P_{i,\text{PV}}^{\text{net}}(t)\right]^{+}} \tag{9.41}$$

where $\lambda_{i,\text{buy}}^{\text{PV}}(t)$ and $\lambda_{i,\text{sell}}^{\text{PV}}(t)$ respectively represent the purchase and sale prices of green electric energy of prosumers; $[*]^{+}$ represents a projection of $*$ in a non-negative quadrant, e.g. $\left[P_{i,k}^{\text{net}}(t)\right]^{+} = \max\left(0, P_{i,k}^{\text{net}}(t)\right)$. Equations (9.40)–(9.41) indicates that the purchase price of green electricity should be the sum of the average value of RTP and FiT and the additional value that prosumers are willing to pay for green electricity. To ensure market balance and fair distribution of trading surplus, the selling price is uniformly set as the average purchase cost of green electricity.

(2) MT electricity price

To ensure the benefits of MT power generation and reduce transaction costs, MT power price is set as the marginal cost of MT operating point, and MT power price is calculated as follows:

$$\lambda_{i,\text{buy}}^{\text{MT}}(t) = \lambda_{\text{M}}^{\text{MT}}(t) + \beta_i^{\text{grid}} \tag{9.42}$$

$$\lambda_{i,\text{sell}}^{\text{MT}}(t) = \frac{\sum_{i\in I} \lambda_{i,\text{buy}}^{\text{MT}}(t)\left[P_{i,\text{MT}}^{\text{net}}(t)\right]^+}{\sum_{i\in I} \left[P_{i,\text{MT}}^{\text{net}}(t)\right]^+} \tag{9.43}$$

where $\lambda_{i,\text{buy}}^{\text{MT}}(t)$ and $\lambda_{i,\text{sell}}^{\text{MT}}(t)$, respectively, represent the purchase and sale prices of the producer's MT electric energy; $\lambda_{\text{M}}^{\text{MT}}(t)$ represents the marginal cost of each MT run point in the LEM.

(3) Power grid price

Since the grid electricity does not circulate within the LEM, the price is not affected by the behavior of each prosumer, the grid electricity price is RTP and FiT.

9.5 Solution Methodology

To protect the privacy information of each prosumer, ADMM is used to solve the problem (Eq. (9.39)), and the subproblem of each prosumer in the first iteration is constructed, which is expressed as follows:

$$\begin{cases} \min \sum_{t\in \boldsymbol{T}} \{\boldsymbol{C}_i(t) + \boldsymbol{U}_i(t) + \sum_{k\in \boldsymbol{K}\backslash\{\text{grid}\}} [\mu_{i,k}^{(v-1)}(t)(\boldsymbol{X}_{i,k}^{(v-1)}(t) - \boldsymbol{Y}_{i,k}^{(v)}(t)) + \\ \frac{\rho}{2}(\boldsymbol{X}_{i,k}^{(v-1)}(t) - \boldsymbol{Y}_{i,k}^{(v)}(t))^2]\} + \\ [\upsilon_i^{(v-1)}(\boldsymbol{Z}_i^{(v-1)} - \boldsymbol{W}_i^{(v)}) + \frac{\rho}{2}(\boldsymbol{Z}_i^{(v-1)} - \boldsymbol{W}_i^{(v)})^2] \\ \text{s.t. 式(1)—(7), (9)—(14), (25), (27)} \end{cases} \tag{9.44}$$

where $\boldsymbol{X}_{i,k}$,$\boldsymbol{Y}_{i,k}$, respectively, represent the vectors formed by the terms at the left and right ends of the Eq. (9.24); $\boldsymbol{Z}_i$, $\boldsymbol{W}_i$, respectively, represent the vector formed by the terms at the left and right ends of the Eq. (9.36); $\boldsymbol{C}_i$ represents the vector composed of the prosumer's MT generation cost, ES loss cost and external power grid purchasing cost; $\boldsymbol{U}_i$ represents the utility corresponding to the prosumer's transaction and energy consumption preference; $\mu_{i,k}$ denotes the dual variable of the Eq. (9.24); υ_i denotes the dual variable of the Eq. (9.36); ρ represents the penalty term coefficient of ADMM.

Each prosumer solves the subproblem (Eq. (9.44)), updates and interacts with the LEM manager on the net demand for transactions:

$$\tilde{\mathbf{X}}_{i.k}^{(\upsilon)}(t) = \mathbf{Y}_{i.k}^{(\upsilon)}(t) \tag{9.45}$$

$$\tilde{\mathbf{Z}}_{i}^{(\upsilon)} = \mathbf{W}_{i}^{(\upsilon)} \tag{9.46}$$

In addition to the subproblem (Eq. (44)), in order to ensure the balance of different types of electricity in the market, the LEM manager subproblem needs to be defined as follows:

$$\begin{cases} \min \sum_{t \in \boldsymbol{T}} \sum_{i \in \boldsymbol{I}} \left\{ C_i^{\mathrm{s}}(t) + \left[\sum_{k \in \boldsymbol{K} \setminus \{\mathrm{grid}\}} \mu_{i,k}^{(\upsilon-1)}(t)(\tilde{\mathbf{X}}_{i.k}^{(\upsilon)}(t) - \mathbf{X}_{i.k}^{(\upsilon)}(t)) \right.\right. \\ \left.\left. + \frac{\rho}{2}(\tilde{\mathbf{X}}_{i.k}^{(\upsilon)}(t) - \mathbf{X}_{i.k}^{(\upsilon)}(t))^2 \right\}\right] \\ + \sum_{i \in \boldsymbol{I}} \left[\upsilon_i^{(\upsilon-1)}(\tilde{\mathbf{Z}}_i^{(\upsilon)} - \mathbf{Z}_i^{(\upsilon)}) + \frac{\rho}{2}(\tilde{\mathbf{Z}}_i^{(\upsilon)} - \mathbf{Z}_i^{(\upsilon)})^2 \right] \\ \text{s.t.} \quad (9.28), (9.37) \end{cases} \tag{9.47}$$

The LEM manager solves the subproblem (Eq. (9.47)) and updates the coupled variables and, which can be expressed as follows:

$$\mu_{i,k}^{(\upsilon)}(t) = \mu_{i,k}^{(\upsilon-1)}(t) + \rho(\tilde{X}_{i.k}^{(\upsilon)}(t) - X_{i.k}^{(\upsilon)}(t)) \tag{9.48}$$

$$\upsilon_i^{(\upsilon)} = \upsilon_i^{(\upsilon-1)} + \rho(\tilde{\mathbf{Z}}_i^{(\upsilon)} - \mathbf{Z}_i^{(\upsilon)}) \tag{9.49}$$

The model (Eq. (9.39)) is a 2-block convex optimization problem, which can ensure the convergence of ADMM [22, 23]. The convergence criterion can be set as follows:

$$\max_{i,k,t} \left\{ \begin{array}{l} \left|\mu_{i,k}^{(\upsilon)}(t) - \mu_{i,k}^{(\upsilon-1)}(t)\right|, \left|\upsilon_i^{(\upsilon)} - \upsilon_i^{(\upsilon-1)}\right|, \\ \left|X_{i.k}^{(\upsilon)}(t) - Y_{i.k}^{(\upsilon)}(t)\right|, \left|Z_i^{(\upsilon)} - W_i^{(\upsilon)}\right| \end{array} \right\} \leq \varepsilon \tag{9.50}$$

where ε indicates convergence accuracy.

The pseudo-code of the model-solving algorithm based on ADMM is shown in Appendix 9.1. In the process of solving the model, the prosumer only needs to interact with the LEM manager on the limited heterogeneous electric energy net demand and global duality variable information of the model, and the DERs parameter and power generation plan will not be transmitted, which realizes the privacy protection of each market member.

9.6 Case Studies

9.6.1 Parameter Setting

The analysis involves 80 smart building prosumers as market participants. The prosumers are categorized as follows: Prosumer 1–20 are labeled as green eco-friendly prosumers; Prosumer 21–40 are labeled as local consumption prosumers; Prosumer 41–60 are labeled as comfort-oriented prosumers; and Prosumer 61–80 have no special transaction preferences. All prosumers are assumed to be located on the same low-voltage bus within a LEM, and network losses, voltage, and flow issues are ignored.

The time step is $\Delta t = 1$ h. The load and PV generation data for each prosumer are adapted from literature [24]. The carbon emission coefficient for MT generation is 0.065 t/GJ [25]. The external REC price and ETS price are 50 RMB/MW [] and 427.35 RMB/t [26], respectively, while the carbon credit price is 0.05 RMB/min. Other DER parameters are detailed in Appendix 9.2, with each building containing 80 TCLs. The model is solved using MATLAB R2020a with Gurobi 9.1. The following scenarios are set up to verify the effectiveness of the model and algorithm:

Scenario 1: In the absence of energy sharing and without considering the carbon credit mechanism, the 8 buildings with no transaction preferences independently arrange their production and consumption plans under external fixed prices.
Scenario 2: In the energy-sharing environment, without considering the carbon credit mechanism, the 8 buildings with no transaction preferences are allowed only internal energy trading.
Scenario 3: In the energy-sharing e9nvironment, but without considering the carbon credit mechanism, the 8 buildings with different preferences are allowed only internal energy trading.
Scenario 4: In the energy-sharing environment, considering the carbon credit mechanism, the 8 buildings with different preferences can engage in both energy and carbon credit trading.
Scenario 5: Building on Scenario 4, gradually increase the number of prosumers to 80, allowing simultaneous energy and carbon credit trading.

Details of the DERs and transaction preferences for the 8 buildings are provided in Appendix 9.3. Scenarios 1–4 are used to demonstrate the effectiveness of the proposed model, while Scenario 5 is used to analyze the solution efficiency of ADMM.

9.6.2 LEM Energy Cost Analysis Under Different Scenarios

Table 9.3 shows the total LEM cost under scenarios 1 to 4. In Scenario 1, energy sharing among LEM prosumers is not considered, and the MTs equipped by prosumers only meet their own electricity needs. Therefore, the overall generation cost is relatively low. Prosumers with insufficient electricity production capacity must purchase electricity from the grid at a higher RTP to meet their energy needs. During peak periods of PV output during the day, prosumers with PV installations experience a net load surplus. Due to the lack of energy-sharing mechanisms, prosumers can only sell the surplus electricity directly to the grid at a lower FiT to generate revenue. As a result, the total LEM cost in Scenario 1 is the highest.

In Scenario 2, the organization of an internal energy market within the LEM reduces the amount of electricity bought and sold to the external grid, incentivized by the price difference between RTP and FiT. The surplus PV electricity is entirely shared with other prosumers, achieving full local consumption of internal PV generation. Additionally, prosumers with MTs can further increase their MT output to sell to other prosumers for profit, resulting in higher overall generation costs. However, the profit margin created by the FiT and RTP price difference is transferred to the prosumers, thereby reducing the total LEM cost. In Scenario 3, due to the bounded rationality of prosumers, they may make decisions that deviate from the LEM's optimal energy management and sharing strategy, leading to a loss of social welfare. Therefore, compared to Scenario 2, the total LEM energy cost in Scenario 3 slightly increases.

Scenario 4 builds on Scenario 3 by considering a carbon credit mechanism to incentivize low-carbon operations within the LEM. Given the limited PV installations and the presence of several carbon-emitting power sources, MT generation bears the environmental responsibility for carbon emissions under the carbon credit mechanism, effectively increasing the cost of MT generation. Consequently, the total LEM energy cost slightly increases. Compared to Scenario 3, MT generation decreases in Scenario 4, and purchasing from the external grid meets the corresponding electricity deficit. Despite the additional carbon credit costs incurred in Scenario 4, the mechanism is expected to promote the substitution of green electricity consumption at the end-user level as the energy supply side transitions further.

Table 9.3 Total cost in different cases

Scenario	Energy cost/yuan			Carbon credits cost/yuan	Sum/yuan
	Generation	Purchase	Sell		
1	2701.1	11,193.5	507.2	0	13,387.4
2	4447.4	7514.1	0	0	11,961.5
3	4638.4	7592.4	0	0	12,230.8
4	3451.3	8795.9	0	291.9	12,539.1

9.6.3 Analysis of Energy Use Characteristics Under Bounded Rational Behavior

As can be seen from Fig. 9.3, there are great differences in the composition of power supply among prosumers with different preferences. During the peak period of PV output in the afternoon, the green electricity produced by the LEM is mainly supplied to green environment-friendly prosumers, and the electricity demand of prosumer 1 and 2 is met by green electricity during 09:00–15:00. As shown in Table 9.8, compared with prosumer 1, prosumer 2 is willing to pay more for green electricity (0.3 yuan/(kW·h)), so prosumer 2 has a higher priority to use green electricity, and still uses green electricity as the main or part of the power supply source during 08:00, 16:00–17:00. In addition, although the PV output is approximately 0 during 04:00–07:00 and 18:00–24:00, producer 2 can still rely on the energy storage capacity of ES equipment in the LEM to obtain part of the green electricity. Figure 9.3a, c, d is taken as an example. Prosumers 1, 5 and 7 all sold varying amounts of green electricity to prosumer 2 through energy storage devices during the above period.

Figure 9.3c, d respectively show the power supply of local prosumers and those with no transaction preference. In the noon period, because the surplus electricity of the producer equipped with PV makes up the electricity shortage within the LEM to a certain extent, the marginal cost of the internal MT is generally less than the external RTP, so that the electricity demand of the prosumers 5 and 7 can be met by local electricity from 11:00 to 14:00. In the rest of the period, compared with prosumer 7, because prosumer 5 is willing to pay extra for local electricity in the LEM (including MT and PV power generation), local electricity will be preferentially supplied to prosumer 5, because the value added by prosumer 5 to local electricity (0.05 yuan /(kW·h)) is less than that added by prosumer 5 to green electricity. Therefore, prosumer 5 has a lower priority to obtain green electricity, and the trading preference of prosumer 5 is mainly satisfied by MT within the LEM except during peak PV output hours. It should be pointed out that night and early morning belong to the low power consumption period, and RTP is already less than the marginal cost corresponding to MT no-load, so MT will not start. During this period, producers and prosumer 5 choose to directly purchase electricity from the external power grid with RTP.

Figure 9.4, respectively, shows the TCL operating power and room temperature changes of prosumers 3, 4, and 7. Prosumer 7 has no special behavior preference, while prosumers 3 and 4 are self-enjoyment type. The black line shows the corresponding TCL set temperature ($T_{set} = 23\,°\mathrm{C}$) under the most comfortable state of each prosumer.

The TCL operating states of prosumers with different requirements for energy comfort are different. As shown in Fig. 9.4, compared with prosumer 3 and 4, prosumer 7 has low requirements for comfort, and TCL's operating power throughout the day is at a relatively low level, which is reflected in TCL's energy-saving operation to reduce costs, and its willingness to actively adjust flexible TCL to improve

Fig. 9.3 Trading results under bounded rational behavior

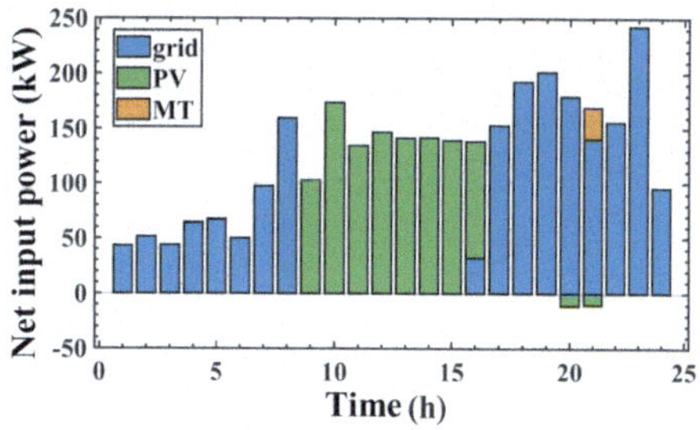

(a) prosumer 1 (green eco-friendly)

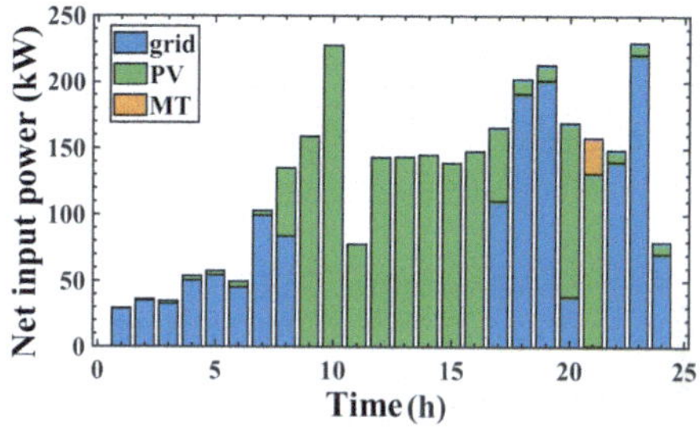

(b) prosumer 2 (green eco-friendly)

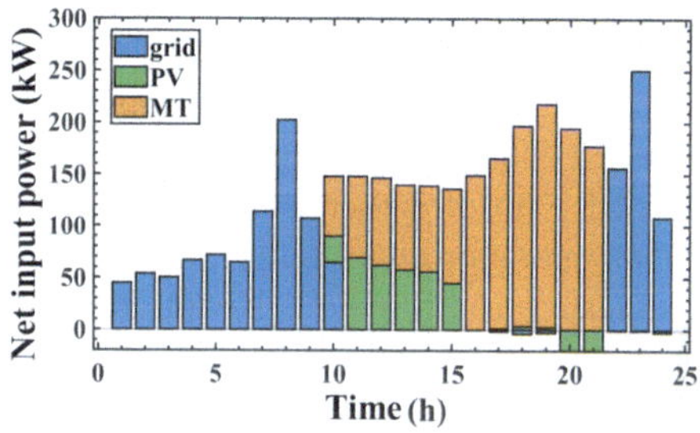

(c) prosumer 5 (local consumption)

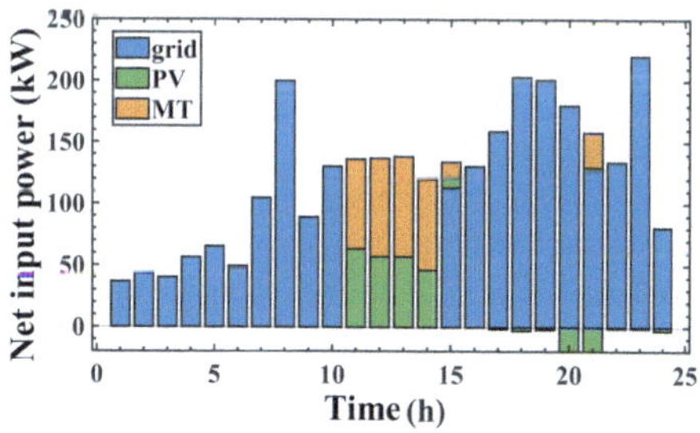

(d) prosumer 7 (no special transaction preferences)

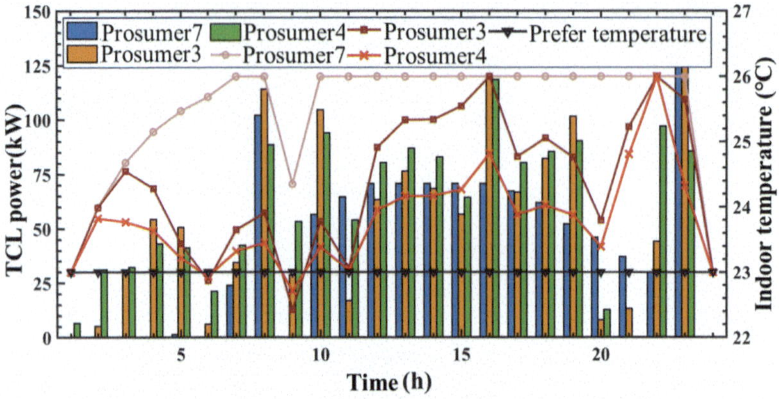

Fig. 9.4 Operating status of TCL under bounded rational behavior

revenue. Therefore, within the tolerable temperature comfort range, the room temperature of the prosumers 7 is basically maintained at the upper limit of the comfort range of 26 °C most of the time.

Prosumers 3 and 4 are enjoyment-type prosumers and have a high demand for energy use comfort, so they avoid sacrificing energy use experience and actively adjusting TCL to a certain extent. As shown in Fig. 9.4, the overall TCL operating power of producer 3 and 4 is significantly higher than that of producer 7, and the indoor temperature is closer to the most comfortable temperature. TCL basically behaves as a conventional rigid load and does not provide any elasticity.

Since ES is not directly related to the bounded rational behavior of each prosumer, the overall operation characteristics of ES of each prosumer are similar, all of which are manifested as arbitrage behaviors based on RTP peak-valley price difference. Different prosumers only have differences in energy storage types due to different preferences for heterogeneous electric energy, which can be seen in Fig. 9.3 and will not be repeated here. Based on the above analysis, it can be found that each prosumer's limited rational behavior preferences are basically satisfied.

9.6.4 LEM Energy Cost Analysis Under Different Scenarios

In Scenario 4, Fig. 9.5 shows the selling price of heterogeneous electric energy, that is, the unified settlement price of the prosumer corresponding to Eq. (9.41) and (9.43) by selling type of electric energy. Based on the different trading preferences of each prosumer, the corresponding purchase price fluctuates up and down based on the selling price. Figure 9.6 shows the power purchased by the LEM from the external grid.

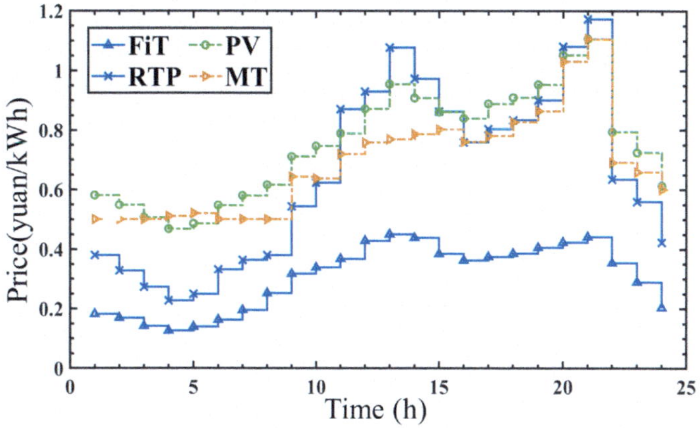

Fig. 9.5 Selling price of heterogeneous electricity in case 4

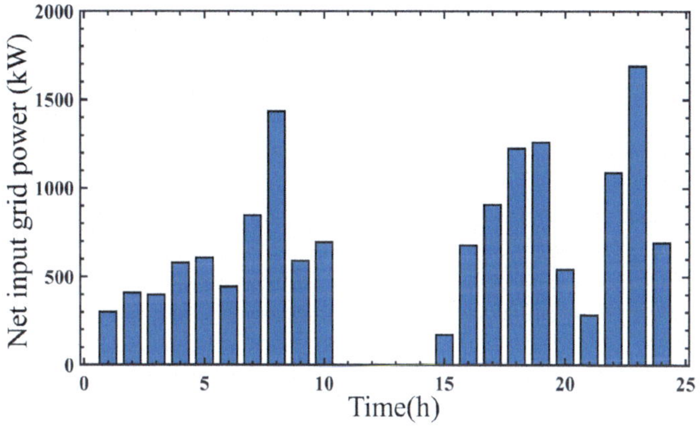

Fig. 9.6 Supplied electricity from external grid in case 4

Because the electricity in the LEM is generally in short supply, the price of all types of electricity is close to RTP. As shown in Fig. 9.5, MT does not start during 01:00–08:00, so the price is basically represented as the no-load marginal cost of MT. In other periods, when RTP is higher than the no-load marginal cost of MT, MT begins to exert its power, and the selling price is jointly determined by its marginal cost and the trading preference of each prosumer. During 11:00–14:00, RTP is higher. Therefore, MT can continuously increase its output to meet the supply and demand balance within the LEM, and the marginal cost is still a big gap with RTP. From 09:00 to 10:00 and 22:00 to 24:00, the marginal cost of MT operation is close to RTP, and the power shortage still needs to be supplied by the external power grid. Meanwhile, MT power generation is preferentially supplied to prosumers who have extra preference for local power. The extra value they are willing to pay also increases the selling price of LEM MT.

The selling price of green electricity is completely dominated by the trading preference of green environment-friendly prosumers, and the price is higher between 01:00 and 08:00, but at this time PV does not generate electricity, and only a very small number of green electricity released by ES is circulating in the market. In other periods, because green prosumers are willing to pay extra for green electricity, there are cases where the price is slightly higher than RTP, and the extra cost is fully borne by green prosumers and will not affect other prosumers.

In addition, combined with Fig. 9.6, it can be found that energy sharing reduces the external net load of the whole LEM during the afternoon peak and evening peak, and improves the overall daily load curve of the LEM to some extent.

Table 9.4 shows the energy costs of each prosumer in scenarios 1, 3, and 4. The cost of scenario 4 includes energy cost and carbon credits cost.

Compared with scenario 1 and scenario 3, only prosumer 1 and 2 pay more extra costs to buy green electricity due to their subjective will, and the energy use cost increases, while other prosumers all reduce their energy use cost due to energy sharing. Scenario 4 further introduces the carbon credits mechanism based on scenario 3. In addition to the carbon credit costs and benefits generated by energy-using behaviors, energy costs also change to some extent. Compared with scenario 3, only the energy cost of prosumer 3, 4, and 8 increases significantly. This is mainly because carbon credits have a certain inhibitory effect on MT power generation, making the marginal operating cost of MT in some periods lower than RTP in Scenario 3, reducing part of the expenditure on MT power generation. Correspondingly, prosumer 3, 4, and 8 have less income from selling MT power generation. In particular, although the prosumer 6 owns MT, it has a high preference for local electricity, so its MT power generation is mainly used for self-use, and it is still mainly reflected as MT power generation prosumers, and its energy cost is also reduced to a certain extent.

Table 9.4 Cost of prosumers in case 1, 3, and 4

Prosumer	Scenarios 1	Scenarios 3	Scenarios 4		
			Energy cost	Carbon credits cost	Sum
1	2216.1	2273.5	2267.1	−51.9	2215.2
2	2227.1	2369.8	2360.1	−79.6	2280.5
3	1827.2	1775.1	1837.5	104.2	1941.7
4	984.0	572.9	646.3	61.5	707.8
5	2285.1	2276.5	2142.1	65.4	2207.5
6	873.4	607.6	567.9	115.9	683.8
7	2091.9	1983.6	1980.5	8.4	1988.9
8	882.5	454.7	527.4	67.9	595.3

In terms of the cost of carbon credits, prosumer 1 and 2 pay higher value for green electricity, and can obtain higher carbon credits when they consume green electricity. The prosumers 3, 4, 6 and 8 who own MT and the local prosumer type prosumer 5 who purchase a lot of MT power generation need to buy carbon credits to compensate for the carbon emissions generated. The prosumer 6 are local prosumers, whose MT is not only for self-use, but also for additional MT power generation to meet consumption preferences, so the cost of carbon credits is the highest. In general, the environmental value within the LEM flows from carbon prosumers to green prosumer.

Compared with the traditional mode of "self-use, surplus electricity online", the energy consumption cost of each prosumer has been reduced to different degrees in general, so the mechanism proposed in this chapter has incentives for most prosumers in the LEM. The energy consumption cost of prosumer 2 increased slightly because of the high subjective compensation willingness for green electricity. Compared with prosumers 4, 6 and 8, prosumer 3 have no PV, and MT works at a higher marginal cost in the traditional self-use mode, which is close to RTP. With the further suppression of MT price by carbon credits, the energy cost of Scenario 4 is basically the same as that of scenario 1, and almost all the cost increment comes from the environmental cost of generating carbon emissions. Therefore, under the current mechanism, prosumers who only have MT as their own power supply may be reluctant to participate in the energy sharing strategy described in this chapter. It should be pointed out that if there are other local prosumer type prosumers with higher preferences in the LEM, or the overall power generation capacity in the LEM is further improved, prosumer 3 still has a large profit margin. In addition, with the deepening of the path of carbon neutrality, it is foreseeable that the rights and responsibilities of carbon emissions will be transferred to multiple entities in more diversified forms [27]. On the one hand, the carbon credits mechanism enables the prosumers to pay for carbon emissions, achieving the goal of a zero-carbon LEM conceptually. On the other hand, carbon credits are shared within the LEM along with MT power generation, so that the carbon emissions cost is equally shared among the prosumers. Therefore, considering the goal of reducing energy cost and building green LEM, the mechanism described in this chapter is effective.

When DERs technical parameters are determined, the cost of carbon credits within the LEM is jointly governed by the price of external REC and ETS. Figure 9.7 shows the impact of changes in external market prices on the overall cost of carbon credits in the LEM.

When the REC price is stable at \$50 /MW, the carbon credits cost increases with the price of the external ETS. Compared with foreign countries, as ETS is still in its infancy and carbon price level is low, the LEM as a whole can still rely on internal PV power generation to obtain part of the income under the current technical conditions. With the deepening of the "dual carbon" path and the further penetration and development of distributed new energy on the distribution side, it is foreseeable that the price of REC will continue to decline, while the price of ETS will continue to rise. As shown in Fig. 9.7, in order to fulfill environmental responsibility, the LEM will pay a higher cost of carbon credits to offset the carbon emissions generated. When the external market develops to a REC price lower than 10 yuan/

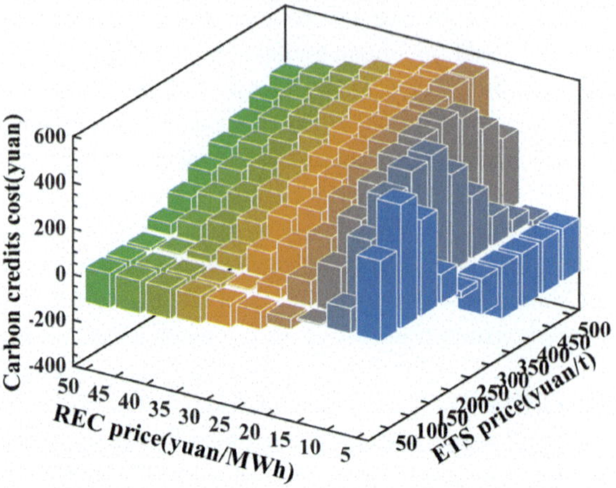

Fig. 9.7 Cost of LEM carbon credit under fluctuant price of REC and ETS

MW and ETS price higher than 250 yuan/t, the overall carbon credits of the LEM are reflected as positive returns, indicating that the MT within the LEM has provided little or completely stopped, and the MT within the LEM is faced with upgrading or transformation, which verifies that the carbon credits mechanism can promote the goal of zero carbon from the economic level to the physical level.

9.6.5 *Model Solving Efficiency Analysis Based on ADMM*

We take $\rho = 0.01$ and $\varepsilon = 10^{-5}$. Figure 9.8 shows the convergence of the residual (9.50) in scenario 4.

The ADMM converges at the 321st iteration with higher accuracy. The maximum solution time of the prosumer problem is 0.11 s, and the maximum solution time of the LEM manager problem is 0.06 s. As shown in Fig. 9.8, the residual decreases rapidly in the early iteration, stabilizes at 0.035 in the 70th iteration, and then

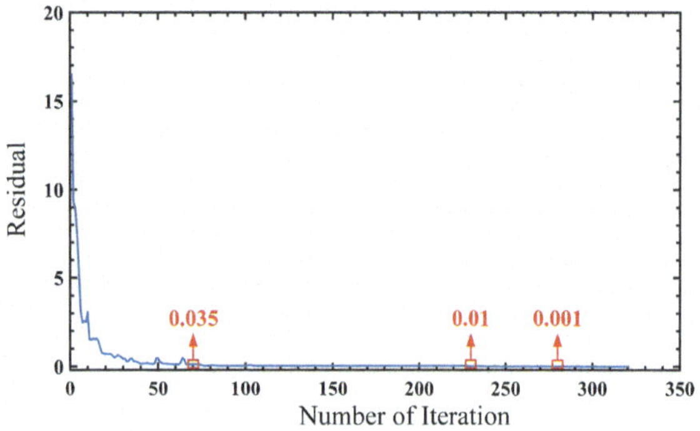

Fig. 9.8 Convergence process of ADMM

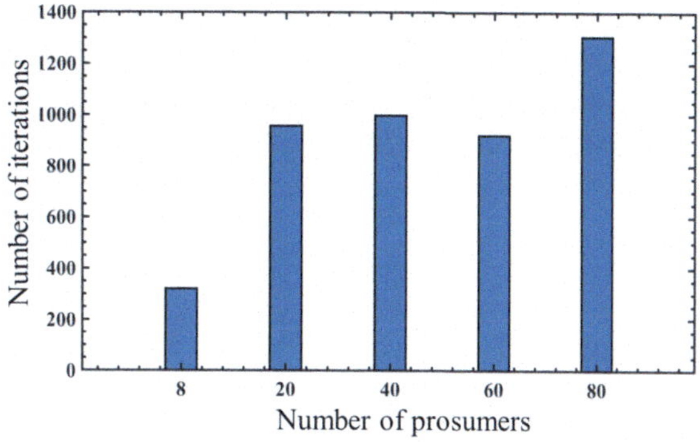

Fig. 9.9 Iterations for different size of prosumers

decreases slowly. Therefore, it is especially suitable for engineering problems where the accuracy requirements are not very strict.

Considering the characteristics of large-scale wide-area distribution of prosumers on the power distribution side in the future, to further verify the algorithm's scalability, the number of prosumers in scenario 4 is gradually increased to 80, and the convergence speed is shown in Fig. 9.9.

Since the prosumer subproblem is solved in parallel mode, the single iteration time does not change significantly with the increase of the number of prosumers. Figure 9.9 shows the change of the number of ADMM iterations with the increase of the number of prosumers. When the number of prosumers is between 20 and 60, the number of iterations tends to be stable. The solution efficiency of the strategy can meet the needs of day-ahead planning at LEM scale.

9.7 Conclusions

The main conclusion of this chapter can be listed as follows:

(1) The proposed market mechanism promotes local electricity trading and on-site renewable energy consumption, thereby reducing the overall LEM energy costs and improving the LEM's overall load profile, achieving optimal resource allocation.

(2) The market mechanism meets each prosumer's individualized transaction demands and energy consumption preferences. Except for prosumers who only have carbon-emitting backup power sources, LEM provides economic incentives for low-carbon energy management and sharing, promoting the transition and upgrade of the distribution-side power structure and the substitution of green electricity consumption at the end-user level.

(3) The ADMM-based solution method accommodates different transaction preferences of prosumers while protecting their privacy. The solution efficiency meets the needs of day-ahead energy management for a large number of prosumers and is effective in scenarios with large-scale prosumers and diverse, heterogeneous DERs.

Appendix 9.1: ADMM Solution Process

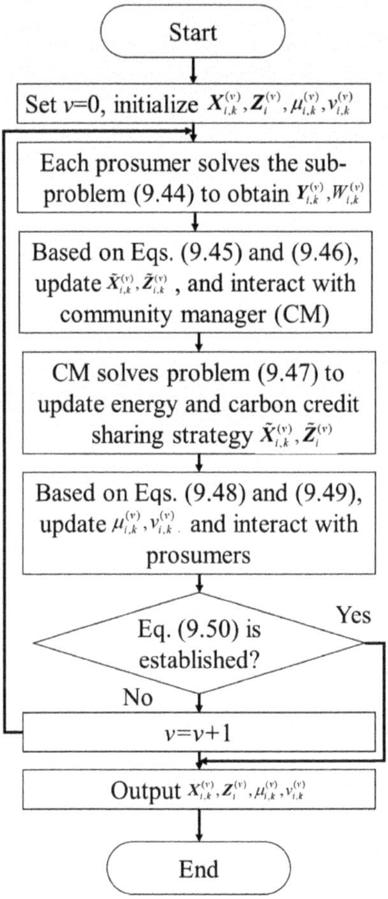

Appendix 9.2: DER Parameters

See Tables 9.5, 9.6, and 9.7.

Appendix 9.3: Parameters of Prosumers

See Table 9.8.

Table 9.5 Parameters of MT

a_i	b_i	c_i
0.00133	0.502	0

Table 9.6 Parameters of TCL

R	C	η	T_{set}	$T_{i,\min}^{\text{in}}$	$T_{i,\max}^{\text{in}}$	$T_i^{\text{in}}(0)$
2	2	2	23	20	26	23

Table 9.7 Parameters of ES

$P_{i,\max}^{\text{ES}}$	$x_{i,\min}^{\text{ES}}$	$x_{i,\max}^{\text{ES}}$	$x_i^{\text{ES}}(0)$	B_i^{ES}	η_i^{ES}
20	0.2	1	0.5	50	0.05

Table 8 Parameters of prosumers

Prosumers	DERs	Behavior marker	Utility coefficient
1	TCL + ES	Environmental protection type	$\beta_i^{\text{PV}} = 0.25$
2	TCL + ES	Environmental protection type	$\beta_i^{\text{PV}} = 0.3$
3	TCL + ES + MT	Enjoyment type	$\beta_i^{\text{TCL}} = 0.35$
4	TCL + ES + MT + PV	Enjoyment type	$\beta_i^{\text{TCL}} = 0.5$
5	TCL + ES	Local consumption type	$\beta_i^{\text{grid}} = 0.05$
6	TCL + ES + PV + MT	Local consumption type	$\beta_i^{\text{grid}} = 0.1$
7	TCL + ES	–	–
8	TCL + ES + MT + PV		–

References

1. Ye, Y., Yuan, Q., Liu, W., Tang, Y., & Goran, S. (2022). Parameter sharing empowered multi-agent deep reinforcement learning for coordinated management of energy communities. In *Proceedings of CSEE, 42*(21), 7682–7695.
2. Melton, R. B. (2013). Gridwise Transactive Energy Framework (draft version) (No. PNNL-SA-22946). Pacific Northwest National Lab. (PNNL), Richland, WA.
3. Hu, J., Wang, K., Ai, X., Han, X., Yang, G., & Yusheng, X. (2019). Transactive energy: An effective mechanism for balancing electric energy system. *Proceedings of the CSEE, 29*, 953–965.
4. Hu, J., Wang, K., Ai, X., Han, X., Yang, G., & Yusheng, X. (2019). Transactive energy: an effective mechanism for balancing electric energy system. *Proceedings of the CSEE, 29*, 953–965.
5. Longo, A., Markandya, A., & Petrucci, M. (2008). The internalization of externalities in the production of electricity: Willingness to pay for the attributes of a policy for renewable energy. *Ecological Economics, 67*(1), 140–152.
6. Zorić, J., & Hrovatin, N. (2012). Household willingness to pay for green electricity in Slovenia. *Energy Policy, 47*, 180–187.
7. Yang, Y., Solgaard, H. S., & Haider, W. (2015). Value seeking, price sensitive, or green? Analyzing preference heterogeneity among residential energy consumers in Denmark. *Energy Research & Social Science, 6*, 15–28.
8. Da Silva, P. G., Karnouskos, S., & Ilic, D. (2012). A survey towards understanding residential prosumers in smart grid neighbourhoods. In *2012 3rd IEEE PES Innovative Smart Grid Technologies Europe (ISGT Europe)* (pp. 1–8).
9. Liu, Y., Gao, S., Wei, X., Huang, T., Zang, T., & Chen, K. (2021). Research on trading model of prosumers considering the social behaviors in the power Internet of Things. *Procedings of CSEE, 41*(9), 3173–3186.
10. Green Power Certificate Subscription Trading Platform of China [EB/OL]. (2022). http://www.greenenergy.org.cn/shop/index.html
11. National Ministry of Ecology and Environment. (2021). *Carbon emissions trading management rules (for trial implementation)* [EB/OL]. https://www.mee.gov.cn/xxgk2018/xxgk/xxgk01/202105/W020210519636657102983.pdf
12. Duan, M., Tian, Z., Zhao, Y., & Li, M. (2017). Interactions and coordination between carbon emissions trading and other direct carbon mitigation policies in China. *Energy Research & Social Science, 33*, 59–69.
13. Böhringer, C., & Rosendahl, K. E. (2010). Green promotes the dirtiest: On the interaction between black and green quotas in energy markets. *Journal of Regulatory Economics, 37*, 316–325.
14. Wang, G., Zhang, Q., Su, B., Shen, B., Li, Y., & Li, Z. (2021). Coordination of tradable carbon emission permits market and renewable electricity certificates market in China. *Energy Economics, 93*, 105038.
15. Feng, C., Xie, F., Wen, F., Zhang, Y., & Hu, J. (2021). Design and implementation of joint trading market for green power certificate and carbon based on smart contract. *Automated Electrical Power System, 45*(23), 1–11.
16. Jun, Y., Jiao, H., & Yongfei, W. (2022). Energy optimization of electricity wholesale market considering carbon emissions trading and green power certificate trading mechanism. *Electric Power, 55*(8), 190–195.
17. Morstyn, T., & McCulloch, M. D. (2018). Multiclass energy management for peer-to-peer energy trading driven by prosumer preferences. *IEEE Transactions on Power Systems, 34*(5), 4005–4014.
18. Yan, M., Shahidehpour, M., Alabdulwahab, A., Abusorrah, A., Gurung, N., Zheng, H., Paaso, E. A., et al. (2021). Blockchain for transacting energy and carbon allowance in networked microgrids. *IEEE Transactions on Smart Grid, 12*(6), 4702–4714

19. Zhang, N., Hu, Z., Dai, D., Dang, S., Yao, M., & Zhou, Y. (2015). Unit commitment model in smart grid environment considering carbon emissions trading. *IEEE Transactions on Smart Grid, 7*(1), 420–427.
20. Ji, Z., Kang, C., Chen, Q., Xia, Q., Jiang, C., Chen, Z., & Xin, J. (2013). Low-carbon power system dispatch incorporating carbon capture power plants. *IEEE Transactions on Power Systems, 28*(4), 4615–4623.
21. Wenhui, G., Hongjun, G., & Haibo, L. (2022). Net zero energy driven power sharing for aggregated buildings: A Peer aggregation model. *Proceedings of CSEE, 42*(24), 8832–8844.
22. Yang, Q., Wang, H., Wang, T., Zhang, S., Wu, X., & Wang, H. (2021). Blockchain-based decentralized energy management platform for residential distributed energy resources in a virtual power plant. *Applied Energy, 294*, 117026.
23. Boyd, S., Parikh, N., Chu, E., Peleato, B., & Eckstein, J. (2011). Distributed optimization and statistical learning via the alternating direction method of multipliers. *Foundations and Trends® in Machine learning, 3*(1), 1–122.
24. Han, D. R. (2022). A survey on some recent developments of alternating direction method of multipliers. *Journal of the Operations Research Society of China*, 1–52.
25. Ratnam, E. L., Weller, S. R., Kellett, C. M., & Murray, A. T. (2017). Residential load and rooftop PV generation: An Australian distribution network dataset. *International Journal of Sustainable Energy, 36*(8), 787–806.
26. Qu, K., Huang, L., Yu, T., & Zhang, X. (2018). Decentralized dispatch of multi-area integrated energy systems with carbon trading. *Proceedings of CSEE, 38*(3), 697–707.
27. The Latest Data on EU ETS Carbon Prices [EB/OL]. (2022). https://ember-climate.org/data/data-tools/carbon-price-viewer/

Chapter 10
Robust Peer-to-Peer Multi-energy Trading Considering Carbon Emission

Abstract In the context of decarbonization, the rise in renewable energy (RE) resources and gas-fired power plants has significantly increased the interdependence between electricity and gas markets. However, centralized market clearing under third-party management raises concerns about data privacy. We propose a peer-to-peer (P2P) multi-energy trading mechanism for integrated electricity and gas distribution systems that incorporates carbon emission limits via a nodal carbon intensity (NCI) model to mitigate this issue. Additionally, we utilize information gap decision theory (IGDT) to build a robust optimization framework that accounts for various uncertainties. The normal boundary intersection (NBI) method is employed to solve the multi-objective optimization problem, ensuring social welfare (SW). This approach considers the impact of uncertainties in RE generation and load demand on decentralized market trading. The model's effectiveness is validated through numerical simulations on a modified IEEE 33-bus power system and a 27-node gas network.

10.1 Introduction

Balancing supply and demand within distributed energy networks under decarbonization targets has become a key research focus [1]. Traditional approaches rely on centralized market structures to optimize joint energy markets within an emission cap, often involving a third-party to manage information exchange. However, this centralized structure can compromise information transparency [2], reduce competition [3], and increase energy prices [4]. Meanwhile, renewable energy (RE) resources have emerged as a crucial strategy for reducing carbon emissions in distributed energy markets [5]. The growing penetration of renewables and diverse energy consumption patterns necessitate market strategies that can hedge against uncertainties to maintain overall energy balance. Thus, we propose a peer-to-peer (P2P) multi-energy trading mechanism enabling direct information exchange between energy buyers and sellers, while considering market uncertainties.

M. Song et al., *Local Energy Markets*, https://doi.org/10.1007/978-981-97-9750-9_10

Integrating gas-fired equipment as replacements for traditional fossil fuel generators, driven by decarbonization efforts, has strengthened the interconnection between gas and electricity networks [6]. A conjectural-variation equilibrium model [7] and a multi-period optimal energy flow model incorporating carbon tax [8] have been employed to manage centralized trading within coupled energy markets. Additionally, an electricity-gas-heating system has been modeled to assess carbon emissions during energy conversion and delivery processes [9]. However, these centralized models face challenges in decentralized systems due to the large number of geographically dispersed participants, administrative costs, and the potential discouragement of prosumers under cap-and-trade or feed-in tariff schemes [10]. Therefore, constructing a decentralized market clearing model that promotes efficient energy allocation while adhering to emission caps is crucial.

Meanwhile, recent studies have focused on modeling energy flows to facilitate simultaneous market clearing in power-gas coupled networks. A steady-state energy flow calculation method based on Energy Circuit Theory [11], combined with existing power system analyses, has been proposed to solve regional power-gas systems [12]. However, these non-linear models are computationally intensive for decentralized market clearing. To address this, linearized models have been introduced, including a second-order cone (SOC)-based flow model [13] that simplifies gas network transmission constraints but reduces accuracy. A linearized power flow model based on the Power Transfer Distribution Factor (PTDF) matrix [14] offers higher accuracy and could be effective for multi-energy market clearing when combined with a linearized gas flow model.

The increasing integration of RE and diverse energy devices introduces significant uncertainty in supply–demand balance and distributed market stability. Variability in energy generation and load demand can disrupt trading and reduce social welfare (SW). Thus, optimizing these uncertainties is essential for ensuring reliable P2P trading and maintaining SW. While opportunity-constrained [15] and stochastic optimization [16] models have been used to address RE uncertainties, they are less suitable for decentralized markets due to the need for precise probability ranges and the inefficiencies associated with complex stochastic problem solving. The Information gap decision theory (IGDT) [17, 18] is more appropriate, as it only requires uncertain intervals and is robust to failure within acceptable loss ranges. IGDT-based robust optimization allows market participants to determine maximum volatility ranges under target SW, aiding in trading decisions.

To our knowledge, the benefits of decentralized market clearing for power-gas networks with emission caps have not been fully explored. Additionally, previous methods for energy flow modeling [11–13] present either computational burdens or accuracy limitations, and existing uncertainty optimization models [15, 16] may yield invalid or inefficient results. Therefore, we propose an IGDT-based market clearing model that ensures trader privacy within coupled energy networks while satisfying emission caps.

Considering the factors mentioned, we propose a practical P2P multi-energy trading mechanism for coupled power-gas networks under a carbon emission cap. Our approach aims to enhance market transparency, improve communication between energy networks, and maximize SW in the presence of uncertainties. The main contributions of our work are as follows:

(1) We introduce a P2P multi-energy trading mechanism using a primal-dual gradient-based Lagrangian method, which models the coordinated interactions in a gas-electricity coupled market. A nodal carbon intensity (NCI) model optimizes and monitors distributed trading within emission constraints.
(2) The proposed trading model incorporates physical constraints on power and gas flows, allowing direct interactions among market participants. Line congestions are detected and managed linearly during Lagrange optimization iterations, utilizing the PTDF and Kirchhoff's laws. This approach accelerates market clearing and ensures feasible P2P multi-energy trading outcomes.
(3) We develop a robust P2P multi-energy trading model based on IGDT to handle uncertainties in renewable generation and load demand. Multi-objective market clearing under uncertainties is converted into a weight-coefficient problem using the Normal Boundary Intersection (NBI) method and solved in a decentralized manner, ensuring the robustness of market clearing against the uncertainties considered.

10.2 Market Structure

The distributed power-gas market structure for the P2P bilateral model is illustrated in Fig. 10.1. This energy network comprises three gas nodes and three power nodes, interconnected by microturbines (MT) and power-to-gas (P2G) equipment for energy conversion and transmission. The gas network consists of gas sources and consumers, enabling P2P multi-energy trading within the gas network and converting surplus energy for trading in the power network. The power network includes prosumers, each node containing energy storage, renewable/conventional generation units (CG), and load. Power nodes can engage in P2P multi-energy trading within the power network and trade with gas nodes via P2G.

10.3 Deterministic Trading Model

10.3.1 Model Formulation

The objective function in the P2P multi-energy trading market is to maximize SW [19], defined as the sum of revenue from buyers RB and sellers RS. For sellers, revenue considers energy trading income and generation costs C, while f1 or buyers, it considers load utility U and energy trading costs. To simplify the modeling, the PTDF [14] is used as a linear approximation of active power flow in the power system, and a simplified branch correlation matrix [20] is employed for the gas system to

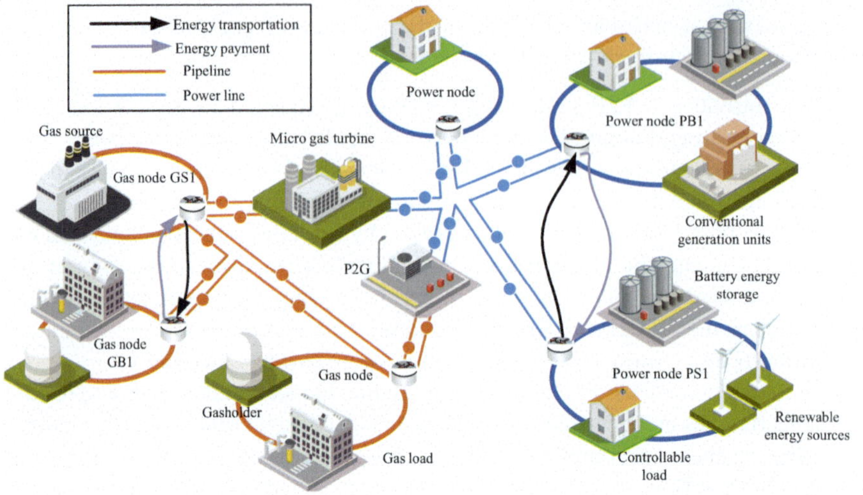

Fig. 10.1 P2P distributed multi-energy market structure diagram

linearize energy variations due to trading.

$$\mathrm{SW} = \mathbf{RS} + \mathbf{RB} \tag{10.1}$$

$$\mathbf{RS} = \sum(\mathbf{x\lambda}) - C(\mathbf{x}) \tag{10.2}$$

$$\mathbf{RB} = U(\mathbf{y}) - \sum(\mathbf{y\lambda}) \tag{10.3}$$

$$\begin{gathered}\boldsymbol{\lambda} \in \left[\lambda_{ij}^{E}, \lambda_{mn}^{G}\right], \boldsymbol{x} \in \left[x_{ij}^{E}, x_{mn}^{G}\right], \boldsymbol{y} \in \left[y_{ji}^{E}, y_{nm}^{G}\right] \\ i \in N_{ES}, j \in N_{EB,} m \in N_{GS}, n \in N_{GB}\end{gathered} \tag{10.4}$$

where $\boldsymbol{x}, \boldsymbol{y}$ refer to the trading volume decision of sellers and buyers, respectively. $\boldsymbol{\lambda}$ represents trading price. Superscripts E and G denote power and gas, respectively. Subscripts i and m refer to sellers; j and n refer to buyers. Further, C [21] and U [22] can be expressed as Eqs. (10.4) and (10.5), where generation cost and load utility are quadratic functions of trading volume of sellers and buyers, respectively.

$$C(\mathbf{x}) = \boldsymbol{\upsilon}(\mathbf{x})^2 + \boldsymbol{\tau}\mathbf{x} \tag{10.5}$$

$$U(\mathbf{y}) = \begin{cases} \boldsymbol{\omega}\mathbf{y} - \boldsymbol{\delta}(\mathbf{y})^2 & \mathbf{y} < \boldsymbol{\omega}/2\boldsymbol{\delta} \\ (\boldsymbol{\omega})^2/2\boldsymbol{\delta} & \mathbf{y} \geq \boldsymbol{\omega}/2\boldsymbol{\delta} \end{cases} \tag{10.6}$$

$$\boldsymbol{\upsilon} \in \left[\upsilon_i^{E}, \upsilon_m^{G}\right], \boldsymbol{\tau} \in \left[\tau_i^{E}, \tau_m^{G}\right], \boldsymbol{\omega} \in \left[\omega_j^{E}, \omega_n^{G}\right], \boldsymbol{\delta} \in \left[\delta_j^{E}, \delta_n^{G}\right] \tag{10.7}$$

where $\boldsymbol{\upsilon}$, $\boldsymbol{\tau}$, $\boldsymbol{\omega}$ and $\boldsymbol{\delta}$ are constant coefficients. Further, the model is subject to:

$$\boldsymbol{x} = \boldsymbol{y} \tag{10.8}$$

$$\boldsymbol{x}^{\min} \le \boldsymbol{x} \le \boldsymbol{x}^{\max} \tag{10.9}$$

$$\boldsymbol{y}^{\min} \le \boldsymbol{y} \le \boldsymbol{y}^{\max} \tag{10.10}$$

$$\sum \boldsymbol{x} = \eta \sum \boldsymbol{y} \tag{10.11}$$

$$-\boldsymbol{F}^{\max} \le \boldsymbol{F} \le \boldsymbol{F}^{\max}, \boldsymbol{F} \in \left[F_\ell^E, F_\ell^G\right] \tag{10.12}$$

$$\boldsymbol{e}_x \boldsymbol{x} \le \boldsymbol{S}_x^{\mathrm{cap}} \tag{10.13}$$

$$\boldsymbol{e}_y \boldsymbol{y} \le \boldsymbol{S}_y^{\mathrm{cap}} \tag{10.14}$$

where Eq. (10.8) restricts the sellers and their corresponding buyers to equal trading volumes. Equations (10.9) and (10.10) restrict the upper and lower limit of participants. Equation (10.11) restricts the energy conversion of coupling nodes, in which η stands for the conversion efficiency of MT/P2G. Equation (10.12) restricts the energy flow limits in transmission lines. Equations (10.13) and (10.14) refer to the emission cap of sellers and buyers. Specifically, $\boldsymbol{e}$ is the NCI matrix [23], which depends on the parameters of all connected generators as well as transmission lines:

$$e = \left(\sum E_G \cdot e_G + \sum F_l \cdot e_l^{\mathrm{line}}\right) \Big/ \left(\sum E_G + \sum F_l\right) \tag{10.15}$$

Here, E_G, e_G, and e_l^{line} refer to energy output of generators, generator CI and CI of transmission lines, respectively. e_l^{line} equals the NCI of its injection bus [24]. $\boldsymbol{S}^{\mathrm{cap}}$ refers to the carbon emission cap, which is directly proportional to the traded energy [8].

Here, E_G, e_G, and e_l^{line} refer to energy output of generators, generator CI and CI of transmission lines, respectively. e_l^{line} equals the NCI of its injection bus [24]. $\boldsymbol{S}^{\mathrm{cap}}$ refers to the carbon emission cap, which is directly proportional to the traded energy [8].

10.3.2 Model Transformation

In combination with Eqs. (10.1–10.3) and (10.6), the model can be transformed into

$$\begin{aligned} & \mathrm{SW} = U(\boldsymbol{y}) - C(\boldsymbol{x}) \\ & \text{s.t.} \quad (10.8) \text{ and } (10.14) \end{aligned} \tag{10.16}$$

For convenience, the PTDF in the power system, defined as linear approximation of the active power flow [14], and simplified branch correlation matrix in the gas system [20] are adopted to linearize the energy variation in transmission lines caused by trading.

$$F^E = \Psi x^E = \Psi y^E \tag{10.17}$$

$$\Psi = B'AC^{-1} \tag{10.18}$$

$$C = A^T B'A \tag{10.19}$$

$$F^G = \left(A^G\right)^{-1} x^G = \left(A^G\right)^{-1} y^G \tag{10.20}$$

where Ψ is the injection sensitivity factor matrix (ISF), which quantifies the impact of nodal load/output on the redistribution of branch power flow [25]. A, B, and A^G represent the node-to-branch incidence matrix of the power network, diagonal branch susceptance matrix and reduced incidence matrix of the gas branch incidence matrix, respectively.

10.3.3 Market Clearing Algorithm

When deterministic optimization functions (Eqs. 10.16–10.20) are solved centrally, participant privacy may be compromised. However, by decomposing the optimization problem, participants can manage their data and computations independently, thus protecting their privacy. The constraints are handled by multiplying the relevant Lagrange multiplier, transforming the original problem into Eq. (10.21), where $\mathcal{L}^P$, $\mathcal{L}^G$, $\mathcal{L}^C$ and $\mathcal{L}^E$ refer to the L-problem of power, gas market, coupling nodes and carbon emission:

$$\begin{aligned} & \mathcal{L}\left(\begin{matrix} x_{ij}^E, y_{ji}^E, \lambda_{ij}^E, x_{mn}^G, y_{mn}^G, \lambda_{mn}^G, \overline{\vartheta}_\ell^E, \underline{\vartheta}_\ell^E, \overline{\vartheta}_\ell^G, \underline{\vartheta}_\ell^G, \overline{\mu}_i^E, \underline{\mu}_i^E \\ \overline{\mu}_j^E, \underline{\mu}_j^E, \overline{\mu}_m^G, \underline{\mu}_m^G, \overline{\mu}_n^G, \underline{\mu}_n^G, \gamma_{in}, \gamma_{mj}, \gamma_{ex}^E, \gamma_{ex}^G, \gamma_{ey}^E, \gamma_{ey}^G \end{matrix}\right) \\ & = \mathcal{L}^P + \mathcal{L}^G + \mathcal{L}^C + \mathcal{L}^E \end{aligned} \tag{10.21}$$

$$\mathcal{L}^P = \sum_{j=1}^{N_{EB}} U_j\left(y_j^E\right) - \sum_{i=1}^{N_{ES}} C_i\left(x_i^E\right) + \sum_{i=1}^{N_{ES}} \sum_{j=1}^{N_{EB}} \lambda_{ij}^E\left(x_{ij}^E - y_{ji}^E\right)$$

$$
\begin{aligned}
&-\sum_{\ell=1}^{N_{E\ell}} \overline{\vartheta}_{\ell}^{E}\left(F_{\ell}^{E}-F_{\ell}^{E,\max}\right)+\sum_{\ell=1}^{N_{E\ell}} \underline{\vartheta}_{\ell}^{E}\left(F_{\ell}^{E}+F_{\ell}^{E,\max}\right)\\
&-\sum_{i=1}^{N_{ES}} \overline{\mu}_{i}^{E}\left(x_{i}^{E}-x_{i}^{E,\max}\right)-\sum_{i=1}^{N_{ES}} \underline{\mu}_{i}^{E}\left(x_{i}^{E,\min}-x_{i}^{E}\right)\\
&-\sum_{j=1}^{N_{EB}} \overline{\mu}_{j}^{E}\left(y_{j}^{E}-y_{j}^{E,\max}\right)-\sum_{j=1}^{N_{EB}} \mu_{j}^{E}\left(y_{j}^{E,\min}-y_{j}^{E}\right)
\end{aligned}
\tag{10.22}
$$

$$
\begin{aligned}
\mathcal{L}^{G}=&\sum_{n=1}^{N_{GB}} U_{n}\left(y_{n}^{G}\right)-\sum_{m=1}^{N_{GS}} C_{m}\left(x_{m}^{G}\right)+\sum_{m=1}^{N_{GS}}\sum_{n=1}^{N_{GB}} \lambda_{mn}\left(x_{mn}^{G}-y_{nm}^{G}\right)\\
&-\sum_{\ell=1}^{N_{G\ell}} \overline{\vartheta}_{\ell}^{G}\left(F_{\ell}^{G}-F_{\ell}^{G,\max}\right)+\sum_{\ell=1}^{N_{G\ell}} \underline{\vartheta}_{\ell}^{G}\left(0-F_{\ell}^{G}\right)\\
&-\sum_{m=1}^{N_{GS}} \overline{\mu}_{m}^{G}\left(x_{m}^{G}-x_{m}^{G,\max}\right)-\sum_{m=1}^{N_{GS}} \underline{\mu}_{m}^{G}\left(x_{m}^{G,\min}-x_{m}^{G}\right)\\
&-\sum_{n=1}^{N_{GB}} \overline{\mu}_{n}^{G}\left(y_{n}^{G}-y_{n}^{G,\max}\right)-\sum_{n=1}^{N_{GB}} \underline{\mu}_{n}^{G}\left(y_{n}^{G,\min}-y_{n}^{G}\right)
\end{aligned}
\tag{10.23}
$$

$$
\mathcal{L}^{C}=\gamma_{in}\left(\sum_{j}^{N_{EB}} x_{ij}^{E}-\sum_{m}^{N_{GB}} y_{nm}^{G}\eta^{MT}\right)+\gamma_{mj}\left(\sum_{n}^{N_{GB}} x_{mn}^{G}-\sum_{i}^{N_{ES}} y_{ji}^{E}\eta^{P2G}\right)
\tag{10.24}
$$

$$
\begin{aligned}
\mathcal{L}^{E}=&-\gamma_{ex}^{E}\left(\sum_{j}^{N_{EB}} e_{x}x_{ij}^{E}-S_{x}^{E\text{cap}}\right)-\gamma_{ex}^{G}\left(\sum_{n}^{N_{GB}} e_{x}x_{mn}^{G}-S_{x}^{G\text{cap}}\right)\\
&-\gamma_{ey}^{E}\left(\sum_{i}^{N_{ES}} e_{y}y_{ji}^{E}-S_{y}^{E\text{cap}}\right)-\gamma_{ey}^{G}\left(\sum_{m}^{N_{GS}} e_{y}y_{nm}^{E}-S_{y}^{G\text{cap}}\right)
\end{aligned}
\tag{10.25}
$$

where λ_{ij}^{E}, λ_{mn}^{G}, $\overline{\vartheta}_{\ell}^{E}, \underline{\vartheta}_{\ell}^{E}, \overline{\vartheta}_{\ell}^{G}, \underline{\vartheta}_{\ell}^{G}$, $\overline{\mu}_{i}^{E}, \underline{\mu}_{i}^{E}, \overline{\mu}_{j}^{E}, \underline{\mu}_{j}^{E}$, $\overline{\mu}_{m}^{G}, \underline{\mu}_{m}^{G}, \overline{\mu}_{n}^{G}, \underline{\mu}_{n}^{G}$, $\gamma_{in}, \gamma_{mj}, \gamma_{ex}^{E}, \gamma_{ex}^{G}, \gamma_{ey}^{E}$ and γ_{ey}^{G} are L-multipliers of corresponding constraints. The distributed optimization update method can be derived by taking the partial derivative of each Lagrange multiplier in Eq. (10.21) using the primal–dual gradient approach [26]. The L-multipliers are then updated according to the process described in [27], and the target energy setpoint is achieved via the inverse gradient method.

The stopping criteria of the model are as follows, where k refers to the iteration times, and ε is a constant coefficient:

$$
\begin{aligned}
&\left|\lambda^{k+1}-\lambda^{k}\right|<\varepsilon_{\lambda}\ \left|\mu^{k+1}-\mu^{k}\right|<\varepsilon_{\mu}\\
&\left|\gamma^{k+1}-\gamma^{k}\right|<\varepsilon_{\gamma}\ \left|\vartheta^{k+1}-\vartheta^{k}\right|<\varepsilon_{\vartheta}
\end{aligned}
\tag{10.26}
$$

10.4 IGDT-Based Robust Trading Model

10.4.1 Uncertainty Model Formulation

IGDT is employed to model and analyze uncertainties in load demand and RE output to maintain robust SW under uncertainty. The model is constructed using the envelope-bound function [28]:

$$B(\mu, b_0) = \{b : b = (1 + \mu)b_0\} \tag{10.27}$$

where b is the uncertain parameter, b_0 is the nominal value, and μ is the extent of the fluctuation. The corresponding load uncertainty characterized by Eq. (10.27) is given in Eq. (10.35).

The IGDT-based robust trading model is formulated to maximize allowable fluctuations in RE output and load demand while ensuring that SW remains at or above a threshold $\delta \mathrm{SW}_0$. Here, δ is a discount coefficient representing performance degradation due to uncertainties, and SW_0 is the optimal value from the deterministic model calculated by Eq. (10.16)

$$\max F = \left[\alpha^2, \beta^2\right] \tag{10.28}$$

$$\mathrm{SW}(\boldsymbol{x}, \boldsymbol{y}) \geq \delta \mathrm{SW}_0 \tag{10.29}$$

$$0 < \delta < 1 \tag{10.30}$$

$$\left(1 - \alpha^2\right)x_i^{E,\min} \leq x_i^E \leq \left(1 - \alpha^2\right)x_i^{E,\max} \tag{10.31}$$

$$\left(1 + \beta^2\right)y_j^{E,\min} \leq y_j^E \leq \left(1 + \beta^2\right)y_j^{E,\max} \tag{10.32}$$

$$-1 \leq \alpha < 1, -1 \leq \beta < 1 \tag{10.33}$$

Over Eqs. (10.8–10.14); (10.17–10.20)
For (α', β'), if there does not exist any (α, β) satisfy

$$\left(\alpha^2 > \alpha'^2\right) \,\&\, \left(\beta^2 > \beta'^2\right) \tag{10.34}$$

Then (α'^2, β'^2) is a Pareto optimal value of the model with two kinds of uncertainties. The market clearing algorithm keeps feasible, as long as the RE output fluctuates within the range of $\left[\left(1 - \alpha^2\right)x_i^{E,\min}, \left(1 - \alpha^2\right)x_i^{E,\max}\right]$ and load demand fluctuates within the range of $\left[\left(1 + \beta^2\right)y_j^{E,\min}, \left(1 + \beta^2\right)y_j^{E,\max}\right]$.

10.4.2 Pareto Optimal Model

Unlike single-objective optimization, the multi-objective problem yields a set of Pareto solutions, commonly modeled by the NBI method [29]. The robust optimization in the P2P multi-energy market is first converted into a single-objective problem where the maximum distance between Utopian lines and the Pareto frontier is the target. This results in a uniformly distributed Pareto solution set, detailed in Eqs. (10.35–10.38).

(1) Single-objective problem solving

First, two single-objective problems are solved independently and the optimal solution is obtained

$$\begin{cases} f_1(\alpha^*) = \max f_1(\alpha) \\ f_2(\beta^*) = \max f_2(\beta) \end{cases} \tag{10.35}$$

where α^*, β^* is the optimal solution when f_1, f_2 is solved separately. The utopia line is obtained by connecting point $(f_1(\alpha^*), f_2(\alpha^*))$ and point $(f_1(\beta^*), f_2(\beta^*))$ on the coordinate plane.

(2) Normalization of objective function

f_1 and f_2 are further normalized through Eq. (10.36):

$$\begin{cases} \overline{f}_1(\alpha) = \frac{f_1(\alpha)-f_1(\alpha^*)}{f_1(\beta^*)-f_1(\alpha^*)} \\ \overline{f}_2(\beta) = \frac{f_2(\beta)-f_2(\beta^*)}{f_2(\alpha^*)-f_2(\beta^*)} \end{cases} \tag{10.36}$$

(3) Formulation of payment matrix

$$P = \begin{bmatrix} \overline{f}_1(\alpha^*) & \overline{f}_1(\beta^*) \\ \overline{f}_2(\alpha^*) & \overline{f}_2(\beta^*) \end{bmatrix} = \begin{bmatrix} 0 & 1 \\ 1 & 0 \end{bmatrix} \tag{10.37}$$

After the normalization of the objective function, we set $m = \begin{bmatrix} -1 & 1 \end{bmatrix}^T$, which represents the unit normal from utopia to the origin. Since any point on the Utopia line can be described as $\mathbf{P}\kappa = \begin{bmatrix} \kappa_1 & \kappa_2 \end{bmatrix}^T$, where $\kappa_1, \kappa_2 \in [0, 1]$ and $\kappa_1 + \kappa_2 = 1$. Therefore, the set of points in the target function space can be represented by $\overline{\Phi}\kappa + \lambda \vec{n}$, $\lambda \in R$.

(4) Solving the single-objective function

When λ reaches the maximum value, it is the Pareto optimal solution of the original multi-objective functions. When κ_1 or κ_2 alters to move the intersection point of normal line $\vec{n}$ and Utopia line, the original multi-objective optimization problem is transformed into a single-objective one:

$$\max \lambda$$
$$\text{s.t.} \begin{cases} \kappa_1 - \lambda = \overline{f}_1(\alpha) \\ 1 - \kappa_1 - \lambda = \overline{f}_2(\beta) \\ 0 \le \kappa_1 \le 1 \end{cases} \tag{10.38}$$

The Pareto curve with uniform distribution, namely Pareto frontier, can be obtained through a series of single-objective optimization solutions.

10.4.3 Market Clearing Under Uncertainty

The objective function then maximizes uncertainty, incorporating SW requirements and revised constraints on RE output/load demand into the optimization, as indicated in Eq. (10.28). The L-method-based distributed optimization model is recalibrated, though details are omitted here due to space constraints.

10.5 Calculation Procedure

The IGDT-based P2P multi-energy trading process is illustrated in Fig. 10.2. Initially, the deterministic market trading optimization function is calculated using the L-method to obtain SW_0. Subsequently, the multi-objective optimization for the P2P trading market addresses the randomness factors α and β, optimizing them to derive robust solutions that meet the required SW. As the P2P market clears in a distributed manner, the L-method is applied again for decentralized updates to the objective function and constraints.

Then, the Pareto frontier of multi-objective optimization is calculated using α^* and β^* and the NBI method, and the maximum distance λ of the normal vector and the weight κ of the objective are obtained. On this basis, the multi-objective optimization is transformed into a single-objective optimization problem with weight, and the transformed model is solved using the L-method. Finally, the P2P multi-energy market is cleared and the trading results of market members are obtained.

Figure 10.3 provides a detailed breakdown of the trading algorithm process for the Lagrangian-based distributed P2P multi-energy market using IGDT, as illustrated in Fig. 10.2. The algorithm is structured into three components: deterministic P2P trading optimization, multi-objective robust optimization, and Pareto optimization.

In Step A, the power and gas networks undergo separate iterative optimizations. The process begins with the initialization of L-multipliers and optimization variables. The L-multiplier is updated first during variable updates, followed by trading volume adjustments. Next, the overall network energy flow is calculated to check for congestion. The process concludes with an evaluation to determine if the stopping criteria are met. In Step B, SW_0 serves as a constraint for robust optimization, with

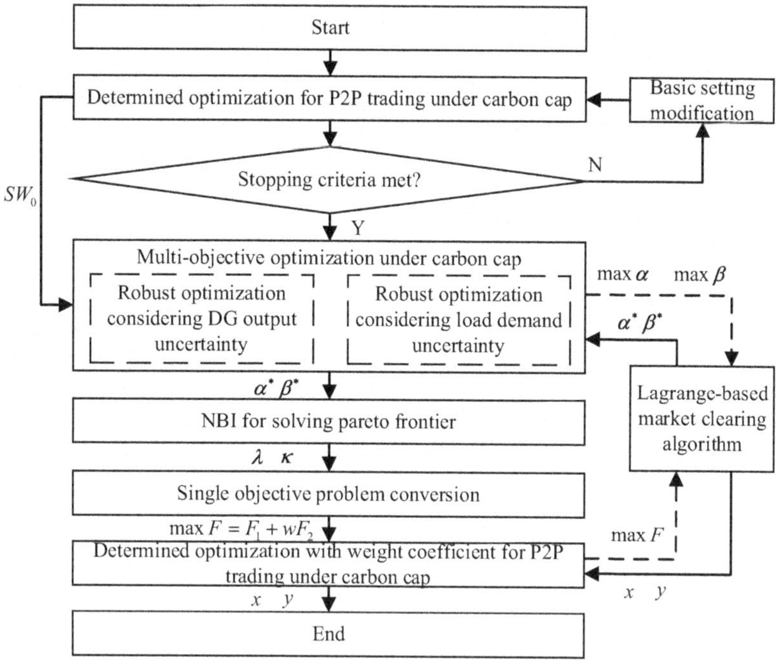

Fig. 10.2 Flowchart of IGDT-based P2P trading

added uncertainty limits. The variables (α and β) and the new L-multipliers (σ and η) are then updated accordingly. Step C applies the NBI method to solve for the Pareto frontier in multi-objective optimization, converting it into a single-objective problem for resolution.

10.6 Case Studies

A case study validates the proposed market trading model within the coupled energy network. First, the deterministic optimization demonstrates iterative convergence, using the clearing price of the electrical network as an example. The proposed market mechanism's performance is compared with existing models, and an analogous coupling point model is introduced for further analysis.

In the second part, the IGDT model is applied for sensitivity analysis of SW, α, and β. The NBI method then calculates the Pareto optimal frontier, verified by sampling points.

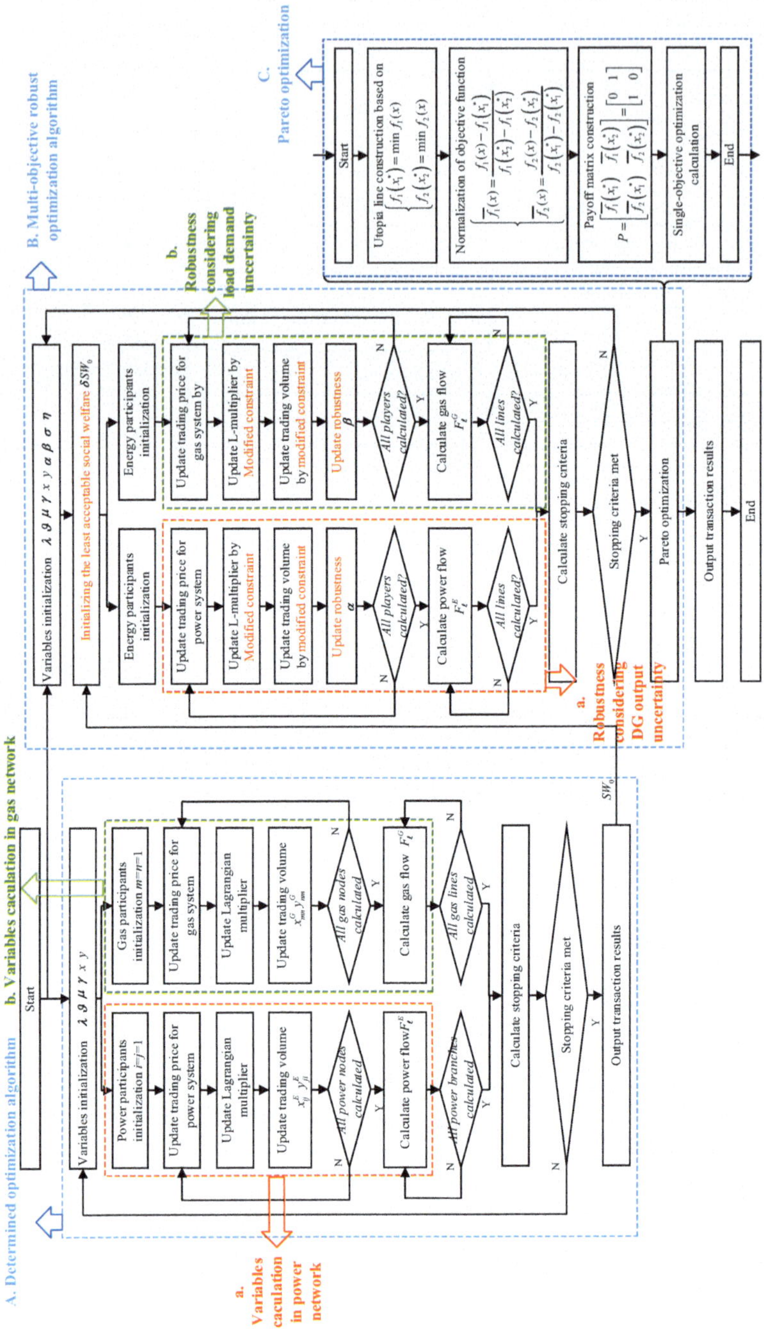

Fig. 10.3 IGDT-based lagrange algorithm for P2P multi-energy market

10.6.1 Basic Configuration

The simulation uses the modified IEEE network topology with 33 power nodes and 27 gas network nodes, as shown in Fig. 10.4. The power network comprises sellers (green) and buyers (yellow), with P1 as the balance bus. Nodes 10, 19, and 33 are conventional generators, while the remaining power nodes are RE. G1 is the gas source, with carbon intensities (CI) of 0.87, 0, and 0.5 $kgCO_2$/kBtu for conventional generators, REs, and gas turbines, respectively. G25 features P2G capabilities. Power-gas networks are interconnected at coupling points: (P4, G25), (P18, G8), (P23, G3), and (P32, G5). Figure 10.5 presents the parameter settings for market participants.

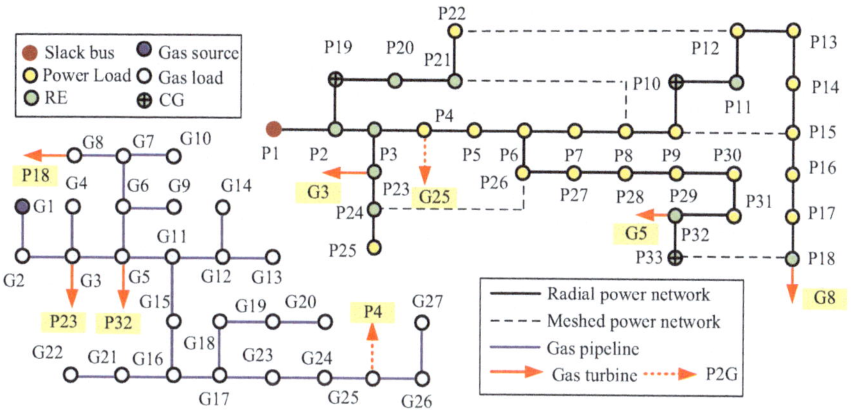

Fig. 10.4 Energy network diagram

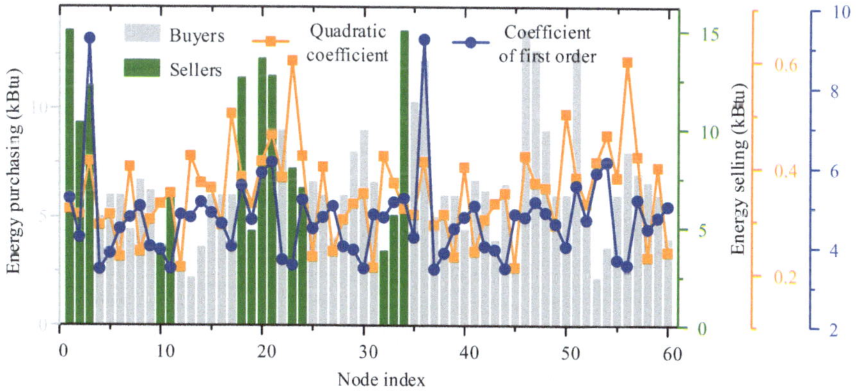

Fig. 10.5 Parameters of market members

10.6.2 Trading in Deterministic Problem

In this section, we analyze the convergence of the proposed clearing model and compare market performance across different trading modes. Additionally, we introduce a coupling node analogy and design four scenarios to calculate clearing results for this analogy.

(1) Convergence analysis

Firstly, constraint Eq. (10.8) is validated by tracking the trading volumes perceived by sellers and buyers during iterations. For example, Fig. 10.6 illustrates the iterative changes in the trading volumes of PS1 and PB1-3. The solid line represents the mean value $(\boldsymbol{X}+\boldsymbol{Y})/2$ of energy trading volume, and the error bar represents the standard deviation. The trend of the curve shows that the error margin decreases from a wide range to zero, indicating convergence of $\boldsymbol{X}$ and $\boldsymbol{Y}$ to a consistent value, which confirms the effective convergence of the proposed P2P trading model.

Secondly, Fig. 10.7 compares the trading price changes between PS1 and all energy-buying nodes during iterations, analyzing the error with the price perceived by PS1 at each step. Without congestion, the grid clearing prices stabilize around \$2.6 with a final error ratio of 0.016, while the gas network clears at approximately \$10.3 with an error ratio of 0.01.

(2) Performance analysis of trading modes

To illustrate the differences between independent networks (mode 1) and coupled energy networks (mode 2), we compare energy utilization and SW. For energy utilization, Fig. 10.8 shows the transacted volume of power sellers with RE equipment under both modes. In mode 2, the final transacted volume aligns with the lower limit of RE

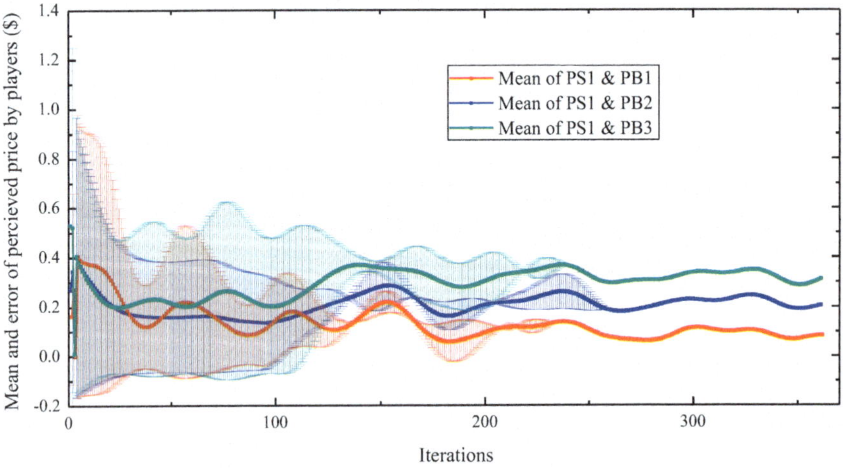

Fig. 10.6 Evolution of traded energy by PS1

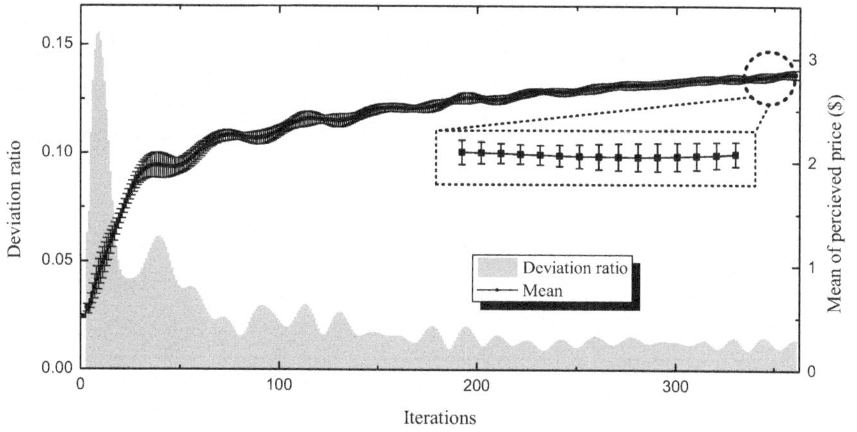

(a)

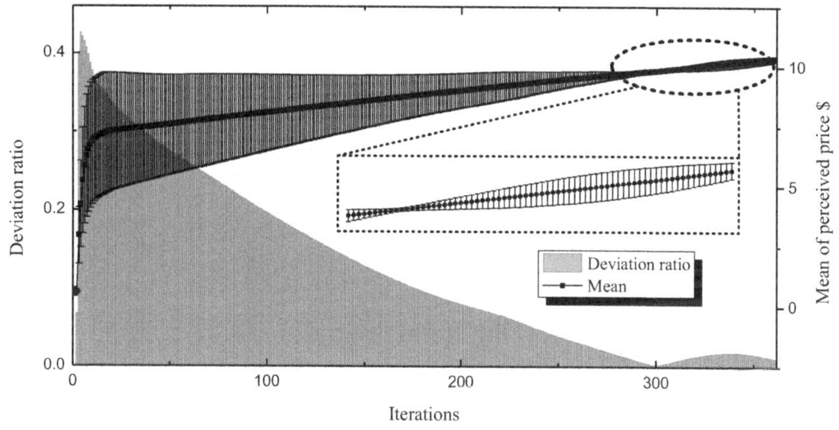

(b)

Fig. 10.7 Energy clearing price in iterations. **a** Power. **b** Gas

output, leading to curtailment, while in mode 1, the volume reaches the upper limit. This difference arises because in mode 2, RE supplies both power and gas loads, unlike in mode 1, where it only serves the power load. Consequently, the power-gas coupled network enhances energy utilization compared to an independent power network.

Next, we compare the proposed trading mechanism with existing models across three modes: power-gas coupled networks (mode 1), independent networks (mode 2 [30]), and centralized trading (mode 3 [31]). Individual and SW are calculated based on (1–3), as depicted in Fig. 10.9. In centralized trading, all traders interact with an energy retailer, where the unit energy buying and selling prices are $8 and

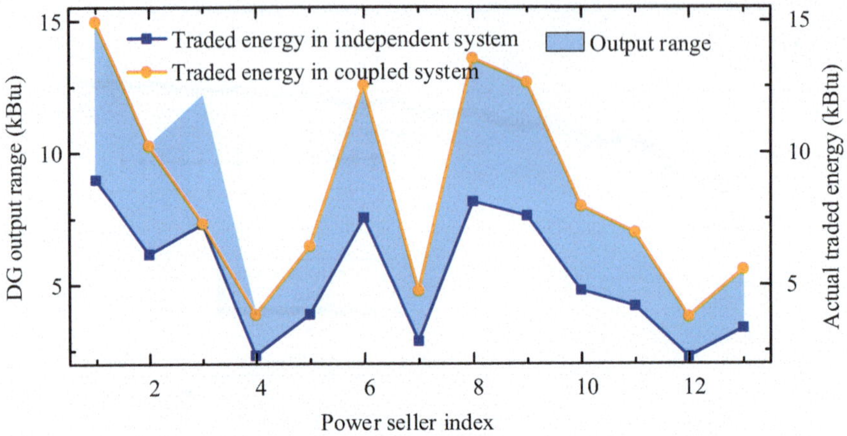

Fig. 10.8 RE consumption comparisons

\$5, respectively. The calculated SW for modes 1, 2, and 3 is \$413 k, \$109 k, and –\$420 k, respectively.

Figure 10.9 demonstrates that the proposed trading mechanism in coupled networks improves both individual and SW compared to independent and centralized trading. The curtailed energy in mode 2 can be converted and sold in the other network, enhancing trading welfare. Conversely, in centralized trading, uniform energy prices for all traders lead to inevitable welfare losses, especially under high generation cost or low load revenue scenarios. Thus, the proposed trading mechanism in power-gas coupled networks effectively increases individual and SW through energy conversion and interaction between gas and power systems.

(3) Emission cap analysis

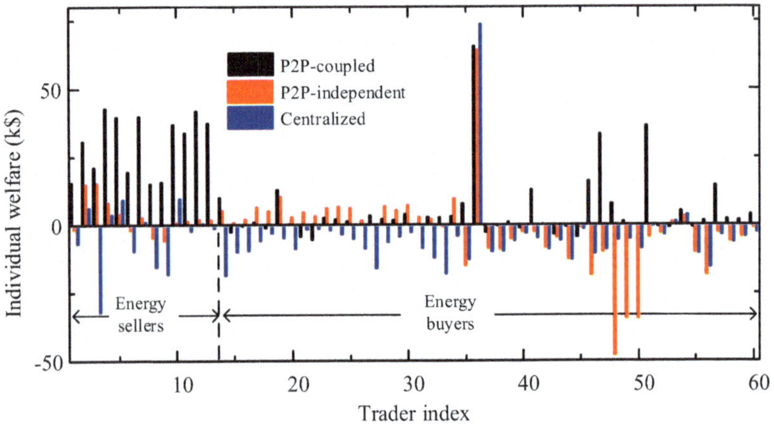

Fig. 10.9 Individual welfare under three trading modes

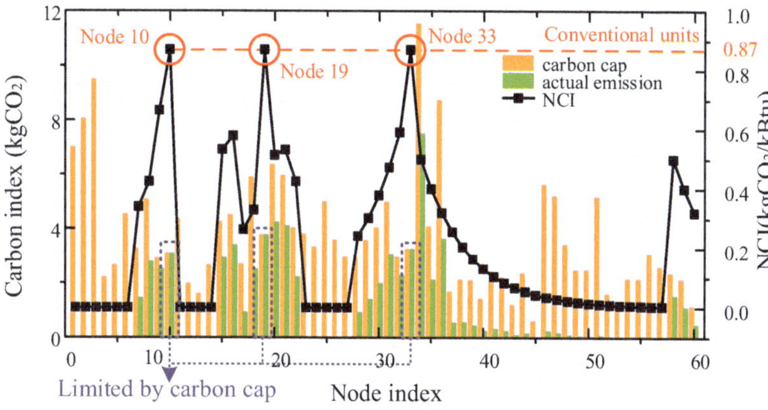

Fig. 10.10 Carbon emission analysis

Lastly, the feasibility of the proposed model under a carbon emission cap is verified by plotting the NCI, carbon cap, and actual emissions according to Eqs. (10.13) and (10.14), as shown in Fig. 10.10. Notably, three peaks are observed in the NCI curve at Nodes 10, 19, and 33, where conventional unit generation has the highest carbon intensity, followed by gas turbines and renewables. As more renewable generations connect, the NCI decreases, explaining the downward trend in CI at these nodes. The nodal actual emission mirrors the NCI trend, particularly at source nodes, where emissions align with the carbon cap due to high NCI. The trading strategy is therefore adjusted to ensure emissions remain within the cap.

10.6.3 Pipelining Analogy of Coupling Points

To evaluate the proposed model under extreme conditions, we conducted an analogy analysis of the coupling node and energy transmission lines, as outlined in Table 10.1. In Case x, when the coupling point lacks a trading volume limit and $\eta < 1$, it resembles an energy transmission pipeline with losses. In Case y, it mirrors a pipeline with a transmission capacity limit, while in Case z, it is analogous to a lossless pipeline.

Figure 10.11 compares the clearing price and SW from Cases x, y, and z with the Base Case. MT and P2G parameters are varied individually, followed by simultaneous changes to all equipment parameters.

First, by comparing SW between Cases x, y, and the Base Case, it is evident that trading limits significantly impact SW. This is because, in the Base Case, η is high, and as η approaches 1, market clearing experiences only minor adjustments.

Second, the influence of MT on clearing prices, compared to P2G, is analyzed by examining λ across different cases. Due to the opposing energy flows of these devices, as η^{MT} increases to 1 in Case y, MT losses decrease compared to the Base Case, resulting in higher gas network prices and lower power prices. Conversely,

Table 10.1 Analogy of coupling nodes

Case	Trading constraints	Conversion efficiency	Analogy
Base case: Case a	Volume $\leq V_{\max}$	$\eta < 1$	Limited transmission line with loss
Case x	Volume ≥ 0	$\eta < 1$	Transmission line with loss
Case y	Volume $\leq V_{\max}$	$\eta = 1$	Transmission line with capacity limits
Case z	Volume ≥ 0	$\eta = 1$	Transmission line without loss

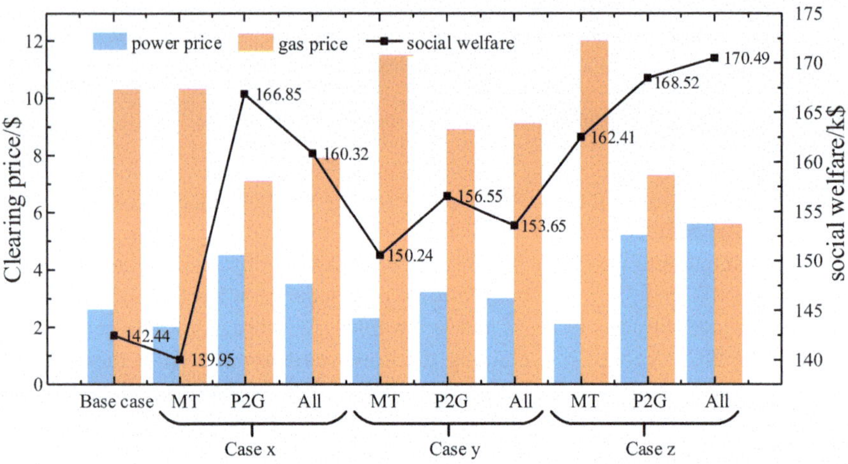

Fig. 10.11 Clearing results of the analogy

when η^{P2G} increases to 1, reduced P2G losses lead to higher grid clearing prices and lower gas network prices.

10.6.4 Trading Under Uncertainties

This section also explores optimization under uncertainties, generating α–δ and β–δ curves for different analogy scenarios. The δ–λ relationship is analyzed under load uncertainty, and the α-β curve is examined under varying δ, confirming the Pareto frontier's validity through sampling points.

(1) Trading under RE output uncertainty

As shown in Fig. 10.12, an inverse relationship between α and δ is observed. As α increases, δ decreases because RE output fluctuations reduce energy supply relative to demand, lowering SW. Additionally, when comparing δ within [0.1, 0.5] and [0.6, 1], α is more sensitive to changes in the lower range, indicating that small

RE output fluctuations have a significant impact on market clearing early on. The α–δ curve varies more in Case a due to the limited energy conversion, making SW more responsive to RE output fluctuations.

(2) Trading under load demand uncertainty

The β–δ curve, depicted in Fig. 10.13, also shows that as β increases, δ decreases. This is due to the unbalanced market caused by rising load demand, which increases prices and decreases SW. Comparing Fig. 10.13 with Fig. 10.12 reveals that the β-δ curve declines faster than the α-δ curve, reflecting SW's higher sensitivity to load demand uncertainty given the buyers' larger utility coefficient.

(3) Trading under both uncertainties

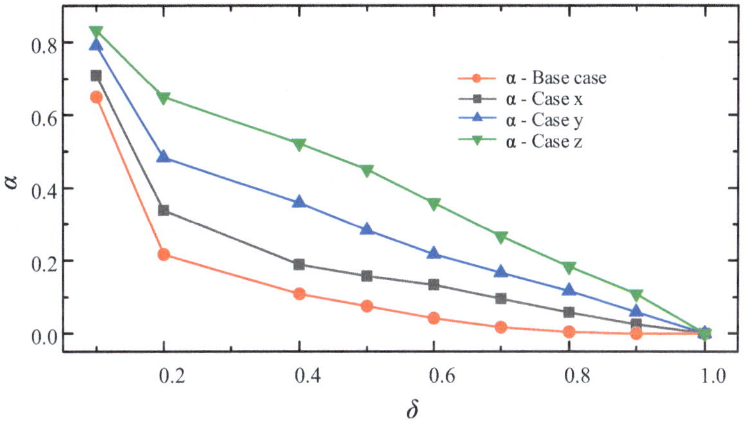

Fig. 10.12 α–δ under different analogy cases

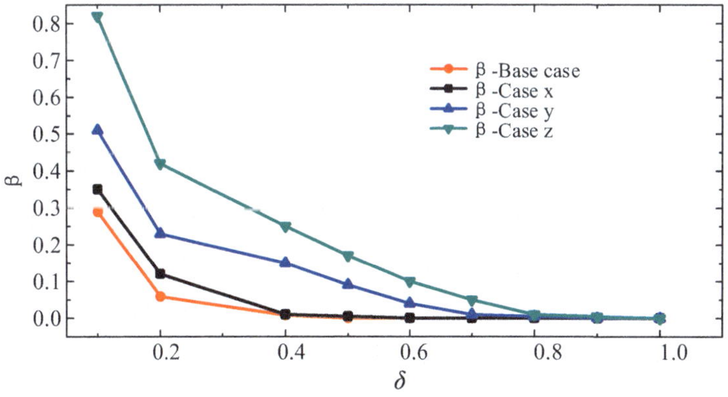

Fig. 10.13 β–δ under different analogy cases

For example, in Case *z*, α–β curves are plotted at δ values of 0.3, 0.7, 0.8, and 0.9, forming the Pareto optimal frontier (Fig. 10.14). These frontiers divide the coordinate space into three regions: A, B, and C. In Region A, any (α, β) combination can achieve 70% of SW_0, in Region B, 30%, while in Region C, achieving the minimum SW_0 is not guaranteed.

To further validate the Pareto frontier, SW was calculated at ten points, as shown in Fig. 10.15. Five points (A: 1, 2; B: 3, 4; C: 5) were selected at $\delta = 0.7$, and another five (A: 6; B: 7, 8; C: 9, 10) at $\delta = 0.3$. Points 3, 4, 5, 9, and 10 failed to meet the required δSW0, as they are located in Regions B and C, beyond the Pareto frontier at δ = 0.7, preventing SW from reaching the target value.

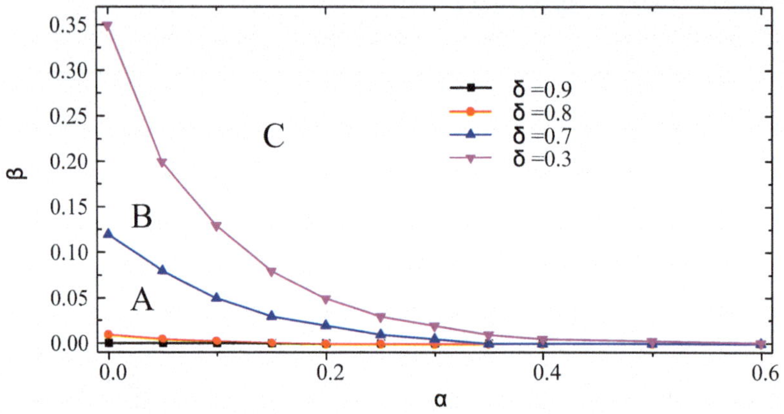

Fig. 10.14 α–β under different δ

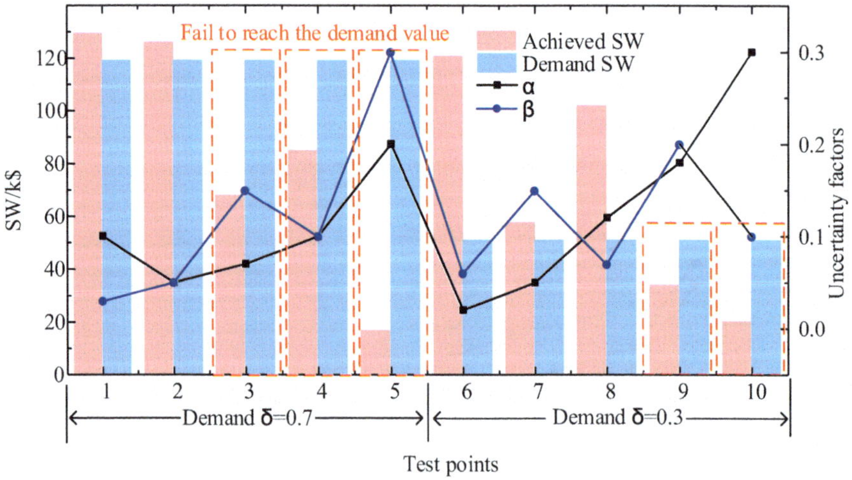

Fig. 10.15 Relationship between α, β and SW at different δ

10.7 Conclusions

A robust P2P multi-energy trading mechanism is proposed for the power-gas-coupled LEM to maximize SW under a carbon emission cap. The PTDF factor and Kirchhoff's law are applied to avoid energy transmission congestion, while the NCI model is integrated to monitor emissions and comply with the carbon cap. The IGDT method robustly optimizes uncertainties in RE output and load demand, defining the uncertainty range while ensuring the required SW. A 37-node power and 27-node gas coupling system validate the model, demonstrating that SW is maintained within the uncertainty range, providing a feasible model for P2P multi-energy trading under a carbon cap.

This work aids policymakers in designing decentralized market mechanisms at the distribution level and lays a theoretical foundation for carbon emission monitoring in power-gas networks. Future work could explore demand response equipment to mitigate uncertainty factors, enabling the practical application of the proposed robust market trading model.

References

1. Qi, F., Shahidehpour, M., Li, Z., Wen, F., & Shao, C. (2019). A chance-constrained decentralized operation of multi-area integrated electricity–natural gas systems with variable wind and solar energy. *IEEE Transactions on Sustainable Energy, 11*(4), 2230–2240.
2. Ullah, M. H., & Park, J. D. (2021). Peer-to-peer energy trading in transactive markets considering physical network constraints. *IEEE Transactions on Smart Grid, 12*(4), 3390–3403.
3. Tushar, W., Saha, T. K., Yuen, C., Smith, D., & Poor, H. V. (2020). Peer-to-peer trading in electricity networks: An overview. *IEEE Transactions on Smart Grid, 11*(4), 3185–3200.
4. Abdella, J., Tari, Z., Anwar, A., Mahmood, A., & Han, F. (2021). An architecture and performance evaluation of blockchain-based peer-to-peer energy trading. *IEEE Transactions on Smart Grid, 12*(4), 3364–3378.
5. Lin, W. T., Chen, G., & Li, H. (2022). Carbon-aware load balance control of data centers with renewable generations. *IEEE Transactions on Cloud Computing, 11*(2), 1111–1121.
6. Wang, C., Wei, W., Wang, J., Wu, L., & Liang, Y. (2018). Equilibrium of interdependent gas and electricity markets with marginal price based bilateral energy trading. *IEEE Transactions on Power Systems, 33*(5), 4854–4867.
7. Chen, S., Conejo, A. J., & Wei, Z. (2021). Conjectural-variations equilibria in electricity, natural-gas, and carbon-emission markets. *IEEE Transactions on Power Systems, 36*(5), 4161–4171.
8. Jiang, T., Deng, H., Bai, L., Zhang, R., Li, X., & Chen, H. (2018). Optimal energy flow and nodal energy pricing in carbon emission-embedded integrated energy systems. *CSEE Journal of Power and energy Systems, 4*(2), 179–187.
9. Cheng, Y., Zhang, N., Wang, Y., Yang, J., Kang, C., & Xia, Q. (2018). Modeling carbon emission flow in multiple energy systems. *IEEE Transactions on Smart Grid, 10*(4), 3562–3574.
10. Peck, M. E., & Wagman, D. (2017). Energy trading for fun and profit buy your neighbor's rooftop solar power or sell your own-it'll all be on a blockchain. *IEEE Spectrum, 54*(10), 56–61.
11. Chen, B., Sun, H., Wu, W., Guo, Q., & Qiao, Z. (2020). Energy circuit theory of integrated energy system analysis (III): Steady and dynamic energy flow calculation. *Proceedings of CSEE, 40*(15), 4820–4831.

12. Wang, W., Wang, D., Jia, H., Chen, Z., Guo, B., Zhou, H. M., & Fan, M. (2017). Steady state analysis of electricity-gas regional integrated energy system with consideration of NGS network status. *Proceedings of CSEE, 37*(5), 1293–1304.
13. Chen, S., Conejo, A. J., Sioshansi, R., & Wei, Z. (2019). Equilibria in electricity and natural gas markets with strategic offers and bids. *IEEE Transactions on Power Systems, 35*(3), 1956–1966.
14. Liu, M., & Gross, G. (2004). Role of distribution factors in congestion revenue rights applications. *IEEE Transactions on Power Systems, 19*(2), 802–810.
15. Guo, Z., Pinson, P., Chen, S., Yang, Q., & Yang, Z. (2020). Chance-constrained peer-to-peer joint energy and reserve market considering renewable generation uncertainty. *IEEE Transactions on Smart Grid, 12*(1), 798–809.
16. Hu, Z., Xu, Y., Korkali, M., Chen, X., Mili, L., & Valinejad, J. (2020). A Bayesian approach for estimating uncertainty in stochastic economic dispatch considering wind power penetration. *IEEE Transactions on Sustainable Energy, 12*(1), 671–681.
17. Song, M., & Sun, W. (2020). Robust distribution system load restoration with time-dependent cold load pickup. *IEEE Transactions on Power Systems, 36*(4), 3204–3215.
18. Ahmadi, A., Nezhad, A. E., Siano, P., Hredzak, B., & Saha, S. (2019). Information-gap decision theory for robust security-constrained unit commitment of joint renewable energy and gridable vehicles. *IEEE Transactions on Industrial Informatics, 16*(5), 3064–3075.
19. Kirschen, D. S., & Strbac, G. (2018). *Fundamentals of power system economics*. John Wiley & Sons.
20. La Scala, M., Bruno, S., Nucci, C. A., Lamonaca, S., & Stecchi, U. (Eds.). (2021). *From smart grids to smart cities: new challenges in optimizing energy grids* (Vol. 2). John Wiley & Sons.
21. Grainger, J. J. (1999). *Power system analysis*. McGraw-Hill.
22. Samadi, P., Mohsenian-Rad, H., Schober, R., & Wong, V. W. (2012). Advanced demand side management for the future smart grid using mechanism design. *IEEE Transactions on Smart Grid, 3*(3), 1170–1180.
23. Kang, C., Zhou, T., Chen, Q., Wang, J., Sun, Y., Xia, Q., & Yan, H. (2015). Carbon emission flow from generation to demand: A network-based model. *IEEE Transactions on Smart Grid, 6*(5), 2386–2394.
24. Wu, T., Wei, X., Zhang, X., Wang, G., Qiu, J., & Xia, S. (2022). Carbon-oriented expansion planning of integrated electricity-natural gas systems with EV fast-charging stations. *IEEE Transactions on Transportation Electrification, 8*(2), 2797–2809.
25. Baldick, R. (2003). Variation of distribution factors with loading. *IEEE Transactions on Power Systems, 18*(4), 1316–1323.
26. Zhu, M., & Martinez, S. (2011). On distributed convex optimization under inequality and equality constraints. *IEEE Transactions on Automatic Control, 57*(1), 151–164.
27. Nedic, A., & Ozdaglar, A. (2010). 10 Cooperative distributed multi-agent optimization. In *Convex optimization in signal processing and communications* (p. 340).
28. Ben-Haim, Y. (2006). *Info-gap decision theory: Decisions under severe uncertainty*. Elsevier.
29. Das, I., & Dennis, J. E. (1998). Normal-boundary intersection: A new method for generating the Pareto surface in nonlinear multicriteria optimization problems. *SIAM Journal on Optimization, 8*(3), 631–657.
30. Li, J., Ye, Y., Papadaskalopoulos, D., & Strbac, G. (2020). Computationally efficient pricing and benefit distribution mechanisms for incentivizing stable peer-to-peer energy trading. *IEEE Internet of Things Journal, 8*(2), 734–749.
31. Yao, Y., Gao, C., Lai, K., Chen, T., & Yang, J. (2021). An incentive-compatible distributed integrated energy market mechanism design with adaptive robust approach. *Applied Energy, 282*, 116155.

Part V
Cutting-Edge Technologies in Local Energy Markets

Chapter 11
Blockchain-Based Local Energy Market for Networked Microgrids

Abstract This chapter introduces a blockchain-based Local Energy Market (LEM) among networked microgrids (MGs). The MGs submit their trading data to the centralized distribution system operator (DSO), which optimizes these transactions to ensure network security. The DSO considers various options, including hourly demand response, on-site MG generation, and trading with the independent system operator (ISO), to maximize MG profits while maintaining network constraints. A cooperative game with externalities models the collaborative behavior among MGs in the market. A two-stage problem is formulated to distribute the grand coalition's payoff among participating MGs. The problem is solved using a column-and-constraint generation (C&CG) method combined with Karush–Kuhn–Tucker (KKT) conditions, offering superior computational efficiency. Blockchain technology ensures secure, transparent settlement, and market operations for the proposed LEM. The approach is tested on a 4-MG network, the IEEE 33-bus system, and the IEEE 123-bus system, demonstrating its effectiveness in encouraging MGs to trade energy and carbon allowances while adhering to network constraints.

11.1 Introduction

In power systems, centralized fossil fuel generators primarily meet energy demands, contributing to significant air pollution and carbon emissions. These excessive CO_2 emissions exacerbate global warming and increase the frequency of severe natural disasters [1]. Microgrids (MGs) offer a promising solution to these environmental challenges by integrating on-site loads and distributed energy resources (DERs) into small, autonomous power systems [2]. When geographically close MGs are networked, they can enhance their benefits further. The LEM facilitates the integration of DERs by encouraging networked MGs to trade energy, promoting low-carbon sustainable development [3].

In parallel, the carbon market, especially the carbon cap-and-trade system, plays a vital role in regulating CO_2 emissions by imposing costs on them. In a cap-and-trade market, participants, including MGs, must operate within their carbon allowances. If

M. Song et al., *Local Energy Markets*, https://doi.org/10.1007/978-981-97-9750-9_11

a MG exceeds its carbon allowance by using fossil fuel sources such as microturbines (MTs) for electricity generation, it must either buy additional allowances or reduce its output to comply.

Integrating carbon trading into the electricity market can motivate participants to limit CO_2 emissions, contributing to a more sustainable society. However, two key issues must be addressed to establish an effective blockchain-based local energy market (LEM): (1) determining trading prices and quantities among MGs while ensuring distribution network constraints are met and incentivizing transactions. This is especially crucial in a combined market where MGs trade both energy and carbon allowances. High energy prices may prompt MGs to purchase carbon allowances to generate on-site energy, while high carbon allowance prices may lead them to buy energy instead; (2) ensuring secure, transparent, and efficient market processes and contract settlements. The large number of active participants in a LEM can overwhelm existing settlement systems, causing delays and a lack of transparency, which may increase electricity prices [4].

Several studies have focused on the first issue within the LEM. References [5, 6] explored centralized optimization methods where the distribution system operator (DSO) collects data from all MGs and clears the market based on distribution locational marginal prices. References [7, 8] proposed decentralized optimization methods that protect individual MG privacy. Gaming approaches are also common in capturing market participants' behaviors. Reference [9] applied a non-cooperative game to highlight competitive MG behaviors in a LEM. A two-level LEM was proposed in [10], where MGs trade energy at a lower level using a non-cooperative Stackelberg game, while the DSO manages distribution network security at the upper level. However, these methods may not be effective if some MGs possess market power, potentially leading to suboptimal trading outcomes and motivating participants to seek more favorable alternatives [11].

Cooperative gaming has also been applied to the LEM, encouraging participants to form coalitions and equitably distribute the resulting payoff. References [12, 13] introduced a cooperative grand coalitional game for community MGs. Reference [14] offered a cooperative coalition formation method allowing MGs to choose their partners flexibly. However, these approaches often neglect power network constraints, which can result in infeasible energy trading decisions. In reality, MG trading decisions are interconnected through distribution network constraints, but traditional cooperative gaming methods do not account for these interdependencies. Some studies focus on the hierarchical relationship between wholesale and LEMs. Reference [15] examined the framework for coordinating wholesale and LEMs. References [16, 17] discussed the hierarchical operation of combined markets, while References [18, 19] demonstrated that incorporating the LEM into the wholesale market can significantly lower electricity prices.

Numerous studies have explored the integration of energy and carbon markets within the transmission system. For instance, Reference [20] examined the equilibrium conditions of these combined markets, while Reference [21] introduced a unit commitment model that incorporates carbon allowance trading.Reference [22] proposed a low-carbon dispatch method that factors in carbon capture technologies, and Reference [23] presented a low-carbon operational approach for multi-energy systems by considering integrated energy-carbon pricing.

Regarding secure settlement of trading contracts, blockchain technology offers a viable solution. Blockchain serves as a decentralized, immutable ledger that records transaction data, which is simultaneously updated and shared across all participants to ensure consistent records. This distributed network enables participants to collaboratively audit and maintain data integrity, creating a trustworthy, secure, and rapid transaction environment without needing a third party [24]. Blockchain's data is traceable, secure, verifiable, and tamper-proof. Additionally, smart contracts within blockchain systems can automate the execution of trading agreements, enhancing the efficiency and reliability of energy and carbon markets. Several references have explored blockchain applications in this context. References [25–27] proposed frameworks for blockchain in LEMs, though they did not delve into specific market mechanisms. Reference [28] focused on blockchain for automated contract settlements in auction-based energy trading. Reference [29] compared the performance of blockchain-based energy markets across different software platforms, and Reference [30] utilized blockchain for energy trading among electric vehicles.

Overall, blockchain introduction into LEM and carbon markets can boost efficiency, reduce operational costs, lower entry barriers, and automate trading processes. In distribution systems, blockchain has been applied to these markets as well. Reference [31] discussed a blockchain-based market for multi-distributed photovoltaic (PV) systems, although it lacked detailed market mechanisms. Reference [32] proposed an auction-based market for MG prosumers, using blockchain to facilitate secure contract settlements, though it did not account for distribution network constraints, which could result in suboptimal outcomes.

When applying cooperative games to combined energy and carbon trading in networked MGs, key issues include: (1) Incorporating distribution network constraints to ensure feasible trading results. (2) Developing a scalable payoff allocation method, as existing methods in cooperative gaming face exponential growth in computational demands with more participants. (3) Creating efficient blockchain-based settlement methods tailored to different market mechanisms, as such solutions are still lacking for cooperative gaming-based LEMs. The differences between the proposed method and existing studies are summarized in Table 11.1.

Table 11.1 Comparative features of previous LEM studies

Reference	Carbon allowance trading	Gaming approach	Blockchain	Network constraints
[5–8]	No	No	No	Yes
[12–14]	No	Yes	No	No
[20–23]	Yes	No	No	Yes
[24–29]	No	No	Yes	No
[31, 32]	Yes	No	Yes	No
Proposed method	Yes	Yes	Yes	Yes

This chapter introduces a blockchain-based LEM for networked MGs using a cooperative game with externalities. In this system, MGs can trade both energy and carbon allowances through a centralized DSO-managed market. The key contributions of this chapter are as follows:

(1) A blockchain-based energy and carbon LEM managed by the DSO is proposed, where networked MGs collaborate within a grand coalition to trade energy and carbon allowances. The cooperative game with externalities models MG behavior while accounting for power distribution network constraints. Demand response is incorporated to maximize MG payoffs while meeting DSO requirements. This is the first known attempt to integrate power network constraints into cooperative gaming.
(2) A two-stage payoff allocation method is developed to distribute the grand coalition's gains among MGs. The proposed solution uses an algorithm combining column-and-constraint generation (C&CG) with Karush–Kuhn–Tucker (KKT) conditions, offering superior computational efficiency.
(3) A blockchain-based energy trading method ensures decentralized, secure, and effective transaction settlements, along with transparent market operations for energy and carbon trading in a cooperative gaming-based market that considers distribution network constraints.

11.2 Blockchain-Based LEM

11.2.1 Cooperative Game with Externalities

Consider a set of players in the game given by $N = \{1, 2, \ldots, n\}$. Each subset of players N can formulate a coalition S. If all players cooperate, they will form a grand coalition N. In the cooperative game, players will first cooperate to maximize the coalition's payoff (i.e., the sum of all players' payoffs in the coalition) rather than their individual payoffs. Then, the coalition's payoff will be allocated fairly to each player.

In practice, players' actions and payoffs are interdependent. In other words, the payoff of coalition S is not only decided by the actions of members in coalition S, but also by the actions of members outside the coalition S. Such interdependencies are known as the externalities in a cooperative game [33]. In a cooperative game with externalities, the payoff of coalition S is defined as $U(S, \rho)$ [34], where ρ is a subset of N with $S \in p$ to represent the actions of members outside the coalition. The payoff $U(S, \rho)$ is obtained by solving the following problem:

$$\begin{aligned} &\max\ U(S, \rho) \\ &\text{s.t.}\quad G(S, \rho) \leq 0 \end{aligned} \tag{11.1}$$

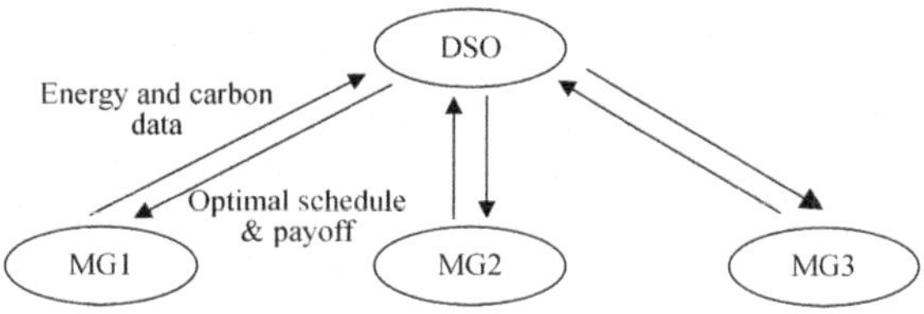

Fig. 11.1 Centralized market structure for transacting energy and carbon

where $G(S, \rho)$ represents the constraints in the cooperative game with externalities. The problem (11.1) is labeled as $J(S, \rho)$.

11.2.2 Local Energy and Carbon Market

In the local energy and carbon market, MGs participate by trading both energy and carbon allowances, operating within a carbon cap-and-trade framework. Each MG has a carbon allowance limiting its CO_2 emissions, and trades are made with peers and the DSO. Given network constraints, the actions of one MG, such as its DER power output, affect the payoffs of others, creating interdependencies modeled by a cooperative game with externalities. All MGs are assumed to cooperate within a grand coalition to trade energy and carbon allowances.

As illustrated in Fig. 11.1, the DSO oversees the market, collecting energy and carbon trading data from MGs and optimizing the trading schedule to maximize the grand coalition's payoff while adhering to distribution network security constraints. The DSO then fairly allocates the grand coalition's payoff among the MGs.

11.2.3 Blockchain for Local Energy and Carbon Market

The DSO handles trade settlements in a centralized market by collecting payments from purchasing MGs and compensating selling MGs. However, this centralized approach has limitations, such as potential single points of failure, and privacy concerns. To address these issues, blockchain technology enables decentralized, secure, and efficient trading settlements.

Blockchain, a shared and distributed ledger, records transaction data transparently. Figure 11.2 illustrates how blockchain facilitates transactions: (1) Participants send trading transactions to others, which are stored as data blocks. (2) These blocks are broadcast to all participants, who collaboratively audit and validate them. (3) Once confirmed, the data blocks are stored in the blockchain, where the transaction is executed. All participants can access the transaction data, ensuring transparency, security, and tamper-proof records without third-party oversight.

In the blockchain network, the DSO and MGs are key participants. MGs trade energy and carbon allowances with their peers through the DSO, acting as market

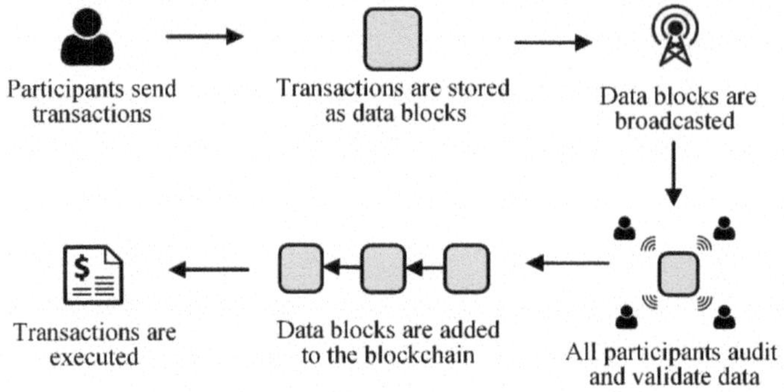

Fig. 11.2 Blockchain process for making transactions

participants. Smart meters within MGs record trading and dispatch information, storing it on the blockchain. The DSO oversees trading events and facilitates communication among MGs, while both the DSO and MGs engage in the validation and consensus processes within the blockchain. Specifically, they can initiate transactions, create data blocks, and store these blocks in the blockchain. Figure 11.3 shows how smart contracts operate in this LEM, with the process including:

(1) Initialization. MGs are each given a unique pair of public and private keys before engaging in blockchain transactions. The public key serves as a participant's identifier in the blockchain network, while the private key acts as a digital signature. Detailed key usage is available in Reference [25].
(2) Trading data submission. MGs submit their trading data to the DSO using smart contracts, which automate this process. These market clearing smart contracts provide an interface for data submission and enforce a schedule for collecting and delivering data to the DSO.
(3) Market clearing. The DSO processes the collected trading data to determine optimal dispatch and payoff allocations while ensuring network security. This process occurs online, with the details provided in Sect. 11.3. Once the market is cleared, the results are automatically broadcast to all MGs by the smart contracts.
(4) Transaction settlement. Another type of smart contract, known as transaction settlement smart contracts, handles payment settlements. These contracts automatically query smart meters for real-time energy and carbon allowance data, comparing it with market clearing results. If MGs meet their commitments, payments are settled automatically; otherwise, penalties are imposed based on pre-set mechanisms detailed in Refs. [35, 36]. These contracts ensure that MGs' account balances are updated accordingly.

In summary, two types of smart contracts are utilized in the proposed market: one for market clearing and another for transaction settlement, both interacting with each other. Transaction settlement contracts query results from market clearing contracts to finalize settlements.

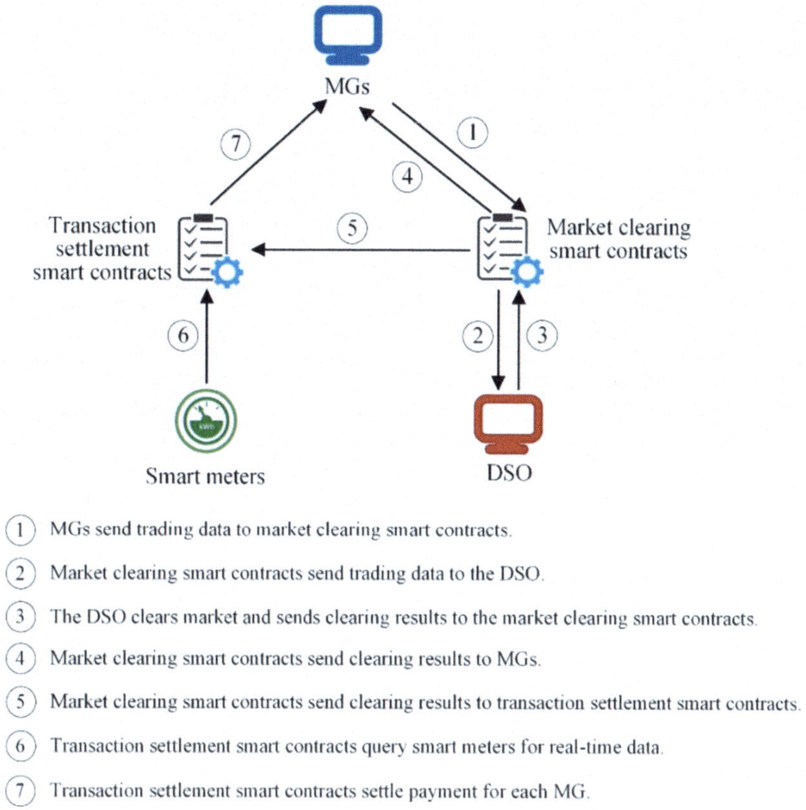

Fig. 11.3 Blockchain for LEM

All data, including trading, market clearing results, and payment data, are validated by participants and stored in the blockchain using a proof-of-work consensus algorithm (see Ref. [24] for details). The blockchain ensures that market clearing results cannot be tampered with and that all participants review and complete transaction settlements, enhancing market operations' transparency.

11.3 Formulation of the Proposed LEM

The LEM based on the cooperative MG game with externalities operates in two stages. First, the DSO determines the optimal schedule for each MG, constrained by the goal of maximizing the grand coalition's total payoff. Second, the DSO distributes this total payoff among the participating MGs according to the optimized schedule.

11.3.1 Mathematical Formulation of the Grand Coalition

In this market model, the DSO functions as a mediator, facilitating the trade of energy and carbon allowances among MGs. The DSO collects energy and carbon allowances from all MGs and may also engage in trades with the wholesale market (i.e., ISO's transmission market) to balance each MG's on-site energy and carbon demands. With all MGs acting within a grand coalition, the DSO's objective is to maximize the overall coalition payoff by optimizing on-site generation through the energy and carbon allowance market and adjusting MGs' loads via demand response.

Each MG is assumed to have adjustable aggregated loads and a mix of MTs and renewable DERs. The total payoff for the grand coalition is calculated as the difference between revenue and cost. Energy trading revenues are derived from selling energy to the DSO/wholesale market and to on-site loads, while costs include purchasing energy from the DSO/wholesale market and generating it on-site through MTs. Similarly, carbon trading payoff is determined by the revenue from selling carbon allowances minus the cost of purchasing them. The formulation is stated as follows:

$$\max \underbrace{\sum_t \sum_i \left(\pi_t^s PS_{i,t} + R_{i,t}(PD_{i,t}) - \pi_t^b PB_{i,t} - \sum_n C(PG_{i,n,t}) \right)}_{\text{Energy}} + \underbrace{\sum_t \left(\beta_t^s CS_t - \beta_t^b CB_t \right)}_{\text{Carbon}} \tag{11.2}$$

$$R_{i,t}(PD_{i,t}) = \lambda_{i,t} PD_{i,t} - \beta_{i,t} (PD_{i,t})^2 \tag{11.3}$$

$$C(PG_{i,m,t}) = a_{i,m} (PG_{i,m,t})^2 + b_{i,m} PG_{i,m,t} + c_{i,m} \tag{11.4}$$

$$\sum_i (PD_{i,t} + PS_{i,t}) = \sum_i \left(\sum_m PG_{i,m,t} + \sum_w PW_{i,w,t} + PB_{i,t} \right) \forall t \tag{11.5}$$

$$\sum_i (C_i^{\max} + CS_i) = \sum_i \left(\sum_t \sum_m CE_{i,m,t} + CB_i \right) \tag{11.6}$$

$$CE_{i,m,t} = \beta_{i,m} PG_{i,m,t} \forall m \in \Omega^M, \forall i, \forall t \tag{11.7}$$

$$CE_{i,m,t} = \beta_{i,m} PG_{i,m,t} - CC_{i,m,t} \forall m \in \Omega^{MC}, \forall i, \forall t \tag{11.8}$$

$$CC_{i,m,t} = \omega PC_{i,m,t} \; \forall m \in \Omega^{MC}, \forall i, \forall t \tag{11.9}$$

$$PG_{i,m,t}^{N} = PG_{i,m,t} - PC_{i,m,t} \forall m \in \Omega^{MC}, \forall i, \forall t \tag{11.10}$$

$$P_{i,t} = PS_{i,t} - PB_{i,t} \forall i, \forall t \tag{11.11}$$

$$\begin{aligned} P_{i,t} &= \sum_{j} PL_{j,t} - \sum_{i} PL_{j,t} \forall i, \forall t \\ Q_{i,t} &= \sum_{j} QL_{j,t} - \sum_{i} QL_{j,t} \forall i, \forall t \end{aligned} \tag{11.12}$$

$$\begin{aligned} P_{i,t} + PB_{i,t}^{ST} - PS_{i,t}^{ST} &= \sum_{j} PL_{ji,t} - \sum_{i} PL_{ji,t} \forall i \in ST, \forall t \\ Q_{i,t} + QB_{i,t}^{ST} &= \sum_{j} QL_{ji,t} - \sum_{i} QL_{ji,t} \forall i \in ST, \forall t \end{aligned} \tag{11.13}$$

$$\begin{aligned} \sum_{i \in ST} PB_{i,t}^{ST} &\geq \sum_{i} PB_{i,t} \forall t \\ \sum_{i \in ST} PS_{i,t}^{ST} &\leq \sum_{i} PS_{i,t} \forall t \end{aligned} \tag{11.14}$$

$$\begin{aligned} PL_{ij,t} &= g_{ij}\left(V_{i,t} - V_{j,t}\right) - b_{ij}\left(\theta_{i,t} - \theta_{j,t}\right) \forall ij, \forall t \\ QL_{ij,t} &= -g_{ij}\left(\theta_{i,t} - \theta_{j,t}\right) + b_{ij}\left(V_{i,t} - V_{j,t}\right) \forall ij, \forall t \end{aligned} \tag{11.15}$$

$$\left(PL_{ij,t}\right)^2 + \left(QL_{ij,t}\right)^2 \leq \left(S_{ij}^{\max}\right)^2 \forall ij, \forall t \tag{11.16}$$

$$V^{\min} \leq V_{i,t} \leq V^{\max} \forall i, \forall t \tag{11.17}$$

$$PD_{i,t}^{\min} \leq PD_{i,t} \leq PD_{i,t}^{\max} \forall i, \forall t \tag{11.18}$$

$$0 \leq PG_{i,m,t} \leq PG_{i,m}^{\max} \forall i, \forall m, \forall t \tag{11.19}$$

$$0 \leq PW_{i,w,t} \leq PW_{i,w,t}^{\max} \forall i, \forall w, \forall t \tag{11.20}$$

Constraint (11.3) defines the revenue for each MG from power consumption, while Constraint (11.4) covers the generation cost for MTs. Constraints (11.5) and (11.6) address the power and carbon balances for MGs, respectively. Constraints (11.7) and (11.8) quantify CO_2 emissions from MTs, both with and without carbon capture devices. The net power output of MTs equipped with carbon capture devices is modeled by Constraints (11.9) and (11.10), where these devices utilize a portion of the generated power for CO_2 capture. Additional constraints for carbon capture devices can be found in [22]. Constraint (11.11) captures the net active power load and generation for each MG, while Constraints (11.12) and (11.13) handle active and reactive power injections into buses, including power exchanges between the DSO and the transmission grid via substation buses (Constraint 11.14). Constraints (11.15) and (11.16) impose limits on line power flows, using a linearized AC optimal power flow model [37], and Constraint (11.17) manages bus voltage magnitudes. Constraint (11.18) regulates the MG load adjustments considering demand response, where the adjustable load $PD_{i,t}$ is optimized in the solution process. Constraints (11.19) and (11.20) set the limits for MT generation capacity and DER output within each MG, respectively. The solution to this problem determines the optimal schedule for MGs, including load and DER output, and establishes the energy trade quantities with the DSO. For example, the energy sold by an MG to the DSO is calculated as the difference between the energy generated by DERs and the load.

11.3.2 Stability of Grand Coalition

In this section, we prove the stability of the grand coalition, i.e., MGs would prefer to cooperate rather than acting individually. We first provide several definitions and lemmas in the cooperative game with externalities.

Definition 1 (*Imputation*)
The vector **x** is considered as the *payoff allocation* representing the payment to each MG. An imputation is defined as a payoff allocation if the following conditions are satisfied:

(1) *Efficiency*: The payoff allocation $\boldsymbol{x}$ is efficient if $\sum_i x_i = U(N)$, indicating that the grand coalition's payoff is allocated to all MGs.
(2) *Individual Rationality*: The payoff allocation $\boldsymbol{x}$ is rational if $x_i \geq U(\{i\}, \rho)\forall i$, indicating that MGs will gain higher payoffs if they join the grand coalition.

Definition 2 (*γ-core*)
γ-core is a set of imputations which satisfy the following condition:

$$RD : \left\{ x | \sum_i x_i = U(N), \sum_{i \in S} x_i \geq U(S, \rho) \forall S \right\} \quad (11.21)$$

Equation (11.21) indicates that the imputations in *γ-core* make sure that MGs can receive higher payoffs by participating in the grand coalition. All MGs will accept the imputations in *γ-core* rather than abandoning the imputations. In *γ-core*, members which do not want to cooperate will act individually. Accordingly, non-cooperating MGs shall only trade energy and carbon allowance with the DSO rather than their counterparts.

Definition 3 (*Superadditivity*)
The cooperative game with externalities is superadditive if the following conditions are satisfied for any $S, G \subseteq N$ and any ρ of $N - S \cup G$:

$$U(S \cup G, \{S \cup G\} \cup \rho) \geq \\ U(S, \{S, G\} \cup \rho) + U(G, \{S, G\} \cup \rho). \tag{11.22}$$

Definition 4 (*Positive homogeneity*)

The cooperative game with externalities is positive homogeneous if the following condition is satisfied for any $S, G \subseteq N$ and ρ of $N - S \cup G$:

$$J(\beta S, \beta \rho) = \beta J(S, \rho) \tag{11.23}$$

where β is a constant.

Definition 5 (*Balanced game*)

A map $\boldsymbol{\omega}$ is defined as a balanced map if $\sum_S \omega_S = 1$. The cooperative game with externalities is a balanced game if the following condition is satisfied for any balanced map:

$$\sum_S \omega_S U(S, \rho) \leq U(N) \tag{11.24}$$

Balanced map is a definition in cooperative game [34]. Any map $\boldsymbol{\omega}$ which satisfies $\sum_S \omega_S = 1$ is defined as a balanced map in the cooperative game.

Definition 6 (*Excess*)

The coalition's dissatisfaction with respect to the imputation $\boldsymbol{x}$ is measured by its excess defined as $\varepsilon(x, S, \rho) = U(S, \rho) - \sum_{i \in S} x_i$.

Lemma 1 *The grand coalition is stable if* γ-core *is nonempty* [33].

Lemma 2 *If a cooperative game with externalities is a balanced game, then* γ-core *is nonempty* [34].

Theorem 1 *The proposed cooperative game with externalities is superadditive and positive homogenous.*

Proof:

See Appendix A in [37].

Theorem 2 *The grand coalition in this local energy and carbon market is stable.*

Proof:

11.3.3 Two-Stage Payoff Allocation Problem

After determining the total payoff for the grand coalition, the DSO will allocate each MG's share fairly. Traditional methods like the nucleolus and Shapley value are commonly used for payoff distribution in standard cooperative games (without externalities). However, these methods are not suitable for cooperative games involving externalities. Therefore, we introduce a two-stage payoff allocation model tailored for this cooperative game, aiming to minimize the worst-case excess across all potential coalitions. This approach ensures that the payoff allocation fits within the *γ-core*. The proposed two-stage payoff allocation process is as follows:

$$\min_{x}\max_{k}\left(U(S,\rho)-\sum_{i\in S}k_i x_i\right) \tag{11.25}$$

$$\sum_{i} x_i = U(N) \tag{11.26}$$

$$U(S,\rho)-\sum_{i\in S}k_i x_i \le 0 \tag{11.27}$$

$$1 \le \sum_{i} k_i \le N-1 \tag{11.28}$$

$$U(S,\rho)=\sum_{t}\sum_{i\in S}k_i\left(U_{i,t}(PD_{i,t})-\sum_{m}C(PG_{i,m,t})+\pi_t^b PB_{i,t}\right.$$
$$\left.-\pi_t^s PS_{i,t}+c_t^b CB_{i,t}-c_t^s CS_{i,t}\right) \tag{11.29}$$

$$PD_{i,t}, PG_{i,n,t}, PB_{i,t}, PS_{i,t}, CB_{i,t}, CB_{i,t} \in$$
$$\arg\left\{\max\sum_{t}\sum_{i}\left(U_{i,t}(PD_{i,t})-\sum_{n}C(PG_{i,n,t})+\pi_t^b PB_{i,t}\right.\right.$$
$$\left.-\pi_t^s PS_{i,t}+c_t^b CB_{i,t}-c_t^s CS_{i,t}\right);$$
$$\text{s.t. Constraints (11.31)} - \text{(11.35)}\} \tag{11.30}$$

$$\sum_{i\in S}(PD_{i,t}+PS_{i,t})=\sum_{i\in S}\left(\sum_{m}PG_{i,m,t}+\sum_{w}PW_{i,w,t}+PB_{i,t}\right)\forall t \tag{11.31}$$

$$\sum_{i\in S}(C_i^{\max}+CS_i)=\sum_{i\in S}\left(\sum_{t}\sum_{m}CE_{i,m,t}+CB_i\right) \tag{11.32}$$

$$PD_{i,t}+PS_{i,t}=\sum_{m}PG_{i,m,t}+\sum_{w}PW_{i,w,t}+PB_{i,t}\forall i\notin S,\forall t \tag{11.33}$$

$$C_i^{\max}+CS_i=\sum_{t}\sum_{m}CE_{i,m,t}+CB_i\forall i\notin S \tag{11.34}$$

$$\text{Constraints (11.3)–(11.4), (11.7)–(11.20)} \tag{11.35}$$

Constraints (11.26)–(11.27) ensure the payoff allocation remains within the *γ-core*. Constraint (11.28) restricts the coalition size. Constraints (11.29)–(11.30) determine the coalition S's payoff, while (11.31)–(11.32) maintain the power and carbon balance for coalition S members, and (11.33)–(11.34) do the same for those outside it. Constraint (11.35) addresses power network limitations. This two-stage allocation model is non-linear, involving quadratic constraints (11.3), (11.4) and a conic constraint (11.16). The quadratic constraints can be linearized using a piecewise linear function [38], and the conic constraint is linearized following the method from [39].

11.4 Solution Techniques

The compact form of the two-stage payoff allocation problem is stated as follow:

$$\begin{aligned} & \min_{x} \max_{y} (A \cdot x + B \cdot z) \\ & \text{s.t. } C \cdot x \leq H,\ D \cdot x + E \cdot y \leq J \\ & z \in \arg\left\{ \max_{z}\ F \cdot z;\ G \cdot y + K \cdot z \leq L \right\} \end{aligned} \tag{11.36}$$

where $\boldsymbol{x}$ donates the set of variables decided at the first stage, while $\boldsymbol{y}$ and $\boldsymbol{z}$ donate the sets of binary and continuous variables decided at the second stage, respectively. $\mathbf{C} \cdot \mathbf{x} \leq \mathbf{H}$ collects the first-stage constraints, while $D \cdot x + E \cdot y \leq J$ donates the second stage constraints. Figure 11.4 provides the flowchart of the proposed solution technique. This two-stage problem cannot be solved directly. We first utilize the C&CG algorithm [38] to decouple the primal problem into a single-level master problem and a two-level subproblem. The KKT condition [40] is further applied to convert the two-level subproblem into a single-level problem.

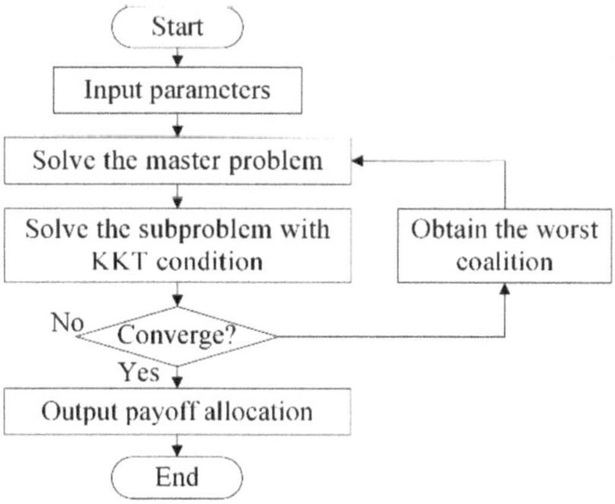

Fig. 11.4 Flowchart of the proposed algorithm

11.4.1 Master Problem

The master problem decides the payoff allocation for each MG, which is illustrated as follow:

$$\begin{aligned} &\min_{x} \alpha \\ &\text{s.t. } C \cdot x \leq H, \alpha \geq \alpha^{\min} \end{aligned} \tag{11.37}$$

where α is an auxiliary variable and $\alpha^{\min}$ is the estimated lower bound for α.

11.4.2 Subproblem

The subproblem finds the coalition in which the excess is the largest with respect to the payoff allocation obtained in the master problem. The subproblem is illustrated as follow:

$$\begin{aligned} &\max_{y}\left(A \cdot x^{*} + B \cdot z\right) \\ &\text{s.t. } D \cdot x^{*} + E \cdot y \leq J \\ &z \in \arg\left\{\max_{z} F \cdot z;\ G \cdot y + K \cdot z \leq L\right\} \end{aligned} \tag{11.38}$$

The subproblem is a two-level problem which cannot be solved by the existing commercial solvers. Therefore, the KKT condition is applied to convert the two-level problem into a single-level problem. The subproblem is reformulated as (11.39) with the introduction of KKT conditions:

$$\begin{aligned} &\max_{y}\left(A \cdot x^{*} + B \cdot z\right) \\ &\text{s.t. } D \cdot x^{*} + E \cdot y \leq J,\ G \cdot y + K \cdot z \leq L \\ &K \cdot \lambda = F,\ (G \cdot y + K \cdot z - L)\lambda = 0 \end{aligned} \tag{11.39}$$

where λ is the dual variable of constraint $F \cdot y + G \cdot z \leq L$.

The bilinear term in (11.39) can be linearized using the big-M approach. Once the subproblem is solved and the optimal solution $\mathbf{z}^{*}$ is calculated, the optimal payoff allocation will be obtained and the algorithm will be terminated if criteria (11.40) is satisfied. Otherwise, we add (11.41) to the master problem for calculating the next iteration:

$$\left|\alpha^{*} - \left(A \cdot x^{*} + B \cdot z^{*}\right)\right| \leq \varepsilon \tag{11.40}$$

$$\begin{aligned} D \cdot x + E \cdot y^{h,*} \leq J, G \cdot y^{h,*} + K \cdot z^h \leq L \\ K \cdot \lambda^h = F, \left(G \cdot y^{h,*} + K \cdot z^h - L\right)\lambda^h = 0 \end{aligned} \tag{11.41}$$

where h is the current iteration and ε is a small number.

11.4.3 Solution Procedure

The proposed solution steps are listed as follow:

1. Input all parameters.
2. Solve (11.37) to calculate x^* and α^*.
3. Set $h = 1$. Solve (11.39) to calculate y^* and z^*.
4. If criteria (11.40) is satisfied, terminate the algorithm and list the optimal payoff allocation. Otherwise, add (11.41) to the master problem (11.37) and repeat steps (2) to (4).

11.5 Case Studies

The proposed local energy and carbon market was tested on a 4-MGs system, the IEEE 33-bus, and IEEE 123-bus testing systems, with a 24-h dispatch period. Each MG can reduce up to 15% of its peak load through demand response, and the carbon allowance per MG during this period is set at 0.2 tons, as detailed in [40]. The mathematical models were solved on a Matlab 2018a platform using Gurobi 9.1, and the blockchain was implemented with Go Ethereum.

11.5.1 Microgrid System

The 4-MGs system is shown in Fig. 11.5. The MG parameters like generator and demand response costs are provided in [10]. MG1 is the slack bus. Six cases are considered here to analyze the proposed local energy and carbon market.

Case I: There is no clearing process by the DSO. MGs trade energy and carbon allowance at the fixed price [40] with the DSO/wholesale market. Demand response is not considered.
Case II: There is a market only for energy. The carbon allowance is traded at a fixed price with the DSO/wholesale market. Demand response is not considered.
Case III: The proposed local energy and carbon market for MGs is applied in which there is a clearing process for both energy and carbon allowance. Demand response is not considered.
Case IV: The same as Case III except the demand response is applied.

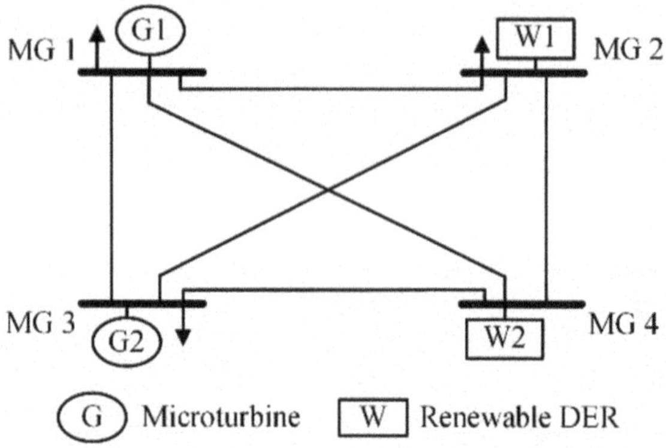

Fig. 11.5 4-MGs system topology

Case V: The same as Case IV except the DSO/wholesale price for carbon allowance is 20% cheaper than that in Case IV.
Case VI: The same as Case V except the DSO/wholesale price for energy is 20% cheaper than that in Case V.

Table 11.2 shows the cost allocation across Cases I-VI. MG2 and MG4, which utilize renewable DERs, sell carbon allowances, while MG1 and MG3, which use MTs, purchase them. In Case I, without inter-MG transactions, MGs buy energy and carbon allowances from the DSO/wholesale market at high prices and sell them back at low prices, resulting in the lowest total payoff. In Case II, an energy-only market encourages MGs to trade energy via the DSO, reducing their purchases from the wholesale market. In Case III, with a market for both energy and carbon allowances, the total MG payoff increases as MGs trade lower-cost carbon allowances through the DSO, leading to less reliance on the DSO/wholesale market for energy and carbon trades.

In Case IV, the introduction of demand response further reduces energy purchases from the DSO/wholesale market, increasing the total MG payoff. In Case V, MGs buy inexpensive carbon allowances from the wholesale market to generate more on-site

Table 11.2 Cost allocations in different cases

Cases	Total payoff ($)	Trading cost for energy with DSO/wholesale market ($)		Trading cost for carbon allowance with DSO/ wholesale market ($)	
		Purchasing	Selling	Purchasing	Selling
I	943.69	501.39	14.23	19.42	8.00
II	1100.63	430.22	0	0	8.00
III	1009.90	406.70	0	0	0
IV	1160.27	114.37	0	0	0
V	1180.78	0	0	51.91	0
VI	1189.88	120.23	0	19.03	0

energy, thereby increasing carbon purchases. Compared to Case V, Case VI shows an increase in cheap energy purchases and a decrease in carbon allowance purchases.

Table 11.3 highlights the optimal solutions for MG3 under various cases. MG3, equipped with MT, typically purchases carbon allowances to generate energy on-site. In Cases I–III, energy consumption remains consistent due to the absence of demand response. In Case I, MG3 buys both energy and carbon allowances at fixed prices from the DSO/wholesale market to meet its load requirements. In Case II, MG3 reduces both energy production and carbon allowance purchases by accessing lower-cost energy through the DSO. In Case III, MG3's energy production increases compared to Cases I and II, as it can acquire low-cost energy and carbon allowances via the DSO, enhancing its on-site energy generation.

In Case IV, MG3 reduces energy consumption and purchases compared to Case III by leveraging demand response and curtailing its load when energy and carbon allowance prices are high. In Case V, with lower carbon allowance prices in the DSO/wholesale market, MG3 increases its carbon allowance purchases to generate more on-site energy, thereby decreasing its energy purchases while boosting energy consumption. In Case VI, MG3 opts to purchase electricity over carbon allowances from the DSO/wholesale market, leading to higher energy purchases and reduced carbon allowance acquisitions.

Figure 11.6 displays the payoffs for each MG in Case VI. MG2 and MG4, with low-cost DERs, supply their on-site loads at minimal cost and without CO_2 emissions, enabling them to sell carbon allowances to maximize their payoffs. Conversely, MG1 and MG3, which rely on MTs, purchase carbon allowances, reducing their payoffs and potentially driving investments in renewable DERs.

Table 11.4 provides the public keys for DSO and MG interchanges within the blockchain network, where smart contracts are executed using the Solidity programming language. The market operates in two phases: initially, MGs trade energy and carbon allowances, followed by payment allocation based on traded quantities. Table V illustrates the payment allocation process via smart contracts, verifying MGs' adherence to transactions. If compliant, the smart contracts facilitate the corresponding payments between MGs. Specifically, MG2 and MG4 sell energy and carbon allowances to MG1 and MG3, prompting smart contracts to transfer payments

Table 11.3 Optimal MG3 solution in different cases

Cases	Energy production (kWh)	Energy consumption (kWh)	Energy purchase (kWh)	Carbon allowance purchase (t)
I	253.78	2640.00	2386.22	0.09
II	114.65	2640.00	2552.36	0.00
III	470.59	2640.00	2169.4	0.30
IV	470.59	1603.20	1132	0.30
V	1635.02	1658.20	23.18	1.25
VI	1003.69	1664.39	660.70	0.75

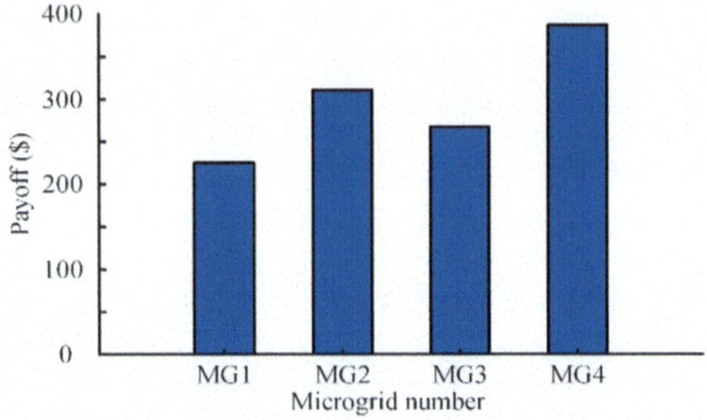

Fig. 11.6 Payoff of each MG in Case VI

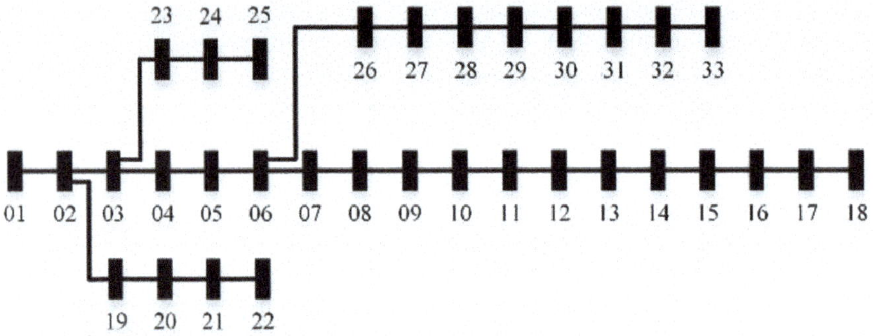

Fig. 11.7 IEEE 33-bus testing system topology

from MG1 and MG3 to MG2 and MG4. As shown in Table 11.4, the payoffs for MG1 and MG3 decrease after the execution of smart contracts, with payments transferred to MG2 and MG4. The resulting payoffs for each MG are consistent with those in Table 11.2 (Table 11.5)

Table 11.4 Public keys of all participants in blockchain network

Participants	Public keys
DSO	0x1A4545135756CC8Fe5301E50c2E1d2E4A7741880
MG1	0x929aB5a6bFf983bC888953664886D01666803D9f
MG2	0x8dA5ad34728805c60aE7d1B4f0fD0145ce75782B
MG3	0x509e76C694D84D74A8736332B1c822045c7cE5A6
MG4	0xC616C75d666B506d356c53FF6D85d68dC18Abd82

Table 11.5 Payment transfer using smart contracts

Cases	MG's payoff ($)			
	MG1	MG2	MG3	MG4
Before smart contracts are executed	327.68	296.56	363.11	302.53
After smart contracts are executed	224.87	310.71	267.48	386.82

Table 11.6 Optimal solutions in TEM and TECM

Cases	Total payoff ($)	CO_2 emission (t)	Energy produced by MTs (kWh)
TEM	25,819.3	28.11	33,019.8
TECM	20,008.1	21.59	25,398.1

11.5.2 IEEE 33-Bus Testing System

The proposed model is tested using the IEEE 33-bus system, featuring 33 MGs and 32 distribution lines, with Bus 01 as the slack bus. Two scenarios are evaluated:

TEM: LEM in which the carbon allowance is not limited.
TECM: The proposed local energy and carbon market.

In TEM, MGs generate significant CO_2 emissions due to the absence of carbon costs, enabling extensive energy production via MTs. In contrast, TECM introduces carbon allowances, reducing the energy generated by MTs and consequently lowering their emissions, as shown in Table 11.6.

Figure 11.8a, b depict the relationship between increased prices for electricity and carbon allowances and the corresponding decreases in total payoff and CO_2 emissions, respectively. As carbon allowance prices rise, MGs reduce emissions by curbing their carbon-intensive activities. However, higher electricity prices lead to increased emissions as MGs opt to generate on-site energy rather than purchase it.

The comparison of TECM with and without network constraints, labeled TECM-I and TECM-II, is illustrated in Fig. 11.9. In TECM-I, unrestricted trades between MGs result in line overloads, compromising network security. TECM-II imposes capacity limits on line flows, ensuring network security. Table 11.7 indicates that while TECM-I achieves higher total payoffs and energy trades, TECM-II emphasizes on-site DER generation to avoid purchasing from other MGs, especially for those with higher CO2 emissions, thereby increasing carbon allowance trades.

11.5.3 IEEE 123-Bus Testing System

The proposed model was tested using the IEEE 123-bus system, which includes 123 MGs. Table 11.8 presents a comparison of optimal solutions between the LEM and the

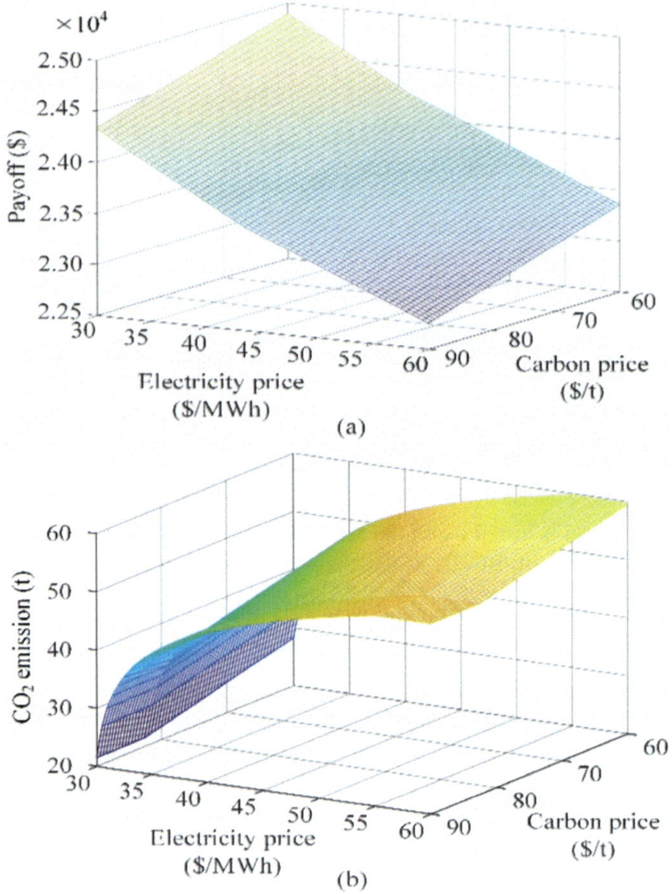

Fig. 11.8 **a** Total payoff and **b** carbon emission as a function of electricity and carbon allowance prices

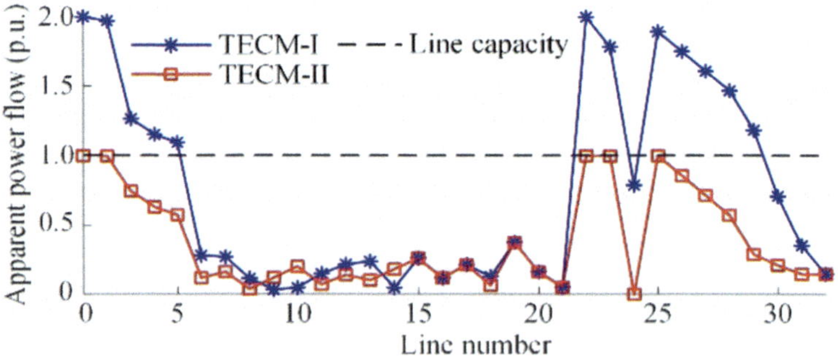

Fig. 11.9 Line flows at hour 9 in TECM-I and TECM-II

Table 11.7 Optimal solutions in TECM-I and TECM-II

Cases	Total payoff ($)	Traded energy (kWh)	Traded carbon allowance (t)
TECM-I	26,582.1	58,232.9	7.81
TECM-II	20,008.1	41,833.0	12.39

local energy and carbon market. In LEM, MGs can freely dispatch MTs to generate energy without CO_2 restrictions, leading to significant emissions. In contrast, local energy and carbon market reduces these emissions by requiring MGs to purchase carbon allowances for CO_2 output.

Table 11.9 outlines the computation time and iterations needed for the proposed payoff allocation method. This method utilizes the C&CG approach to break down the original two-level problem into a master problem and a subproblem, solving the latter using KKT conditions. The total computation time is 7359 s, with 437 s for the master problem and 6922 s for the subproblem. The C&CG algorithm converges in 104 iterations, significantly fewer than the 2123-1 iterations typically required by existing methods. The computational load is primarily due to solving the subproblem, where the big-M method is employed. To improve performance, the method suggested in Ref. [41] could be implemented.

Table 11.11 demonstrates the blockchain's performance in contract settlement. A private blockchain network was established for the participating MGs using Go Ethereum on a personal computer, allowing flexible adjustment of consensus times. The average consensus time for transaction verification is approximately 5 min, with a cost of about 0.001 ETH. To further reduce consensus time, the method could be adapted to other platforms like Hyperledger, which offers better computational efficiency.

Table 11.8 Optimal solutions in TEM and local energy and carbon market

Cases	Total payoff ($)	CO_2 emission (t)	Energy produced by MTs (kWh)
Local energy market	96,214.2	78.31	91,905.70
Local energy and carbon market	87,461.9	69.85	82,170.32

Table 11.9 Computational performance of the proposed payoff allocation method

Solution time (s)			Iterations
Total	Master problem	Subproblem	
7359	437	6922	104

Table 11.11 Performance of the proposed blockchain contract settlement system

Time to achieve consensus	Cost for achieving consensus
About 5 min	About 0.001 ETH

11.6 Discussion

Accurately forecasting load and renewable DER output is challenging, leading to potential load imbalances during real-time operations. To manage this variability, the proposed energy market can be extended into a combined energy and reserve market, where MTs, renewable DERs, and loads can serve as reserves [42]. The mathematical formulation for the grand coalition in this combined market is outlined as follows:

$$\max UR(N) \tag{11.42}$$

$$0 \leq PG_{i,m,t} + RG_{i,m,t} \leq PG_{i,m}^{\max} \forall i, \forall m, \forall t \tag{11.43}$$

$$0 \leq PW_{i,w,t} + RW_{i,w,t} \leq PW_{i,w,t}^{\max} \forall i, \forall w, \forall t \tag{11.44}$$

$$PD_{i,t}^{\min} \leq PD_{i,t} + RD_{i,t} \leq PD_{i,t}^{\max} \forall i, \forall t \tag{11.45}$$

$$\text{Constraints } (11.2)-(11.17) \tag{11.46}$$

where $UR(N)$ represents the total payoff for the grand coalition N for the combined energy and reserve market. The detailed expression for $UR(N)$ is provided in [43]. $RG_{i,m,t}$, $RW_{i,w,t}$, and $RD_{i,t}$ represent the reserve provided by MTs, renewable DERs, and loads. After solving the above problem consisting of constraints (11.42)–(11.46), the grand coalition's payoff is obtained. Then, the grand coalition's payoff can be allocated to each MG by using the method provided in Sect. 11.3.3. The reserve in this market can accommodate the potential uncertainties. It should be noted that the payoff allocation for each MG includes the payoffs for both energy and reserve.

11.7 Conclusions

This chapter introduces a blockchain-based local energy and carbon market for interconnected MGs. The DSO gathers data from all MGs and oversees energy and carbon allowance trades to maintain network security. The MGs' cooperative behaviors, influenced by the power distribution network, are modeled using a cooperative game with externalities. A two-stage payoff allocation problem, along with the corresponding solution techniques, is proposed to distribute the coalition's payoff to each MG within a reasonable time frame. The use of asymmetric encryption (public and private keys) within the blockchain ensures secure and transparent transactions. Smart contracts enhance operational efficiency by automating market processes, including transaction settlements. Simulations demonstrate the proposed local energy and carbon market's effectiveness, computational performance, and scalability. Test results show a 76.8% reduction in total carbon emissions, validating the method's effectiveness in reducing emissions. Additionally, this approach motivates MGs to trade energy and carbon allowances, as their total payoff increases by over 25%. Future research will explore the LEM for energy hubs and test the method on various blockchain platforms.

References

1. Li, Z., Shahidehpour, M., Aminifar, F., Alabdulwahab, A., & Al-Turki, Y. (2017). Networked microgrids for enhancing the power system resilience. *Proceedings of the IEEE, 105*(7), 1289–1310.
2. Rahimi, F. A., & Ipakchi, A. (2012). Transactive energy techniques: Closing the gap between wholesale and retail markets. *The Electricity Journal, 25*(8), 29–35.
3. Adeyemi, A., Yan, M., Shahidehpour, M., Bahramirad, S., & Paaso, A. (2020). Transactive energy markets for managing energy exchanges in power distribution systems. *The Electricity Journal, 33*(9), 106868.
4. Molyneaux, L., & Head, B. (2020). Why information and transparency about electricity matter: Fragmentation of governance and accountability under New Public Management. *Australian Journal of Public Administration, 79*(1), 143–164.
5. Du, Y., & Li, F. (2018). A hierarchical real-time balancing market considering multi-microgrids with distributed sustainable resources. *IEEE Transactions on Sustainable Energy, 11*(1), 72–83.
6. Bai, L., Wang, J., Wang, C., Chen, C., & Li, F. (2017). Distribution locational marginal pricing (DLMP) for congestion management and voltage support. *IEEE Transactions on Power Systems, 33*(4), 4061–4073.
7. Liu, W., Zhan, J., & Chung, C. Y. (2018). A novel transactive energy control mechanism for collaborative networked microgrids. *IEEE Transactions on Power Systems, 34*(3), 2048–2060.
8. Wang, Y., Huang, Z., Shahidehpour, M., Lai, L. L., Wang, Z., & Zhu, Q. (2019). Reconfigurable distribution network for managing transactive energy in a multi-microgrid system. *IEEE Transactions on Smart Grid, 11*(2), 1286–1295.
9. Liu, W., Gu, W., Wang, J., Yu, W., & Xi, X. (2018). Game theoretic non-cooperative distributed coordination control for multi-microgrids. *IEEE Transactions on Smart Grid, 9*(6), 6986–6997.
10. Yan, M., Shahidehpour, M., Paaso, A., Zhang, L., Alabdulwahab, A., & Abusorrah, A. (2020). Distribution network-constrained optimization of peer-to-peer transactive energy trading among multi-microgrids. *IEEE Transactions on Smart Grid, 12*(2), 1033–1047.
11. Han, L., Morstyn, T., & McCulloch, M. (2018). Incentivizing prosumer coalitions with energy management using cooperative game theory. *IEEE Transactions on Power Systems, 34*(1), 303–313.
12. Feng, C., Wen, F., You, S., Li, Z., Shahnia, F., & Shahidehpour, M. (2019). Coalitional game-based transactive energy management in local energy communities. *IEEE Transactions on Power Systems, 35*(3), 1729–1740.
13. Li, J., Ye, Y., & Strbac, G. (2020). Stabilizing peer-to-peer energy trading in prosumer coalition through computational efficient pricing. *Electric Power Systems Research, 189*, 106764.
14. Tushar, W., Saha, T. K., Yuen, C., Azim, M. I., Morstyn, T., Poor, H. V., & Bean, R. (2020). A coalition formation game framework for peer-to-peer energy trading. *Applied Energy, 261*, 114436.
15. Mukherjee, M., Marinovici, L. D., Hardy, T. D., & Hansen, J. (2020). Framework for large-scale implementation of wholesale-retail transactive control mechanism. *International Journal of Electrical Power & Energy Systems, 115*, 105464.

16. Renani, Y. K., Ehsan, M., & Shahidehpour, M. (2017). Optimal transactive market operations with distribution system operators. *IEEE Transactions on Smart Grid, 9*(6), 6692–6701.
17. Marinovici, L., Hardy, T., & Hansen, J. (2019, August). Hierarchical architecture for transactive wholesale-retail market integration. In *2019 IEEE Power & Energy Society General Meeting (PESGM)* (pp. 1–5). IEEE.
18. Hansen, J., Hardy, T., & Marinovici, L. (2019, February). Transactive energy: Stabilizing oscillations in integrated wholesale-retail energy markets. In *2019 IEEE Power & Energy Society Innovative Smart Grid Technologies Conference (ISGT)* (pp. 1–5). IEEE.
19. Ghamkhari, M. (2019, June). Transactive energy versus demand response in cutting wholesale electricity prices. In *2019 3rd International Conference on Smart Grid and Smart Cities (ICSGSC)* (pp. 182–187). IEEE.
20. MacCormack, J., Zareipour, H., & Rosehart, W. D. (2012). Long-term market equilibrium model with strategic, competitive, and inflexible generation. *IEEE Transactions on Power Systems, 27*(4), 2291–2292.
21. Zhang, N., Hu, Z., Dai, D., Dang, S., Yao, M., & Zhou, Y. (2015). Unit commitment model in smart grid environment considering carbon emissions trading. *IEEE Transactions on Smart Grid, 7*(1), 420–427.
22. Ji, Z., Kang, C., Chen, Q., Xia, Q., Jiang, C., Chen, Z., & Xin, J. (2013). Low-carbon power system dispatch incorporating carbon capture power plants. *IEEE Transactions on Power Systems, 28*(4), 4615–4623.
23. Cheng, Y., Zhang, N., Zhang, B., Kang, C., Xi, W., & Feng, M. (2019). Low-carbon operation of multiple energy systems based on energy-carbon integrated prices. *IEEE Transactions on Smart Grid, 11*(2), 1307–1318.
24. Adeyemi, A., Yan, M., Shahidehpour, M., Botero, C., Guerra, A. V., Gurung, N., & Paaso, A. (2020). Blockchain technology applications in power distribution systems. *The Electricity Journal, 33*(8), 106817.
25. Shahidehpour, M., Yan, M., Shikhar, P., Bahramirad, S., & Paaso, A. (2020). Blockchain for peer-to-peer transactive energy trading in networked microgrids: Providing an effective and decentralized strategy. *IEEE Electrification Magazine, 8*(4), 80–90.
26. Hamouda, M. R., Nassar, M. E., & Salama, M. M. A. (2020). A novel energy trading framework using adapted blockchain technology. *IEEE Transactions on Smart Grid, 12*(3), 2165–2175.
27. Li, Z., Kang, J., Yu, R., Ye, D., Deng, Q., & Zhang, Y. (2017). Consortium blockchain for secure energy trading in industrial internet of things. *IEEE Transactions on Industrial Informatics, 14*(8), 3690–3700.
28. AlAshery, M. K., Yi, Z., Shi, D., Lu, X., Xu, C., Wang, Z., & Qiao, W. (2020). A blockchain-enabled multi-settlement quasi-ideal peer-to-peer trading framework. *IEEE Transactions on Smart Grid, 12*(1), 885–896.
29. Abdella, J., Tari, Z., Anwar, A., Mahmood, A., & Han, F. (2021). An architecture and performance evaluation of blockchain-based peer-to-peer energy trading. *IEEE Transactions on Smart Grid, 12*(4), 3364–3378.
30. Li, Y., & Hu, B. (2019). An iterative two-layer optimization charging and discharging trading scheme for electric vehicle using consortium blockchain. *IEEE Transactions on Smart Grid, 11*(3), 2627–2637.
31. He, H., Luo, Z., Wang, Q., Chen, M., He, H., Gao, L., & Zhang, H. (2020). Joint operation mechanism of distributed photovoltaic power generation market and carbon market based on cross-chain trading technology. *IEEE Access, 8*, 66116–66130.
32. Hua, W., Jiang, J., Sun, H., & Wu, J. (2020). A blockchain based peer-to-peer trading framework integrating energy and carbon markets. *Applied Energy, 279*, 115539.
33. Chander, P., & Tulkens, H. (1997). The core of an economy with multilateral environmental externalities. *International Journal of Game Theory, 26*, 379–401.
34. Hafalir, I. E. (2007). Efficiency in coalition games with externalities. *Games and Economic Behavior, 61*(2), 242–258.
35. Ghorani, R., Fotuhi-Firuzabad, M., & Moeini-Aghtaie, M. (2019). Main challenges of implementing penalty mechanisms in transactive electricity markets. *IEEE Transactions on Power Systems, 34*(5), 3954–3956.

36. Du, E., Zhang, N., Kang, C., Kroposki, B., Huang, H., Miao, M., & Xia, Q. (2016). Managing wind power uncertainty through strategic reserve purchasing. *IEEE Transactions on Power Systems, 32*(4), 2547–2559.
37. Yang, J., Zhang, N., Kang, C., & Xia, Q. (2016). A state-independent linear power flow model with accurate estimation of voltage magnitude. *IEEE Transactions on Power Systems, 32*(5), 3607–3617.
38. Yan, M., Zhang, N., Ai, X., Shahidehpour, M., Kang, C., & Wen, J. (2018). Robust two-stage regional-district scheduling of multi-carrier energy systems with a large penetration of wind power. *IEEE Transactions on Sustainable Energy, 10*(3), 1227–1239.
39. Zhou, Y., Shahidehpour, M., Wei, Z., Li, Z., Sun, G., & Chen, S. (2019). Distributionally robust co-optimization of energy and reserve for combined distribution networks of power and district heating. *IEEE Transactions on Power Systems, 35*(3), 2388–2398.
40. Yan, M., He, Y., Shahidehpour, M., Ai, X., Li, Z., & Wen, J. (2018). Coordinated regional-district operation of integrated energy systems for resilience enhancement in natural disasters. *IEEE Transactions on Smart Grid, 10*(5), 4881–4892.
41. Yan, M., Shahidehpour, M., Alabdulwahab, A., Abusorrah, A., Gurung, N., Zheng, H., & Paaso, E. A. (2021). Blockchain for transacting energy and carbon allowance in networked microgrids. *IEEE Transactions on Smart Grid, 12*(6), 4702–4714.
42. Guo, Z., Pinson, P., Chen, S., Yang, Q., & Yang, Z. (2020). Chance-constrained peer-to-peer joint energy and reserve market considering renewable generation uncertainty. *IEEE Transactions on Smart Grid, 12*(1), 798–809.
43. Huang, G., Wang, J., Chen, C., Qi, J., & Guo, C. (2017). Integration of preventive and emergency responses for power grid resilience enhancement. *IEEE Transactions on Power Systems, 32*(6), 4451–4463.

Chapter 12
Reinforcement Learning-Based Local Energy Trading

Abstract In this chapter, we examine the role of emerging energy brokers in a local energy market (LEM), facilitating indirect local energy trading. This chapter proposed a reinforcement learning-based local energy trading strategy that aims to diversify the energy ecosystem at the edge of distribution networks rather than replace existing energy services or become the best trading model. The trading mechanism provides additional options for customers and prosumers to participate occasionally in the retail electricity market, supplementing existing utility services. Considering customer behavior characteristics, it improves trading efficiency and encourages local power balance. The energy trading process will be structured as a Markov decision process (MDP), incorporating reinforcement learning and data-driven methods. Meanwhile, economic concepts like search friction, related to typical search costs in this trading model, are also discussed.

12.1 Introduction

In today's retail electricity market, customers have limited options for choosing different energy services. Despite the widespread installation of distributed energy resources (DERs) in many regions, most customers and prosumers with local energy generation and surplus can only trade with utility companies. They buy energy from or sell surplus back to the utilities, often facing a price gap. Consumer-centric business models and localized energy trading platforms need to be developed to provide more freedom in energy trading and foster open innovation. The research community is working on these ideas to create a future local energy market (LEM) where all customers can equally participate directly [1].

Many studies have proposed future market mechanisms with significant potential, such as prosumer grid integration [2], peer-to-peer (P2P) models [3, 4], and prosumer community groups [5, 6]. Direct integration of DERs into the main grid and participation in the electricity market through aggregators is widely accepted due to its straightforward applicability and manageability. The work in [2] analyzes the optimal planning and operation of aggregated DERs participating in the electricity market.

M. Song et al., *Local Energy Markets*, https://doi.org/10.1007/978-981-97-9750-9_12

However, to diversify the customer-side energy ecosystem and promote a deregulated market, a local energy trading model with an eBay-like market platform is also frequently discussed and gaining research interest, though it remains largely conceptual [7–9]. The e-commerce-like retail activity of energy trading constitutes only a small fraction of the overall energy business, particularly the wholesale market or bilateral contracts. The trend of gamification of energy activities [10] and the design of new business models is being actively pursued to develop a new energy ecosystem and market reform. For example, the concept of a prosumer community is used in [5, 11] for local energy sharing and internal trading. In [3], a matching mechanism-based market model is proposed to enable individual prosumers to conduct bilateral trades. In [4], a scalable and modular P2P system is designed, along with a prototype, to demonstrate on-site building energy trading between prosumers. Additionally, the same group of researchers proposed a novel decentralized digital currency called NRGcoins to encourage prosumers to trade their excess energy locally and make payments using NRGcoins [12].

This chapter examines the role of emerging energy brokers in a LEM to facilitate indirect local energy trading. Many customers seek greater "energy choice" to share or exchange self-generated electricity locally. Local energy trading can complement existing market mechanisms, reduce distribution losses from long-distance exchanges, and improve resource allocation efficiency by considering attitudinal and contextual factors [13]. However, small-scale electricity consumers and prosumers often cannot afford the time to find specific trading partners, making pure P2P modes impractical. In contrast, localized adaptive services can gather, synthesize, and explore distinct renewable energy trading characteristics and size-related patterns [14]. In LEM, trades involve buyers, sellers, and multiple intermediaries [15]. Classical economic approaches, like competitive equilibrium analysis, often overlook the role of these middlemen. Therefore, understanding middlemen's optimal strategies when choosing among potential opportunities of varying quality is necessary, especially considering the costs of delaying decisions. This chapter's contributions are:

(1) Proposing a reinforcement learning-based paradigm for indirect local energy trading in a LEM;
(2) Introducing the role of the retail energy brokers (REBs) to facilitate market operations with specific actions;
(3) Using reinforcement learning techniques to benefit all market participants through a smart strategy with learning capabilities;
(4) Providing insights into the proposed market mechanism based on simulation results.

12.2 Reinforcement Learning-Based Market Architecture

A LEM operates at the distribution level to facilitate retail energy trading with the aid of REBs. This market mechanism provides additional trading opportunities alongside existing utility services, allowing customers to participate directly in local energy transactions with REBs acting as intermediaries. This reinforcement learning-based LEM, similar to [16], functions as a backup trading platform based on local requirements and seasonal events, such as high solar irradiance or regional energy shortages, rather than being open year-round. The future retail electricity market, including the LEM, is illustrated in Fig. 12.1.

In this system, energy entities like buildings, campuses, and communities can join the local trading platform as sellers or buyers based on their energy surplus or deficit forecasts. The LEM model is structured as a Markov decision process (MDP) [17] from the REB's perspective. The REB manages the market, gaining trading experience and developing strategies through reinforcement learning techniques. This approach demonstrates that both customers and traders can benefit simultaneously by employing intelligent methods and considering customer behavior characteristics. Additionally, economic concepts such as search friction [18], related to the costs involved in this market model, are discussed in subsequent sections, offering insights into improving energy market efficiency.

The chosen market model for a LEM is not selected for its superior performance but for two primary reasons: (1) Deregulation, though often cited, remains crucial in transforming traditional energy frameworks. Historical examples from the airline and telecommunications industries illustrate consumer benefits, as seen with platforms like Priceline and Uber. These industries showcase how deregulation and innovative platforms can redefine business models and ecosystems within an existing market.

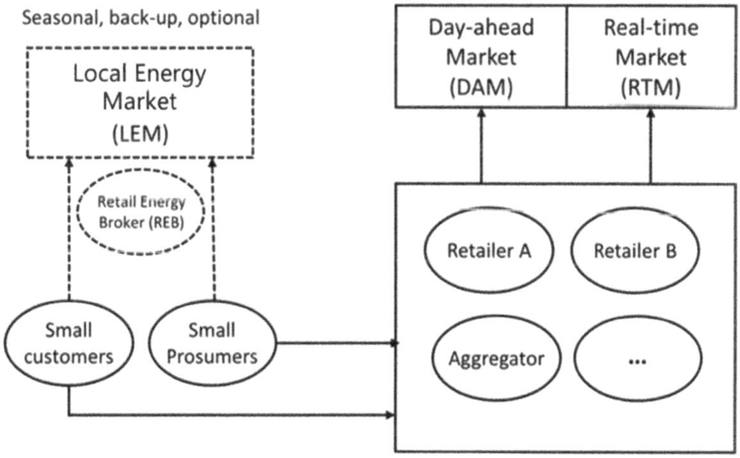

Fig. 12.1 Future retail electricity market

Conversely, despite being one of the largest globally, the electricity sector is often viewed as the least innovative due to stringent regulations; (2). There is a demand for new business model designs and gamification within the energy ecosystem. If effectively implemented, these models could introduce numerous new players (e.g., REBs) into a more competitive retail market. Instead of passively adapting to emerging energy trading trends, we can enhance customer comprehension of the future energy landscape. For instance, mobile Internet made up only 5% of all Internet connections before smartphones, but now it accounts for half. We should create similar enthusiasm for local energy trading. Therefore, the LEM aims to boost regular retail efficiency by diversifying current energy business models and offering energy trading opportunities at the distribution network's edge, close to energy demand sources.

In our holistic market model (Fig. 12.2), we outline the primary functions of each entity. The LEM will support short-term or immediate local energy transactions. The DSO or DNO will oversee market regulation, ensuring reliability and security. Electric utilities will continue to serve customers and offer various long-term retail plans. Customers, meanwhile, can create their own bidding/offering strategies on this local energy trading platform using different energy devices (e.g., batteries, DGs, flexible loads). Local energy transactions will be carried out using existing distribution lines and smart meters for billing and payment.

This chapter does not aim to cover every detail of the market model. Instead, it focuses on the new role of the REB, highlighted in Fig. 12.2. We assume that local participants are geographically close, likely connected to the same distribution feeder, as shown in Fig. 12.3. Thus, we do not model the AC power flow constraint

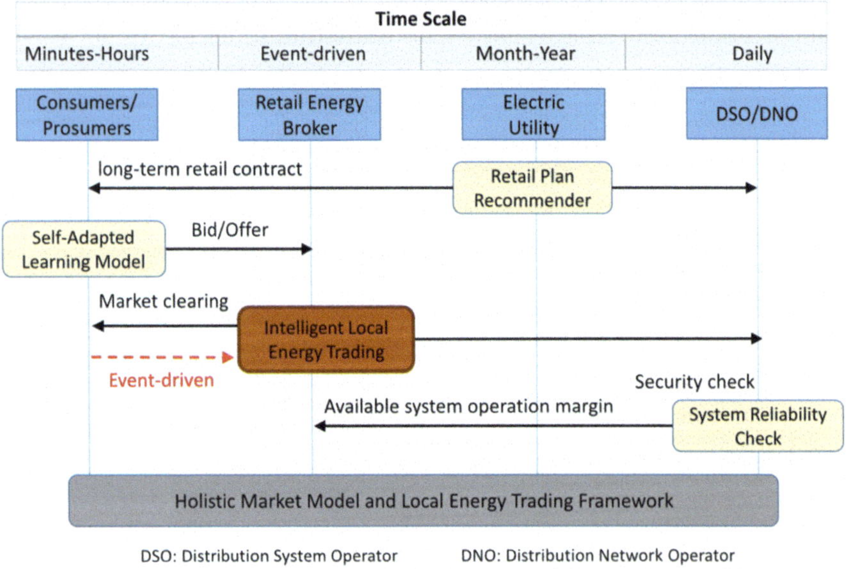

Fig. 12.2 Holistic market model design

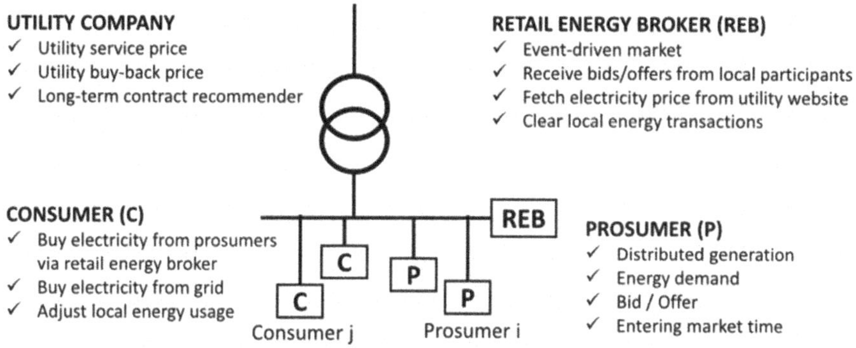

Fig. 12.3 Physical connection of consumer/prosumer and REB

of distribution networks here. Future work will explore the impact of these physical constraints on the proposed LEM.

12.3 System Model

We examine a localized market model including an REB, consumers (buyers) $\mathcal{I}^b$, and prosumers (sellers) $\mathcal{I}^s$. The market operates with day-ahead bidding and time slots of equal duration, allowing the REB to make decisions at each interval. The REB's actions in each time slot aim to clear the market and satisfy both sellers and buyers. Consequently, the market model will be formulated as a MDP, enabling multi-scenario analysis of the market mechanism dynamics and characteristics.

12.3.1 Market Operation

The market mechanism is designed based on a day-ahead bidding process, similar to the wholesale electricity market day-ahead market [19]. Each day is divided into n time slots, allowing market participation from sellers or buyers during these intervals, as shown in Fig. 12.4. Each time interval T will be equal to 24 $hours/n$. Sellers (buyers) can stay in the market for up to H consecutive intervals after entering. The REB can take various actions based on the current market conditions for each time interval. Sellers and buyers typically enter the market on different schedules. Renewable energy supply, such as solar power, does not usually coincide with peak demand periods. Buyers tend to enter the market early morning or late afternoon, coinciding with peak consumption, while sellers prefer noon or midnight, aligning with peak solar or wind energy production. Without energy storage systems (ESS), this temporal discrepancy requires the REB to adopt strategic decision-making,

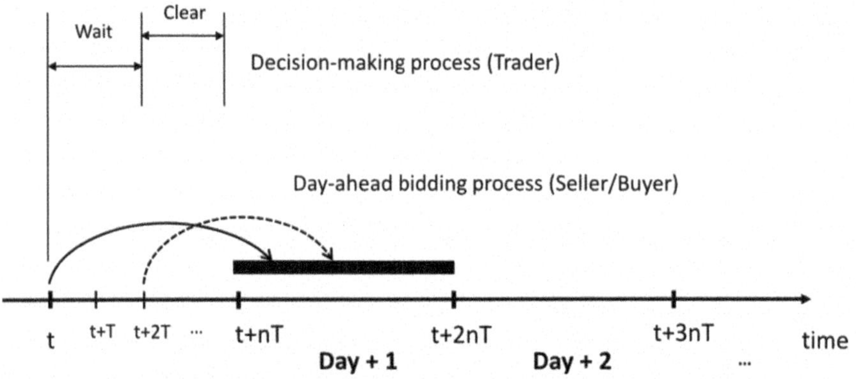

Fig. 12.4 Timeline of the proposed market model

depending on whether it is a buyer's or seller's market. Moreover, decision-making delays or waiting times, akin to opportunity costs, can lead to varied market clearance outcomes due to search friction.

This timeline design facilitates future exploration of customer behaviors, such as withdrawing or modifying bids or offers in consecutive time slots. However, this chapter focuses on market dynamics from the REB's perspective, limiting customers to one-time bidding or offering. Additionally, we introduce a remaining time counter $h_{i,t}^s \in \mathcal{H}_t^s$ and $h_{j,t}^b \in \mathcal{H}_t^b$ associated with each seller and buyer to record their staying status in this market with an initial value of $h_{i,t_i^s}^s = H$ and $h_{j,t_j^b}^b = H$ respectively, where the subscript $t_i^s \in \mathcal{T}^s (t_j^b \in \mathcal{T}^b)$ indicates the market entering time of each seller (buyer). As the time evolves $t \leftarrow t+1$, the remaining time counter decreases accordingly after each time interval: $h_{i,t}^s \leftarrow h_{i,t}^s - 1$, $h_{j,t}^b \leftarrow h_{j,t}^b - 1$. If the remaining time counter approaches empty, $h_{i,t}^s = 0$ or $h_{j,t}^b = 0$, the seller (buyer) will quit this market and exercise direct transaction with the utility company.

12.3.2 *Sellers and Buyers*

Consumers and prosumers participate as sellers or buyers in the localized retail electricity market based on their power demand and generation surplus. Prosumers, regardless of their specific generation sources (e.g., wind, PV, biomass), are considered uniformly for simplification. Upon obtaining predictions for the next hourly interval, seller i provides the REB with bidding information, including s selling price p_i^s and supply energy block e_i^s as price-quantity pairs. Meanwhile, it is essential to ensure that

$$p^{ub} \le p_i^s \le p^u - _p, \quad \forall i \in \mathcal{I}^s \tag{12.1}$$

where p^{ub} is the buy-back price of utility if sellers directly sell their energy surplus to the utility company, and p^u is the regular electricity service price of the utility if buyers directly purchase energy from the utility company. The lowerbound p^{ub} of a seller's bidding price p_i^s is intuitive because the only motivation for sellers' participation in this market is the fact that the seller can sell energy at a higher price than through direct trading with the utility. However, the upperbound $p^u - \epsilon_p$is implicitly established because any bidding price higher than p^u will force buyers to prefer direct trading with the utility, pushing them away from participating in this localized market. Furthermore, item ϵ_p is a small speculation margin controlled by the REB to guarantee fairness and enhance the seller/buyer's pairing chance (no exactly equivalent price to the utility's is allowed to participate, otherwise all customers will bid the same as the utility's price). Similarly, the same principle also applies to buyers and their offering information, with demand energy block e_j^b and purchasing price p_j^b as:

$$p^{ub} + \epsilon_p \leq p_j^b \leq p^u, \ \forall j \in \mathcal{I}^b \tag{12.2}$$

Therefore, buyers can benefit from purchasing electricity at a lower price than the utility's regular service price p^u while leaving enough price margin to attract sellers, and vice versa. These price-quantity pairs of both sellers and buyers can be contained in sets, $\mathcal{P}^s = \{p_i^s\}$, $\mathcal{P}^b = \{p_j^b\}$, $\varepsilon^s = \{e_i^s\}$, $\varepsilon^b = \{e_j^b\}$. These sets, along with sets of the remaining time counters at time t, will be reformulated again as follows:

$$\mathcal{P} = \mathcal{P}^s \times \mathcal{P}^b, \quad \mathcal{E} = \mathcal{E}^s \times \mathcal{E}^b, \quad \mathcal{H}_t = \mathcal{H}_t^s \times \mathcal{H}_t^b. \tag{12.3}$$

We also use $\mathcal{I}_t = \mathcal{I}_t^s \times \mathcal{I}_t^b$ to denote the set of sellers (buyers) that are still present in the market at current time t with $h_{i,t}^s, h_{j,t}^b > 0$. In this way, the market status at current time t will be totally determined by all the current customers' bidding parameters, the current timestamp and the status of the remaining time counters associated with current customers. Thus, the transition probability of the market status, consisting of the status of sellers and buyers, from state $s_t = (\mathcal{H}_t, \mathcal{I}_t)$ to state $s_{t+1} = (\mathcal{H}_{t+1}, \mathcal{I}_{t+1})$ with the REB's given action a_t can be represented as:

$$p_m(s_{t+1}|s_t, a_t) = \prod_{i \in \mathcal{I}_t^s} p_s(h_{i,t+1}^s|s_t, a_t) \times \prod_{j \in \mathcal{I}_t^b} p_b(h_{j,t+1}^b|s_t, a_t) \tag{12.4}$$

where $\mathcal{I}_t$ implicitly contains all the information associated with parameter sets $\mathcal{P}$ and ε, and $\mathcal{H}$ explicitly indicates the possible action candidates available to the energy broker.

12.3.3 Retail Energy Broker

In the proposed market model, the REB determines the market open rate when events or requests are detected. Acting as an intermediary between sellers and buyers, it facilitates local energy transactions in a LEM indirectly (i.e., indirect local energy trading). This eliminates the need for sellers or buyers to search for specific trading partners during the pairing process.

The REB's actions depend on the current time t and the market situation, based on the price-quantity pairs from current seller bids or buyer offers. Essentially, the REB determines the mapping function $a_t : \mathcal{R}_+^{|\mathcal{S}|} \to \mathcal{A}$ as follows:

$$a_t(\mathcal{H}_t, \mathcal{I}_t, t) = a_i, \quad i = 1, 2, \ldots, A. \tag{12.5}$$

From there, it decides how to process the market-clearance at current time-slot t where $|\mathcal{S}|$ denotes the cardinality of market status tuple set $\mathcal{S} = \mathcal{H}_t^s \times \mathcal{H}_t^b \times \mathcal{I}_t^s \times \mathcal{I}_t^b$ and $\mathcal{A}$ denotes finite action set $\{a_1, a_2, \ldots, a_A\}$ as follows (let $A = 4$; four types of actions are designed here):

(1) *Waiting*: Do not accept any bid or offer price-quantity pairs, and delay market clearance until the next time slot, $t + 1$, anticipating more speculative bids or offers from additional sellers or buyers. This action must consider search friction, as the REB may miss opportunities for greater or immediate profit. The remaining time counters for sellers and buyers will be updated to $h \leftarrow h - 1$ for the next interval.

(2) *Retention*: Do not accept any bid or offer price-quantity pairs and delay the market-clearance until the next time slot, $t + 1$; however, in contrast to *waiting*, the REB will issue a retention bonus ξ to all the sellers/buyers present in the current market to compensate for the delay of the market-clearance. Moreover, the seller and buyer's remaining time counters will be updated by $h \leftarrow h - 1$ for the next time interval.

(3) *Retention*: Similar to waiting, do not accept any bid or offer price-quantity pairs and delay market clearance until the next time slot, $t + 1$; However, unlike *waiting*, the REB issues a retention bonus ξ to all current sellers and buyers to compensate for the delay. The remaining time counters for sellers and buyers will also be updated to $h \leftarrow h - 1$ for the next interval.

(4) *Partial-Clearance*: Accept the highest bid p_{high}^s and lowest offer p_{low}^b associated with a price-quantity pair to lock the instant profit so far, setting the chosen seller/buyer's remaining time counter $h \leftarrow 0$, meanwhile delaying the rest of the market-clearance until the next time-slot, with all the other sellers/buyers' remaining time counters updated by $h \leftarrow h - 1$.

(5) *Clearance*: Accept all the bids and offers belonging to price-quantity pairs to exercise the standard double-auction market (action $a_t = a_4$ in Eq. (12.7) and Fig. 12.5), and set all the chosen sellers/buyers' remaining time counters $h \leftarrow 0$ to force them to quit the market. The rest of the customers that fail to be selected

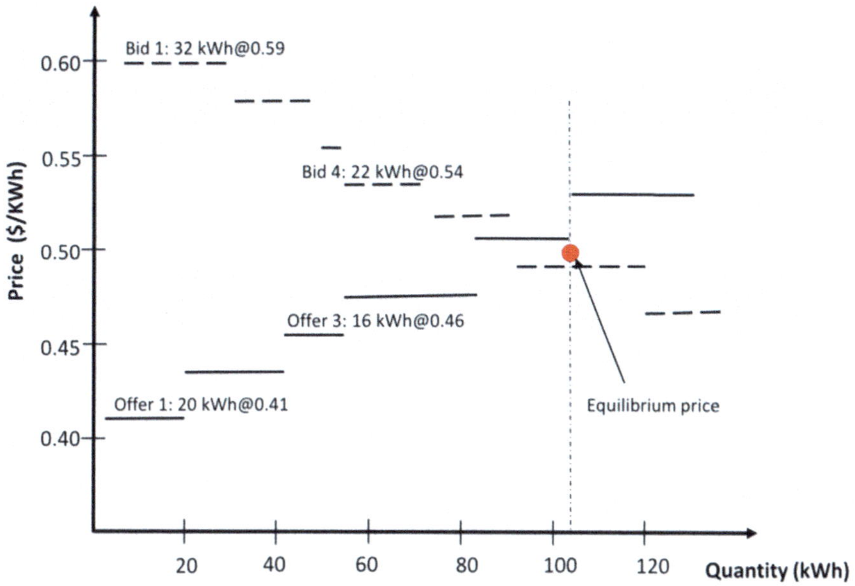

Fig. 12.5 Market-clearance of the double-auction model

continue to stay in the market for future chances, with their remaining time counters updated by $h \leftarrow h - 1$.

The goal of the REB in this LEM is to choose a series of actions to maximize its profit with constant consideration of some opportunity cost when many a seller/buyer frequently enters and quits the market. The optimal policy π^* can be defined as: (1) the best action is always correctly chosen for maximizing long-term expected profit (reward), and (2) the decision is always made at the most suitable time without any delay. Thus, the expected discounted revenue of the REB is defined as in the following MDP problem P:

$$\mathrm{P}: \max_{\pi:\mathcal{S}\rightarrow\mathcal{A}} E[\sum_{t=0}^{nT} (\gamma)^t r_t(s_t, \pi(s_t))] \tag{12.6}$$

where $0 \leq \gamma \leq 1$ is the discount factor, which represents the relative importance of the future market profit compared with that of the present market profit. The reward function $r_t(s_t, \pi(s_t))$ is calculated according to different chosen actions, as follows:

$$r_t(s_t, a_t) = \begin{cases} 0 & a_t = a_1 \\ -\xi & a_t = a_2 \\ p^s_{\text{high}} e^s_{\text{high}} - p^b_{\text{low}} e^b_{\text{low}} & a_t = a_3 \\ z^* & a_t = a_4 \end{cases} \tag{12.7}$$

where a_1, a_2, a_3 and a_4 indicate the actions of *waiting*, *retention*, *partial-clearance* and *clearance*, $p^s_{\text{high}} = \max\{p^s_i | i \in \mathcal{I}^s_t\}$, and $p^b_{\text{low}} = \max\{p^b_j | j \in \mathcal{I}^b_t\}$. The numerical value $z^\star$ corresponding to $a_t = a_4$ in Eq. (12.7) is obtained as an optimal value through solving the following optimization problem Pa, in Eqs. (12.8–12.11), which is also a double-auction problem, as shown in Fig. 12.5. It follows a pay-as-bid model, in which all the price-quantity pairs left at the equilibrium price will be cleared.

$$Pa : \max_{x^s_i, x^b_j} z = \sum_{j \in \mathcal{I}^b_t} p^b_j x^b_j - \sum_{i \in \mathcal{I}^s_t} p^s_i x^s_i \tag{12.8}$$

$$\text{s.t.} \quad \sum_{j \in \mathcal{I}^b_t} x^b_j - \sum_{i \in \mathcal{I}^s_t} x^s_i = 0 \tag{12.9}$$

$$0 \le x^b_j \le e^b_j, \quad \forall j \in \mathcal{I}^b_t \tag{12.10}$$

$$0 \le x^s_i \le e^s_i, \quad \forall i \in \mathcal{I}^s_t \tag{12.11}$$

It is also worth commenting that in a reinforcement learning framework, the defined actions can be low-level controls, such as assigning a particular numerical value to a variable, or high-level decisions, such as whether or not to solve a subdecision problem [20] like ours. Additionally, retention bonus ξ in *retention* explicitly indicates the fact that decision delay is costly, and some other implicit *search cost* is also contained in *waiting* because delaying the decision-making by waiting may often introduce some regret or opportunity cost, which will be quantitatively measured in Sect. 12.5.4.

12.3.4 Real-World Implementation Issues

To implement the local energy trading process, it is assumed that every participant can join the LEM using a home energy management system (HEMS) [21] with embedded trading functionality. The trader must also access a regional central market operation system responsible for collecting local customer subscriptions to this optional energy service. Over time, as trading experience grows, both customers and the trader will adapt to daily, weekly, or seasonal trading patterns. The cost of communication hardware for this service is expected to be low, with minimal hardware modification required, as the intelligent trading platform eliminates subjective negotiations, and localized energy services tend to be stable within neighborhoods or local communities.

12.4 Markov Decision Process and Modified Q-Learning Algorithm

An MDP is usually described by state space S, action space $\mathcal{A}$, reward function set $\mathcal{R}$, state transition probability matrix $\mathcal{P}$ and a discount factor $\gamma \in [0, 1].[0, 1]$. It is also assumed that an MDP remains the Markov property, which implies the transition probabilities of a state are only affected by the previous one step with no memory [20]. As our market model and problem P is built as an MDP (Fig. 12.6), it can be supposed to be solved by any standard reinforcement learning techniques, such as temporal-difference learning.

However, practical issues arise, such as handling the continuous state space. For example, the market status S includes price information, energy quantity, customer entry times, and various combinations of customer statuses, all of which are continuous variables and difficult to enumerate. One solution is to approach this as a multi-armed bandit problem, specifically a contextual bandit problem [22]. In this approach, the timestamp t is the only state space, with other information $\{\mathcal{P}, \mathcal{E}, \mathcal{T}^s, \mathcal{T}^b, \mathcal{H}_t, \mathcal{I}_t\}$ serving as reward parameters or external signals, assuming certain parameter distributions. This method limits the decision-maker's learning ability since it does not utilize past experience for future actions. In a bandit problem, each action affects only the immediate reward. If actions impact the subsequent situation and reward, it becomes a full reinforcement learning problem [20]. In our market model, the choice of action (e.g., *waiting, retention, partial-clearance, clearance*) affects immediate profit and the market situation and profit potential for the next interval. Therefore, we use an episode-dependent modified Q-learning algorithm, based on temporal-difference learning and theoretical analysis [23], to address this issue. This algorithm, shown in Algorithm 1, underpins our smart strategy for market operation.

The function $\phi(\tau)$ in Algorithm 1 is based on the tabular non-stationary reward structure described in [23]. In our market model, non-stationarity has two aspects: (1) the reward distribution is influenced by the market status, which changes across episodes (days) and may follow certain probability distributions, with one sample generated per episode; and (2) the action taken at each step of an episode affects the potential reward in subsequent steps. These mechanisms introduce interesting

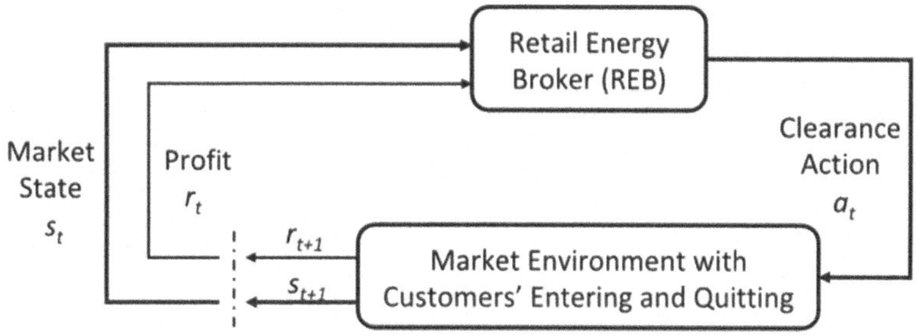

Fig. 12.6 Market model as an MDP with agent-environment interaction

market dynamics, which are demonstrated through various case studies in the next section. For more theoretical details, readers can refer to [23–25].

Similar to the basic Q-learning algorithm, the optimal stationary policy π^* can be well defined by using the optimal *action-value function* $Q^* : \mathcal{S} \times \mathcal{A} \rightarrow \mathcal{R}$, which satisfies the following Bellman optimality equation:

$$Q^*(s, a) = r(s, a) + \gamma \sum_{s' \in \mathcal{S}} p(s'|s, a) V^*(s') \tag{12.12}$$

where $V^*(s')$ is the optimal *state-value function* [20], which is defined as

$$V^*(s') = \min_{a \in \mathcal{A}} Q^*(s', a), \ \forall s \in \mathcal{S}. \tag{12.13}$$

Since $V^*(s')$ is the expected discounted system cost, with action a in state s, we can obtain the optimal stationary policy as

$$\pi^*(s) = \operatorname*{argmin}_{a \in \mathcal{A}} Q^*(s, a) \tag{12.14}$$

In solving problem P, any temporal-different learning-based method, unlike dynamic programming methods [26], can provide the solution without acquisition of the state transition probabilities $p_s(s_{t+1}|s_t, a_t), \quad \forall s \in \mathcal{S}$ a priori [27]. In this way, reinforcement learning techniques based on an MDP will allow the REB, as a trader, to take the LEM as a pure data-driven model to build its strategy intelligently with increasing experience.

12.5 Case Studies

This section provides numerical results to showcase observations regarding the proposed LEM model and evaluate our market learning algorithm's performance. Unless otherwise specified, each day consists of 72 time intervals, each lasting 20 min. Simulations were run in Python 3.6 on a desktop with an Intel i7 and 16 GB RAM.

12.5.1 Smart Strategy with Learning

In the first case study, we compare the results of the smart strategy with learning against a dummy strategy without learning. The dummy strategy means the REB clears the market immediately without using historical trading data, aiming to maximize current profit. In contrast, the smart strategy uses learning to choose actions (waiting, retention, clearance, partial-clearance) to maximize profit over the entire day. The parameters of the market model are defined in Table 12.1.

Table 12.1 Parameters of the market model

Parameter sets	Values
$\ominus^s, \ominus^b$	$\ominus^s \sim \ominus(39, 12^2), \ominus^b \sim \ominus(54, 12^2)$
$\ominus^s, \ominus^b$	$u(0.4, 0.6)$
$\varepsilon^s, \varepsilon^b$	$u(20, 40)$
	$\epsilon_p = 0, \lvert\mathcal{I}^s\rvert = 100, \lvert\mathcal{I}^b\rvert = 110, H = 3, n = 72, \xi = 0.5$

This market model is entirely data-driven, relying on accumulated trading experience and independent of distribution choice. The parameter sets $\mathcal{T}^s$ and $\mathcal{T}^b$ are assumed to be normally distributed for convenience, but they could also follow Weibull, uniform, or any other distribution. Sensitivity analysis in subsequent case studies will show similar results with hybrid distributions or random data. Some parameters related to the modified Q-learning algorithm are also given as follows: $\epsilon = 0.1, \alpha = 0.5, \gamma = 1$. The algorithm runs 500 episodes in total and provides the last 400 episodes' results while averaging over 20 independent runs in Fig. 12.7. The learned optimal policy with chosen actions is also shown in Fig. 12.8. The action code is defined to be the same as in Eq. (12.7): $1 \rightarrow waiting$, $2 \rightarrow retention$, $3 \rightarrow partial - clearance$, $4 \rightarrow clearance$.

From Fig. 12.7, it is evident that the smart strategy initially performs poorly due to its learning phase involving trial and error. However, as more episodes are run and more experience is gained, the smart strategy starts to adapt to market characteristics, optimizing its policy [20]. Over time, it surpasses the dummy strategy, which lacks

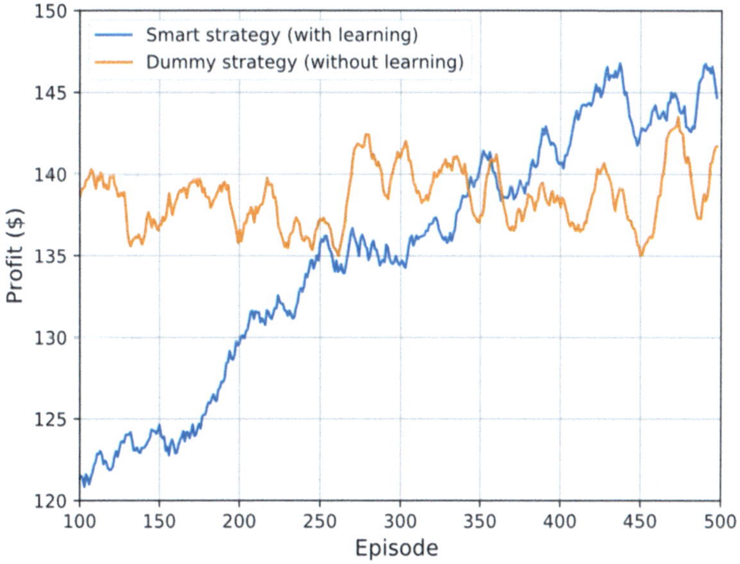

Fig. 12.7 Smart strategy (with learning) vs dummy strategy (without learning)

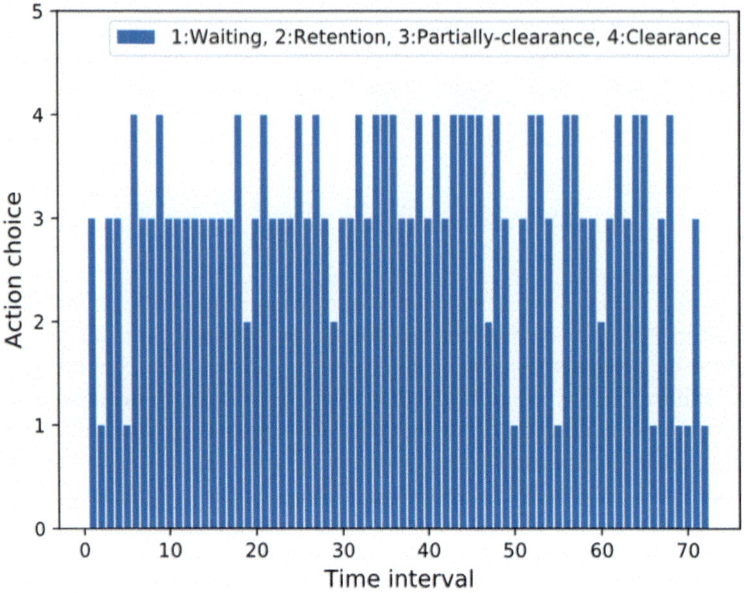

Fig. 12.8 Optimal policy with chosen actions

learning capability. To illustrate this long-term improvement, we ran the algorithm for about 2000 episodes and presented the results from the first episode in Fig. 12.9.

Interestingly, the smart strategy with learning improves quickly and consistently outperforms the dummy strategy in the long term. The optimal policy adjusts with increased experience. Figures 12.8 and 12.9a show that while the smart strategy generally outperforms the dummy strategy, this is not always the case for every episode due to the non-deterministic nature of the results. Occasionally, the smart strategy may perform worse depending on specific market conditions. Statistically, however, the smart strategy with learning is superior in the long term.

Figure 12.10 analyzes a benchmark optimization model similar to Eqs. (12.8–12.11), considering all time intervals and perfect future forecasts, to compare with the reinforcement learning method. The objective function is given as follows:

$$\max \quad Z_{\text{opt}} = \sum_{j \in \mathcal{I}^b} p_j^b x_j^b - \sum_{i \in \mathcal{I}^s} p_i^s x_i^s \tag{12.15}$$

where $\mathcal{I}^b = \bigcup_{t=1}^{nT} \mathcal{I}_t^b$ and $\mathcal{I}^s = \bigcup_{t=1}^{nT} \mathcal{I}_t^s$. We can see that because of the assumption that perfect forecast is available and given as a priori, the benchmark optimization actually has the best performance with maximum profit. However, this perfect forecast that removes all the uncertainty characteristics of customers' trading behavior is rarely the case in reality. Considering its adaptivity to volatile local market situations, it is

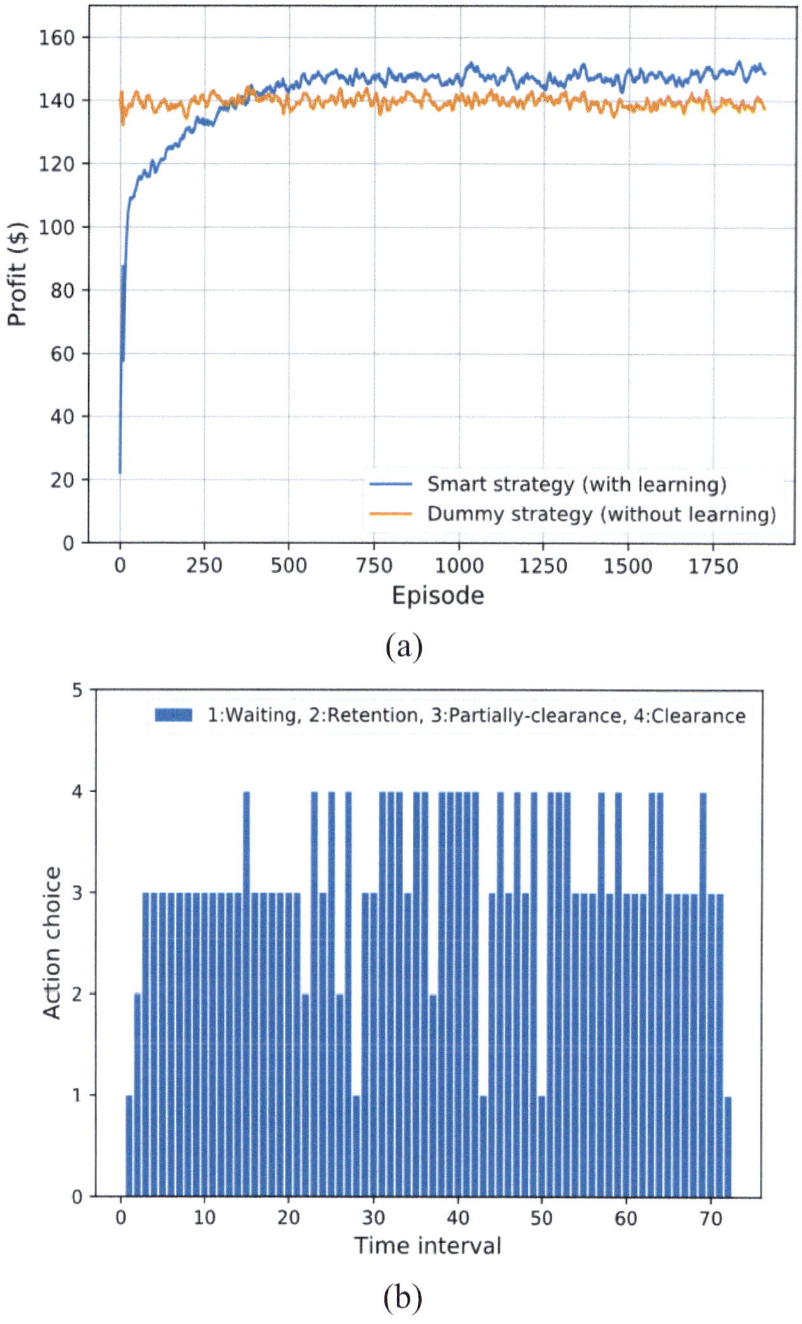

Fig. 12.9 Results of running for a longer time with more experience accumulated. **a** Profit with more experience. **b** Optimal policy with more experience

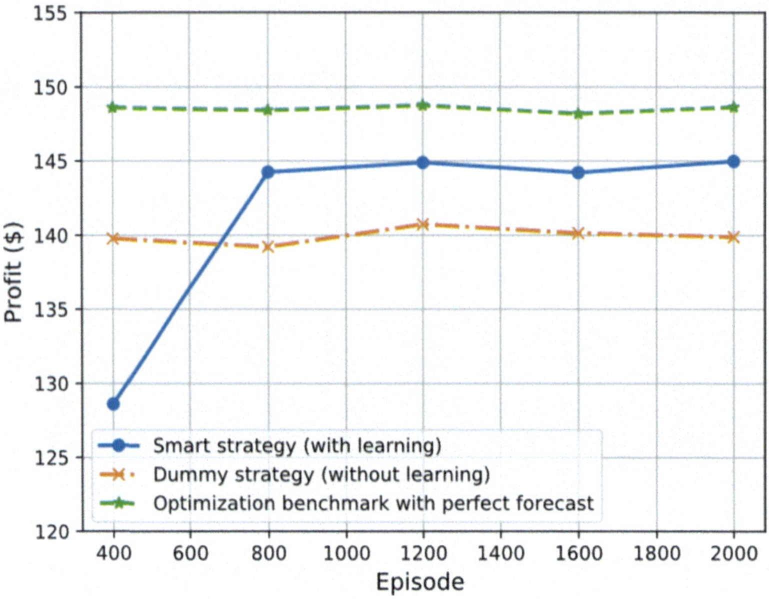

Fig. 12.10 Optimization benchmark with perfect forecast

still reasonable to accept reinforcement learning as a well performing method for the proposed market model.

12.5.2 *Advantages of LEM*

To highlight the advantages of the LEM model over existing energy services without local energy trading, we present the total benefit for all participants at different participation levels in a one-day market operation in Table 12.2. Sellers' economic benefits are shown by the increase in revenue from bidding, while the cost savings from energy purchases reflect buyers' benefits.

Interestingly, the total benefit at medium participation levels (~400 sellers and buyers) is more than double that at low participation levels (~200 sellers and buyers).

Table 12.2 Benefit ($) of a LEM for all participants

Participation level	Sellers		Buyers		Traders	
	Mean	Std.	Mean	Std.	Mean	Std.
Low (~200)	240.4	16.8	265.2	12.4	121.1	2.4
Medium (~400)	501.2	27.8	552.6	21.5	276.8	4.2
High (~1000)	1307.9	49.0	1441.9	39.2	698.9	9.1

This is because increased participation leads to more energy transaction opportunities due to a wider range of bid/offer prices. Additionally, we also present the average annual benefit per participant at different participation levels in the LEM, assuming occasional market openings throughout the year, using a one-time average benefit (12.16).

$$\overline{R_{\text{seller}}} = \frac{1}{3N_r}\left[\frac{1}{N_{\text{low}}^S}\sum_{k=1}^{N_r}\sum_{i=1}^{N_{\text{low}}} R_{\text{low}}^S(i,k) + \frac{1}{N_{\text{med}}^S}\sum_{k=1}^{N_r}\times\sum_{i=1}^{N_{\text{med}}^S} R_{\text{med}}^S(i,k) + \frac{1}{N_{high}^S}\sum_{k=1}^{N_r}\sum_{i=1}^{N_{high}^S} R_{high}^S(i,k)\right] \tag{12.16}$$

where $\overline{R_{\text{seller}}}$ indicates the average revenue of a seller for one-day participation in different scenarios, where $N_{\text{low}}^s, N_{\text{med}}^s, N_{\text{high}}^s$ are the number of customers in different participation level, respectively. N_r is the number of algorithm running times. The annual revenue of a seller in a LEM can be calculated by $365 \times r_o \times \overline{R_{\text{seller}}}$, where r_o indicates the occasional market open rate. The average annual cost of a buyer and the average annual profit of a trader can be calculated in a similar way. The overall annual benefit effect (per capita, with $r_o = 20\%$) of introducing a LEM in addition to an existing utility's energy service is shown in Fig. 12.11.

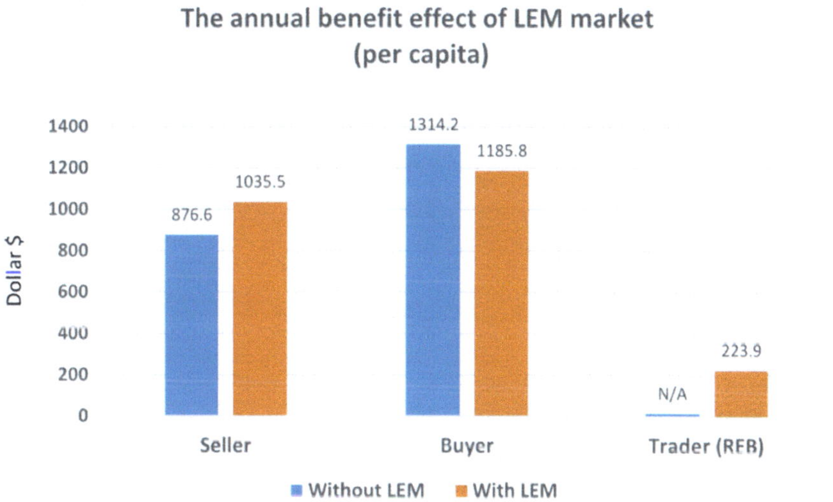

Fig. 12.11 Annual benefit effect (per capita) on participants in a LEM (values in dollar are revenue for seller and REB, and cost for buyer)

12.5.3 Sensitivity Analysis

In this market model, a lot of market parameters and learning parameters are involved, and many of them are significant in the REB's decision-making process. Here, some important parameters are studied during the sensitivity analysis with many interesting results and phenomenon observed. In the following cases, we use a uniform distribution for market entering time and a bid/offer price with $|\mathcal{I}^s| = 200$, $|\mathcal{I}^b| = 220$ other parameters the same as in Sect. 12.5.1 if without explicit illustration.

(1) *Time Intervals*: The number of time intervals per day affects the frequency at which customers can enter the market. With a fixed remaining time counter (H), fewer time intervals mean customers spend a larger portion of the day in the market, increasing their chances of being selected for market-clearance. This also provides the REB more opportunities for profit, as shown in Fig. 12.12. Conversely, increasing time intervals effectively reduces the remaining time counters, raising the trader's opportunity cost. This relates to the concept of search friction, discussed in Sect. 12.5.4.

(2) *Exploration Ratio*: In reinforcement learning, there is a balance between exploration and exploitation. The agent, such as the REB, must use its existing knowledge to gain rewards while exploring new actions for better future choices [20]. The exploration ratio determines how often the agent chooses actions with non-maximum rewards. To observe its long-term effect, we ran the algorithm for 1000 episodes, with results shown in Fig. 12.13.

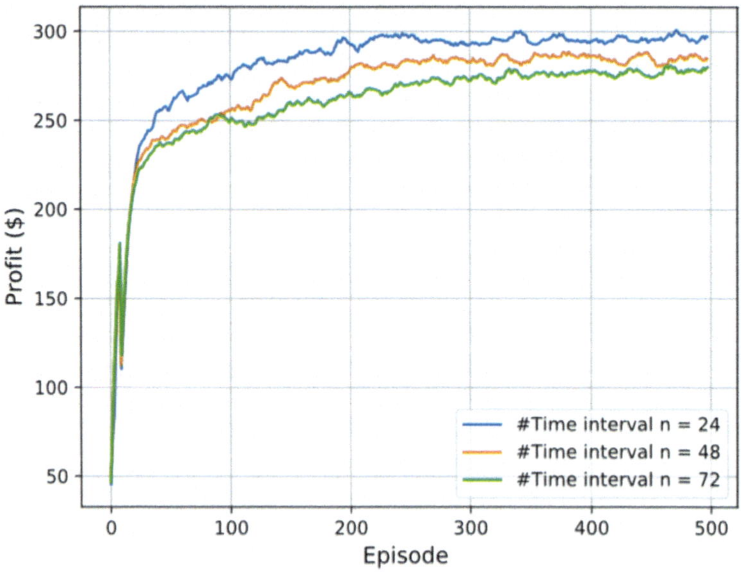

Fig. 12.12 Impact of the number of time intervals

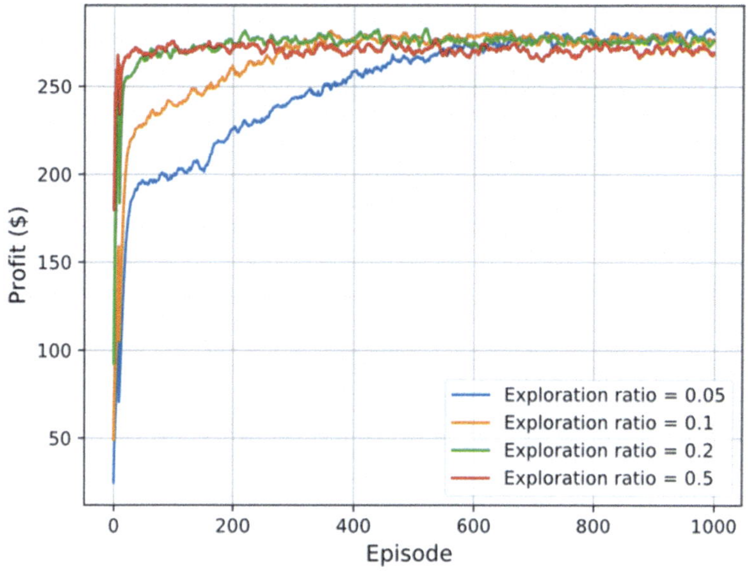

Fig. 12.13 Obtained profit with different exploration ratios

We can observe that a very low exploration ratio $\epsilon = 0.05$ introduces very low profit at the beginning; however, the profit will gradually converge to a normal level, similar to that of other choices, in the long term. In contrast, a very high exploration ratio $\epsilon = 0.05$ provides higher profit at the beginning but will suffer some loss in the long term because of the inefficient effort on random exploration.

(3) *Price Margin*: The price margin ϵ_p, discussed in Sect. 12.3.2, sets a price threshold ensuring fair trading for customers, who must consider their trading partners' benefits. The REB can set this margin to 0, allowing customers to bid/offer at utility service prices, though they risk not being selected. A reasonable increase in the price margin boosts REB's profit and market efficiency by encouraging quick and frequent trades. Since the price margin statistically affects price samples, we compared results with different margins ϵ_p using a violin plot in Fig. 12.14, showing that a slightly higher margin shifts the total profit distribution toward larger numbers.

12.5.4 The Impact of H and Search Friction

Market operations are event-driven and open occasionally during specific conditions, such as high solar irradiance or main grid supply shortages. Participant activity during these scenarios influences market efficiency, as long wait times for trading opportunities may cause some customers to leave. The anxiety from waiting and searching for suitable trades can be described qualitatively and quantitatively. In this

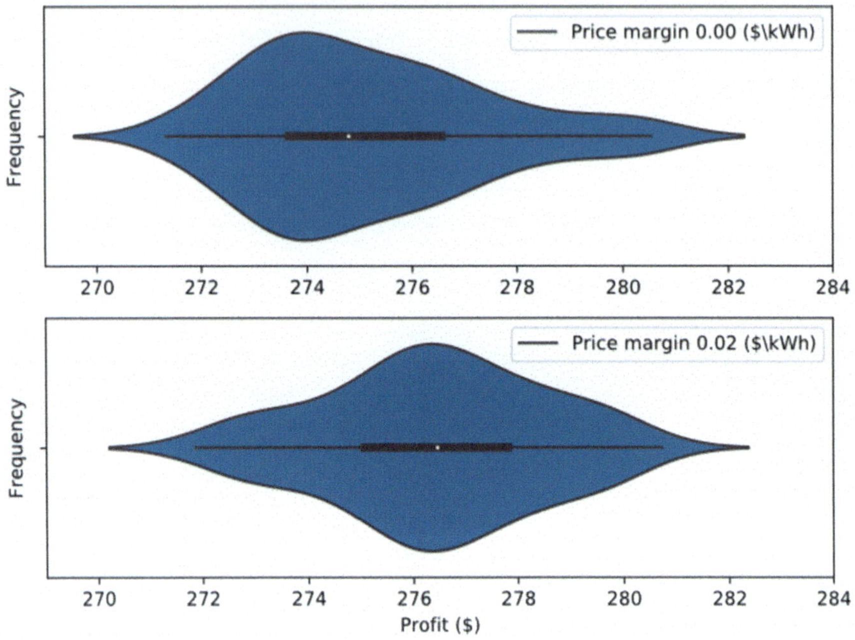

Fig. 12.14 Effect of price margin ϵ_p

test case, we will try different values of parameter H, which is used to describe the remaining time counter, $h^s_{i,t^s_i}, h^b_{j,t^b_j} = H$, or how many chances each customer (seller or buyer) still has before they must quit the market, to see the impact on the final profit of the REB for the whole day. Some other parameters are the same as in the first case study with more customers,$|\mathcal{I}^s| = 200$ and $|\mathcal{I}^b| = 210$, distributed more intensively over all the time intervals. The result is averaged over 10 independent runs and shown in Fig. 12.15. It matches the intuition that the more chances to enter or the longer customers are allowed to stay in the market, the more profit the REB can obtain by delaying its decision-making until the overlap of many a suitable bid and offer.

In microeconomics, search friction arises during a search process, such as sequentially sampling a population X, where samples x_t can be energy prices, service quality, bids, etc. [18]. The 2010 Nobel Prize in Economics recognized the analysis of markets with search frictions [28]. The optimal stopping problem, a common application of search friction, involves choosing the best time to act to maximize reward or minimize cost [29]. In our energy market model, each time interval t presents a decision problem for the REB. Early decisions may miss better offers, while late decisions may incur opportunity costs if customers exit the market. This issue is quantitatively analyzed in our previous work [30], assuming a probability distribution of customer quality and entry times. The optimal stopping result is estimated by Eq. (12.17), with P_{dummy} the baseline profit from the dummy strategy, and Eq. (12.18) showing the

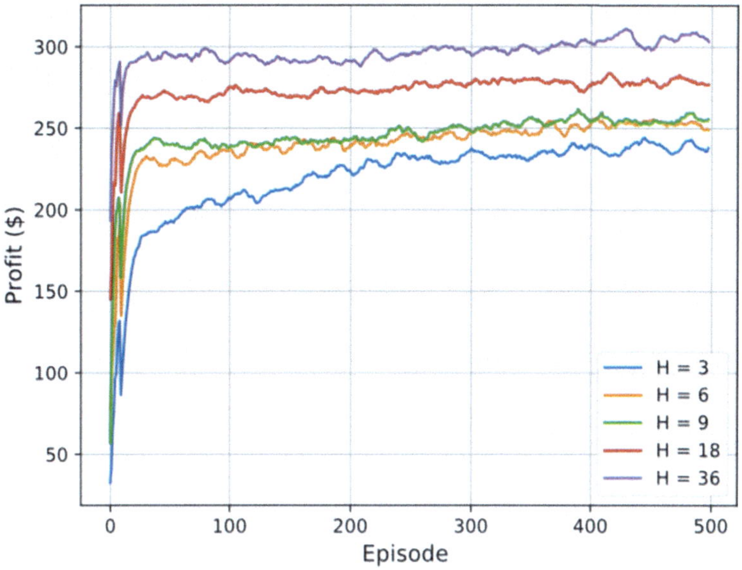

Fig. 12.15 Impact of parameter H

normalized profit in an optimal stopping problem, assuming equal profit distribution in each interval [31]. The reduced search friction from the learning method is obtained by averaging the profit gap in Fig. 12.15. Table 12.3 summarizes results related to search friction analysis. Actions like retention and partial-clearance are designed to reduce search friction, locking in partial profit and increasing decision-making opportunities

$$C_{\text{friction}} = n \times \min_{c=2,\ldots,H} [V_T(c) - V_T(c-1)] \times P_{\text{dummy}} \tag{12.17}$$

$$V_T(c) = \sum_{t=c}^{H-1} [\prod_{w=c}^{t-1} (\frac{w-1}{w})](\frac{1}{t+1}) + 0.5[\prod_{w=c}^{H-1} (\frac{w-1}{w})] \tag{12.18}$$

Table 12.3 Reduced search friction by learning method (daily)

Remaining time counter	Reduced cost	Analytical cost estimation
$H = 3$	\$5.96	\$23.33
$H = 6$	\$17.32	\$46.67
$H = 9$	\$19.12	\$62.22
$H = 18$	\$40.25	\$81.67
$H = 36$	\$63.45	\$97.22

12.6 Conclusions

This chapter explores how the REB facilitates indirect local energy trading using reinforcement learning. The localized reinforcement learning based trading model offers additional energy trading options, promoting retail electricity market deregulation and potentially reshaping the energy business landscape. This local energy trading strategy aims to diversify and gamify the energy ecosystem at the distribution network edge and near end-users, without replacing existing energy services. Most local participants are assumed to be geographically close, likely on the same distribution feeder, so this chapter does not model distribution networks' AC power flow constraint. Future research will consider more realistic conditions and might employ virtual experience or experience replay-based deep reinforcement learning techniques to enhance market operator intelligence. Next-generation distribution networks are anticipated to see more energy business models and local trading platforms, revolutionizing the energy ecosystem.

References

1. Su, W. (2014). The role of customers in the us electricity market: Past, present and future. *The Electricity Journal, 27*(7), 112–125.
2. Calvillo, C. F., Sánchez-Miralles, A., Villar, J., & Martín, F. (2016). Optimal planning and operation of aggregated distributed energy resources with market participation. *Applied Energy, 182*, 340–357.
3. Sikdar, S., & Rudie, K. (2013). Microgrid level competitive market using dynamic matching. In *2013 IEEE Electrical Power & Energy Conference* (pp. 1–6). IEEE.
4. Mihaylov, M., Jurado, S., Avellana, N., Razo-Zapata, I. S., Van Moffaert, K., Arco, L., Nowé, A., et al. (2015). SCANERGY: A scalable and modular system for energy trading between prosumers. In *AAMAS* (Vol. 15, pp. 1917–1918).
5. Liu, N., Yu, X., Wang, C., Li, C., Ma, L., & Lei, J. (2017). Energy-sharing model with price-based demand response for microgrids of peer-to-peer prosumers. *IEEE Transactions on Power Systems, 32*(5), 3569–3583.
6. Yuan, C., Illindala, M. S., & Khalsa, A. S. (2017). Co-optimization scheme for distributed energy resource planning in community microgrids. *IEEE Transactions on Sustainable Energy, 8*(4), 1351–1360.
7. Morstyn, T., Teytelboym, A., & McCulloch, M. D. (2018). Bilateral contract networks for peer-to-peer energy trading. *IEEE Transactions on Smart Grid, 10*(2), 2026–2035.
8. Zhang, C., Wu, J., Long, C., & Cheng, M. (2017). Review of existing peer-to-peer energy trading projects. *Energy Procedia, 105*, 2563–2568.
9. Chen, T., Alsafasfeh, Q., Pourbabak, H., & Su, W. (2017). The next-generation US retail electricity market with customers and prosumers—A bibliographical survey. *Energies, 11*(1), 8.
10. Johnson, D., Horton, E., Mulcahy, R., & Foth, M. (2017). Gamification and serious games within the domain of domestic energy consumption: A systematic review. *Renewable and Sustainable Energy Reviews, 73*, 249–264.
11. Zafar, R., Mahmood, A., Razzaq, S., Ali, W., Naeem, U., & Shehzad, K. (2018). Prosumer based energy management and sharing in smart grid. *Renewable and Sustainable Energy Reviews, 82*, 1675–1684.

12. Mihaylov, M., Jurado, S., Avellana, N., Van Moffaert, K., de Abril, I. M., & Nowé, A. (2014). NRGcoin: Virtual currency for trading of renewable energy in smart grids. In *11th International conference on the European energy market (EEM14)* (pp. 1–6). IEEE.
13. Xu, X., Maki, A., Chen, C. F., Dong, B., & Day, J. K. (2017). Investigating willingness to save energy and communication about energy use in the American workplace with the attitude-behavior-context model. *Energy Research & Social Science, 32*, 13–22.
14. Brock, W. A., & Evans, D. S. (1989). Small business economics. *Small Business Economics, 1*, 7–20.
15. Rubinstein, A., & Wolinsky, A. (1987). Middlemen. *The Quarterly Journal of Economics, 102*(3), 581–593.
16. Park, S., Lee, J., Hwang, G., & Choi, J. K. (2017). Event-driven energy trading system in microgrids: Aperiodic market model analysis with a game theoretic approach. *IEEE Access, 5*, 26291–26302.
17. Papadimitriou, C. H., & Tsitsiklis, J. N. (1987). The complexity of Markov decision processes. *Mathematics of Operations Research, 12*(3), 441–450.
18. Kohn, M. G., & Shavell, S. (1974). The theory of search. *Journal of Economic Theory, 9*(2), 93–123.
19. Contreras, J., Candiles, O., De La Fuente, J. I., & Gomez, T. (2001). Auction design in day-ahead electricity markets. *IEEE Transactions on Power Systems, 16*(1), 88–96.
20. Sutton, R. S., & Barto, A. G. (1999). Reinforcement learning: An introduction. *Robotica, 17*(2), 229–235.
21. Han, D. M., & Lim, J. H. (2010). Design and implementation of smart home energy management systems based on zigbee. *IEEE Transactions on Consumer Electronics, 56*(3), 1417–1425.
22. Yue, Y., Broder, J., Kleinberg, R., & Joachims, T. (2012). The k-armed dueling bandits problem. *Journal of Computer and System Sciences, 78*(5), 1538–1556.
23. Shibuya, T., & Yasunobu, S. (2011, October). Reinforcement learning with nonstationary reward depending on the episode. In *2011 IEEE international conference on systems, man, and cybernetics* (pp. 2145–2150). IEEE.
24. Rana, R., & Oliveira, F. S. (2014). Real-time dynamic pricing in a non-stationary environment using model-free reinforcement learning. *Omega, 47*, 116–126.
25. Feinberg, E. A., & Shwartz, A. (Eds.). (2012). *Handbook of Markov decision processes: Methods and applications* (Vol. 40). Springer Science & Business Media.
26. Bertsekas, D. (2012). *Dynamic programming and optimal control: Volume I* (Vol. 4). Athena Scientific.
27. Watkins, C. J. C. H. (1989). *Learning from delayed rewards.*
28. Albrecht, J. (2011). *The 2011 nobel memorial prize in search theory*. Department of Economics, Georgetown University. http://www9.georgetown.edu/faculty/albrecht/SJESurvey.pdf
29. Jacka, S. D. (1991). Optimal stopping and the American put. *Mathematical Finance, 1*(2), 1–14.
30. Chen, T., Su, W., & Chen, Y. S. (2017). An innovative localized retail electricity market based on energy broker and search theory. In *2017 North American Power Symposium (NAPS)* (pp. 1–5). IEEE.
31. Bearden, J. N. (2006). A new secretary problem with rank-based selection and cardinal payoffs. *Journal of Mathematical Psychology, 50*(1), 58–59.

The manufacturer's authorised representative in the EU is Springer Nature Customer Service Centre GmbH, Europaplatz 3, 69115 Heidelberg, Germany. If you have any concerns regarding our products, please contact ProductSafety@springernature.com

Printed and bound by CPI Group (UK) Ltd, Croydon, CR0 4YY
01/10/2026
02225926-0015